T

09

Third Edition

Principles of
TOXICOLOGY

Karen E. Stine
Department of Biology
Auburn University at Montgomery
Montgomery, Alabama, USA

Thomas M. Brown
GeneCTAr Com LLC
Whitefish, Montana, USA

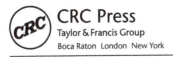

CRC Press
Taylor & Francis Group
Boca Raton London New York

CRC Press is an imprint of the
Taylor & Francis Group, an **informa** business

10 0742 1774

CRC Press
Taylor & Francis Group
6000 Broken Sound Parkway NW, Suite 300
Boca Raton, FL 33487-2742

© 2015 by Taylor & Francis Group, LLC
CRC Press is an imprint of Taylor & Francis Group, an Informa business

No claim to original U.S. Government works

Printed on acid-free paper
Version Date: 20150311

International Standard Book Number-13: 978-1-4665-0342-7 (Hardback)

Visit the Taylor & Francis Web site at
http://www.taylorandfrancis.com

and the CRC Press Web site at
http://www.crcpress.com

I dedicate this to my parents, Betty and Dale Stine, my children, Melinda Jacobs and Danny Jacobs, and my husband, Steve Ballard, as well as to the teachers, colleagues, and students with whom I have been privileged to work and who have taught me so much along the way.

—**K.E.S.**

I dedicate my contributions to this book to my daughter, Annie May Brown, to my father, Alexander Musgrove Brown, who was proprietor of Brown's Corner Drugstore, formerly of Philipsburg, Pennsylvania, and to the nature of Montana which motivates me.

—**T.M.B.**

Contents

Preface

Principles of Toxicology has evolved (through three editions now) from a course originally developed and taught by us at Clemson University beginning in the spring of 1987. In searching for a textbook to accompany this new introductory course on the basic principles and concepts of toxicology, we discovered that, unfortunately, many of the available books were more appropriate as reference materials for more advanced students, while others were too narrowly focused on a single topic (such as biochemical toxicology or ecological toxicology) for our purposes.

Thus, we developed this book. The purpose of the book is to serve as a comprehensive, yet readable, textbook for a first course in toxicology at the undergraduate or beginning graduate level. We have chosen to cover the broad and interdisciplinary field of toxicology through a "levels of organization" approach, with initial chapters focusing on molecular and cellular toxicology, the middle of the book emphasizing physiological toxicology, and a later chapter introducing ecological toxicology. Additional chapters cover a few of the many applied areas within the field.

Each of these chapters combines background material in the appropriate discipline (to help students review and remember basics) with new information on toxicology in a manner that stresses key principles and concepts. We have also included a selection of case studies in which these principles and concepts are applied to specific real-world issues, and an extensive cross-referencing system to help students tie it all together.

We would like to thank all of our colleagues (and students) who have contributed suggestions, comments, and support to this project over the years, as well as the editors and staff at CRC Press/Taylor & Francis Group who have helped make this book a reality.

Authors

Karen E. Stine, PhD, is a professor in the Department of Biology at Auburn University at Montgomery and a former dean of the School of Sciences there. She was formerly a professor of biology/toxicology and director of the toxicology program in the Department of Biology/Toxicology at Ashland University in Ashland, Ohio. Dr. Stine holds a BS in physics and biology from the College of William and Mary in Virginia, an MS in environmental sciences from the University of Virginia, and a PhD in toxicology from the University of North Carolina at Chapel Hill. She is a member of the Society of Toxicology, the Society of Environmental Toxicology and Chemistry, and the Cell Stress Society. At Clemson University, Dr. Stine co-developed and co-taught a Principles of Toxicology course that was open to both undergraduate and graduate students. At Ashland University and Auburn University at Montgomery, she taught an undergraduate Principles of Toxicology course, along with numerous other courses in the areas of toxicology and biology. She has also authored or co-authored several research publications in the field of toxicology. Her research interests are in the area of mechanisms of toxicity and focus on the role of stress proteins in cellular function and dysfunction. She is also interested in the evolution of toxins.

Thomas Miller Brown, PhD, is president of Genectar Com LLC in Whitefish, Montana, which conducts research in toxicology, genetics, and genomics, now focusing on pigment cell development in the common wood nymph butterfly as a melanoma model. Genectar participates in the Montana BioScience Alliance and, by affiliation, the Washington Biotechnology & Biomedical Association. Formerly a professor, he taught toxicology of insecticides, principles of toxicology, and insect biotechnology. He led the discovery of an actively transposing short interspersed nuclear element (*Insect Science* 2009, 16:219–226). He holds a BS from Adrian College and a PhD from Michigan State University, where he was introduced to the study of toxicology by Ronald E. Monroe and Anthony William Aldridge Brown. He has published papers on the biochemical toxicology of organophosphorus compounds and on the mechanisms of insecticide resistance in insects. He was program chairman of the American Chemical Society Special Conference, "Molecular Genetics and Ecology of Pesticide Resistance." He has conducted research at Nagoya University and Tsukuba Science City in Japan.

1 Measuring Toxicity and Assessing Risk

INTRODUCTION

Toxicology is the science of poisons and has as its focus the study of the adverse effects of chemicals on living organisms. Although almost any substance in sufficient quantities (even water) can be a poison, toxicology focuses primarily on substances that can cause these adverse effects after exposure to relatively small quantities. Knowledge of the relative toxicity of substances is fundamental to all applications of toxicology, from development of a new drug to the modeling of the effects of an environmental pollutant. This chapter describes approaches used by toxicologists to determine the toxicity of a substance. We consider principles of the dose versus response relationship, methods used to evaluate toxicity in laboratory animals, and subsequent statistical analyses for quantitation of toxicity. We also discuss the use of toxicity data in assessing the risks of exposure to potentially hazardous substances.

CHEMISTRY OF TOXICANTS

Knowledge of the chemistry of a poison is of primary importance because it is a major determinant in the solubility and reactivity of the substance. Chemistry can dictate the *vehicle* used to administer the substance in testing and can predict to some extent the duration of the test, as well as any potential products of biotransformation (metabolism). For small molecules, knowledge of chemistry is relatively simple to gain by examining the chemical formula and performing tests such as an octanol/water partitioning experiment to find the relative lipid solubility. However, the coming generation of drugs will include many large polymeric compounds such as proteins and nucleic acids, which are more challenging to study chemically. There is also a trend to return to botanical products, which are often highly complex mixtures with multiple active ingredients. This can complicate the administration and interpretation of testing for toxicity.

TOXICITY TESTING METHODS

There are a variety of different methods for gathering information on toxicity. First of all, observation of a chemical structure can give a strong indication as to the behavior of a chemical in a living system. Most directly, chemicals can be compared to compounds of similar structure for which toxicity has already been observed and measured. The relationship between a chemical structure and toxicity is known as the *structure–activity relationship*, and the ability to accurately use structure to predict toxicity has the potential to greatly expedite the identification of chemicals requiring further testing for potential toxic effects.

Another option for toxicity testing is testing *in vitro*. *In vitro*, literally meaning "in glass" in Latin, means that testing takes place outside of a living organism, typically in a test tube or a petri dish. Examples of toxicity testing *in vitro* would include the use of cell cultures to identify carcinogens (an assay that will be discussed in a later chapter) or testing the effects of toxicants on isolated, perfused organs (such as liver or kidney). These tests can give valuable insight into mechanisms of toxicity, but yet are by nature limited in their scope.

To gain the greatest insight into the potential effects of a chemical on human health, animal bioassays are currently the most effective model. The choice of species, of course, depends on the desired endpoint in terms of information. If the focus is on human health, species are often chosen on the basis of similarities of structure and function between the species and the human physiological system(s) of greatest interest. If the studies are focused on ecological health, organisms representing key species within an ecosystem may be chosen. Of course, pragmatically, smaller and simpler model organisms are typically easier and less expensive to culture and easier to handle in the laboratory setting. Recently, *transgenic* models are increasingly available for testing. These are animals whose genome has been deliberately modified, often in such a way as to produce a model for at least some aspects of human disease. One example of this is the *knockout* animal, in which a particular gene has been deliberately disabled. There are now many knockout mouse lines available, each with a different gene knocked out (more on this in Chapter 5).

TESTING IN ANIMALS

A wealth of information can be gathered from toxicity testing by carefully observing animals during and following exposure. These data may provide evidence for the mode of action of the substance and clues as to which physiological systems, organs, and tissues could be affected.

Although specific protocols for toxicity testing have been developed by various regulatory agencies (such as the Food and Drug Administration [FDA] and the Environmental Protection Agency [EPA]), they share many characteristics in common. It is important to test effects on both male and female organisms, and often, testing across a range of ages is appropriate. Of course, in any study, the handling and treatment of animals must be humane and must be the same for all animals in the study, whether they are in treated or control groups. Animals may be tagged for identification, using either simple numbered, metal *ear tags*, or more sophisticated devices such as *electronic transponding implants*. Animals must be housed in clean, comfortable conditions with access to adequate food and water. Typically, animals are housed in conventional, box-type cages, although in some cases specialized cages such as metabolism cages may be used. These cages are equipped with a separator for collection and measurement of urine and feces, so that consumption of food and water as well as alteration of the test substance can be measured accurately.

In a typical animal test, *body weight* of the test animals is measured either daily or periodically. Animals are observed for *behavior* (comparing behavior of treated with control animals) and *symptomology* (such as tremors or convulsions, for example). During the exposure period, animals are monitored closely for symptoms of poisoning, as well as timing of the appearance of those symptoms, which might suggest the mechanism of poisoning of the substance. A slow onset of poisoning, for example, might suggest a *bioactivation* of the substance to a more toxic *metabolite*, or product, which accumulates as the *parent substance* is converted.

Following an exposure period, the animals are sacrificed and *necropsy* is performed. This is a procedure in which the treated and control animals are dissected and organs are weighed and examined for toxic effects on gross morphology and physiology. Sections of tissue samples may be sliced on a *microtome* and examined under the microscope for evidence of *histopathology* (which is any abnormality in cell or tissue). Tissue samples may also be analyzed for the presence of biochemical indicators of pathology.

HUMAN DATA AND EPIDEMIOLOGY

Of course, in some cases (in which the risk of harm is low, for example), toxicity studies may be carried out using human subjects. Information on toxicity can also be obtained from *epidemiological studies*, which assess the relationship between exposure to a toxicant (usually accidental or voluntary exposure) and either disease *incidence* (the rate at which new cases of the disease appear in a human population) or disease *prevalence* (the number of existing cases of the disease at a particular point in time). These types of studies will be discussed in more detail later in this chapter.

FACTORS TO BE CONSIDERED IN PLANNING TOXICITY TESTING

There are several questions that must be answered in designing laboratory studies to determine the toxicity of a chemical substance. Among these are

1. Through what physiological route does exposure occur (in other words, how does the substance enter the body)?
2. How much of the substance is necessary to produce toxicity?
3. Over what period of time does exposure occur?

Toxicity testing attempts to answer these questions and thus provides practical information about the risks involved in exposure to potentially toxic compounds.

ROUTES OF EXPOSURE

Various means of administration or *routes of exposure* are used in toxicity testing. Oral toxicity is of primary concern when considering a substance that might be ingested in food, such as a pesticide or food additive, or that might be taken orally as

Routes of Exposure

See also:
 Absorption
Chapter 2

a drug. Dosing through the mouth is technically described as the *peroral* or *per os* method. In some cases, the substance to be tested may be added directly to the animal's food or water. Alternatively, it may be dissolved in water, vegetable oil, or another *vehicle* (depending on the solubility of the test substance) and introduced directly into the esophagus or stomach through the use of a curved needle-like tube (a process called *gavage*). Dermal administration may be considered for a substance that might be handled by workers, such as paints, inks, and dyes, or for cosmetics applied to the skin. The test substance is painted onto the skin, covered with a patch of gauze held with tape, and plastic is wrapped around the body to prevent ingestion of the substance. Finally administration though the respiratory system should be considered in testing industrial solvents or cosmetics applied in an aerosol spray.

Toxicity may also be assessed by direct injection of the substance, using a syringe and a needle. *Intraperitoneal injections* are made into the body cavity; *intramuscular injections* are placed into a large muscle of the hind leg; *subcutaneous injections* are placed just beneath the skin; and *intravenous injections* are made directly into a large vein. Data derived from these injections are especially useful in estimating doses for the investigation of drugs that eventually may be injected by an analogous method in human patients.

DETERMINING THE RESPONSES TO VARYING DOSES OF A SUBSTANCE

Several terms are used to describe levels of exposure to toxicants. The terms *dose* and *dosage* have been used nearly interchangeably, although dose commonly refers to the amount of a chemical administered and dosage refers to the amount of a chemical administered per unit body weight of the recipient. Thus, the dose of a drug might be expressed in milligrams, whereas the dosage would be expressed as milligram per kilogram of body weight. In toxicity testing, most chemical amounts are calculated and administered as dosages, which allows better standardization of the amount of chemical received and allows a better basis for the comparison of effects between individuals and species of widely varying body size. In respiratory exposures, exposure levels are usually measured by the concentration of the substance in the air (in parts per million).

Quantitative toxicology involves challenging test animals with the substance to be evaluated, which is applied in an ordered series of doses. The dose is controlled by the toxicologist; therefore, it is considered to be the independent variable. Response of the animals may be measured in many different ways and is generally dependent on the dose applied (i.e., it is the dependent variable).

Responses can be scored and related to dose in order to determine the dose versus response relationship. One response considered in toxicology is the death of the animal. This is scored at a *quantal value*, alive (no response: 0) or dead (response: 1), and recorded at a predetermined time as *mortality*, for which an objective criterion, such as lack of heartbeat, is observed. A dose producing mortality is a lethal dose of the substance. In other experiments, the observed response may be a *continuous variable* that can be measured in each subject. Examples of continuous variables include consumption of oxygen, time to onset of convulsions, degree of inhibition of an enzyme, and loss of weight.

A basic principle of toxicology is that response varies proportionally to a geometric, not arithmetic, increase in dose. This means that to test a substance that produces responses in a small proportion of animals at 1–2 mg/kg, a geometric dosing range (1, 2, 4, 8, and 16 mg/kg) would be used rather than an arithmetic range (1, 2, 3, 4, and 5 mg/kg). Because of this, graphs relating dose and response are generally plotted with the response value on the y-axis and the logarithm of the dose on the x-axis.

TIMING OF EXPOSURE

Often, the first of many considerations in toxicity testing is to assess the acute toxicity of the chemical. *Acute toxicity* is the toxicity that results from a single exposure to the substance. Typically, animals are dosed with a single dose and then observed for up to 14 days. One example of an acute toxicity test is the LD_{50}, which will be discussed next in this chapter. *Subacute toxicity testing* measures the response to substances that are delivered through repeated or continuous exposure over a period that generally does not exceed 14 days, and *subchronic toxicity testing* involves repeated or continuous exposure over a period of 90 days. The final category of exposure is *chronic toxicity testing*, which refers to repeated or continuous exposures that last for more than 90 days. To ensure sufficient challenge, animals are often exposed to the maximum tolerated dose, the greatest dose that neither kills nor causes incapacitating symptoms. While very high doses are used in order that any chronic toxicity of the test compound will be observable, some experts consider that effects seen at large doses may be due to massive physical damage or *mitogenesis* (regeneration due to cell death) and thus may not accurately predict a substance's toxicity at lower doses.

THE LD$_{50}$ (MEDIAN LETHAL DOSE) EXPERIMENT

TESTING

Traditionally, the *median lethal dose* has been used as a general measure of acute toxicity of any substance. This is the predicted dose at which one half of the individuals in a treated population would be killed. The median lethal dose is determined by exposing groups of uniform test animals to a geometric series of doses of the substance of interest under controlled environmental conditions. The abbreviation LD_{50}, for lethal dose 50, is often used for the median lethal dose. The standard laboratory animal commonly used in this test is the white Norway rat, *Rattus norwegicus*.

The dose is expressed in dosage units of milligrams of active ingredient of the test substance administered per kilograms of body weight of the test organisms (mg/kg). The highest dose administered in a typical LD_{50} experiment is chosen so that 90% or more of the animals in the highest dose group will be killed. The choice of the highest dose can be estimated from previous results with chemically similar substances, or by a pilot, range-finding experiment in which a smaller number of animals are exposed to a wide range of dilutions. Then *serial dilutions* of that dose are used to produce a gradient of intermediate responses over four or five doses. Typically, at least 10 animals (ideally, five males and five females) are exposed at each of the six doses, and there is always a negative control group exposed only to the vehicle.

In LD_{50} determinations, the test substance is usually applied as the *technical grade*, the practical grade as manufactured for sale and use (usually of approximately 95% purity). In some cases, an agency might require additional tests with the *analytical grade* (which is greater than 99% pure). If necessary, a sample can be tested by gas chromatographic or high-performance liquid chromatographic analysis to verify the purity. A vehicle and route of administration must be chosen, based in part on the physical properties of the test substance, including solubility in water, organic solvents, and corn oil; melting point, boiling point, and vapor pressure; and color and odor.

To perform the dosing, animals of similar weight are chosen, fasted overnight, numbered, and then assigned to treatment groups using a table of random numbers. Doses are applied beginning with the negative control, which is vehicle only, and then increasing doses of the test substance. (In this way, the same syringe can be used with no risk of contaminating a lower dose with the residue of a higher dose.) For each dose, a solution is prepared so that a small volume of vehicle (e.g., no more than around 2.0 mL) will contain the intended dose when administered to the animal of average size. For example, if a dose of 10 mg/kg is intended and the average rat weighs 200 g (0.2 kg), then the concentration of the test substance in oil solution should be 1.0 mg/mL. If the subject rat weighs 200 g, then 2.0 mL administered will result in the intended dose of 10 mg/kg. If the next rat weighs only 190 g, then the amount administered can be reduced to 1.9 mL to maintain the desired dose of 10 mg/kg.

ANALYSIS

In any population, a very small proportion of individuals are very *susceptible*, whereas another very small proportion of individuals are very *tolerant* to the same dose of the same poison. Some variability is actually *experimental error* due to factors such as the precision in administering the dose, environment of the animal, and condition of the animal, such as fasting and handling of the animal. True heterogeneity is due to the genetic variability in physiological characteristics of the animals (although it is expected that inbred strains of rodents used in laboratory experiments would be much less heterogeneous in response than wild populations of the same species).

This variation in response is observed as the long tails of the dose versus response histogram (Figure 1.1). If tolerances of individuals are *normally distributed* (values are symmetrical around the mean, with 68% of the values falling within one standard deviation of the mean and 98% of the values falling within two standard deviations of the mean) in the population of rats tested (as commonly assumed), then a *sigmoid curve* would describe the accumulated percentage mortality plotted versus the logarithm of the dose (Figure 1.2). The increasing mortality observed at each higher dose is the result of accumulating responses of less tolerant to more tolerant individuals in each group.

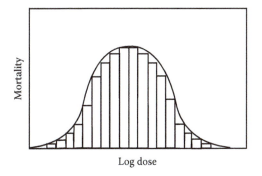

FIGURE 1.1   Histogram showing a normally distributed response to a toxic substance.

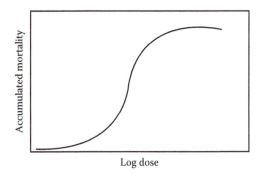

FIGURE 1.2   Graph showing the sigmoid curve characteristic of accumulated responses.

The median lethal dose is often calculated by transforming the accumulated percentage mortality at each dose to a *probit* mortality score. This is a type of probability transformation in which one probit unit is defined as being equal in magnitude to one standard deviation unit of the response. Probit mortality scores plotted versus the logarithm of the dose will produce a straight line from which the median lethal dose can be predicted (Figures 1.3 and 1.4). The slope of the probit plot line

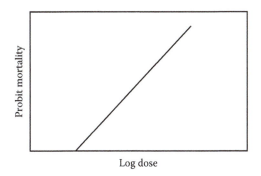

FIGURE 1.3   Log dose versus probit-transformed responses.

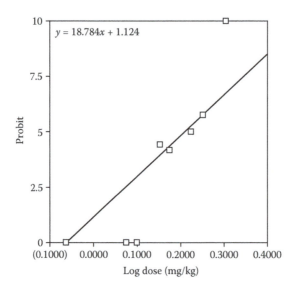

FIGURE 1.4 Mortality of CF1 mice exposed to 4-nitrophenyl methyl(phenyl)phosphinate. (Reprinted from *Toxicol. Appl. Pharmacol.*, 84, Joly, J.M. and Brown, T.M., Metabolism of aspirin and procaine in mice pretreated with O-4-nitrophenyl methyl(phenyl)phosphinate or O-4-nitrophenyl diphenylphosphinate, 523, Copyright 1986, with permission from Elsevier.)

is related to the uniformity of response within the animal population. If the slope of the response were 2.0, a rather typical value, then approximately 68% of the population would be expected to respond to a 10-fold range in doses centered at the median lethal dose. In contrast, if the slope of the response was 6.0, then more than 99.6% of the population would be expected to respond to a 10-fold increase in doses centered at the median lethal dose. As an example, an extremely homogeneous response was observed in CF1 mice exposed to an organophosphorus agent; there was a change of 18.8 probit units of response per 1.0 $\log_{10}$ increase in dose (Figure 1.4).

ALTERNATIVE TESTS

Although the median lethal dose is traditionally a very important value for toxicologists to use when considering the toxicity of a substance, this experiment has been criticized increasingly because of the large number of animals needed to gain a rigorous estimate. If a less precise estimate of the median lethal dose is acceptable, then a substitute called the up-and-down method, which requires fewer animals, can be used. This method was developed to find optimum mixtures of explosive materials with fewer trials and less waste and hazard. In this method, one animal is exposed to one dose of the substance. If it survives, then a second animal is exposed to a higher dose. If the second animal survives, another higher dose is administered until mortality is observed. Following mortality, the next animal is exposed to a lower dose, and following survival, the next animal is given a higher dose, until an equilibrium is observed.

For fewer than 10 animals, the population median lethal dose can be estimated by the up-and-down method with accuracy similar to that of the full-scale LD_{50}

experiment exposing more than 60 animals. One deficiency of this test, however, is that the variability of response cannot be estimated. Another disadvantage is the extra time required in waiting to score one animal before determining the dose for the next animal; however, this problem can be partially solved by reducing the observation period. Despite these problems, for most routine purposes of comparing the toxicity of poisons, the up-and-down method will provide sufficient information while sacrificing far fewer animals.

Biotechnology is also being applied to provide more sensitive test animals to supplement conventional testing. Lethality can be highly dependent on biotransformation of the substance of interest. This can be controlled by engineering for a high or low capacity of biotransformation. Also, tumor suppressor capacity or DNA repair capability can be negated in the animal to increase sensitivity in detecting carcinogens. For example, genetically modified mice in development include the p53+/−, Tg.AC, and Tg.rasH2 assays; they are to be used to supplement information gained in the conventional inbred strains and not to replace them.

CATEGORIES OF TOXICITY

A somewhat arbitrary system of toxicity ranking has evolved on the basis of the median lethal dose of a substance. A substance with a median lethal dose of less than 1 mg/kg is considered to be extremely toxic, and various definitions of highly toxic, moderately toxic, and slightly toxic have been proposed. Generally, highly toxic substances have a median lethal dose of less than 50 mg/kg, moderately toxic have a median lethal dose of less than 500 mg/kg, and slightly toxic have a median lethal dose of greater than 500 mg/kg and up to approximately 5 g/kg, which approaches the practical limit of most dosing techniques.

NO OBSERVED ADVERSE EFFECT LEVELS

Another common approach for comparing toxicity is to measure the *"no observed adverse effect level"* (NOAEL) for a compound. The definition of an NOAEL is the highest dosage tested at which the response measured is not significantly different statistically from the control (more about statistics later). Another measure, the *"lowest observed adverse effect level,"* or LOAEL is the lowest dosage tested at which the response is significantly different from the control. Of course, this approach also has its limitations, including the fact that careful selection of dosages to be tested is necessary to effectively determine the threshold between no effect and effect.

MIXTURES

Mixtures of poisons can be more toxic or less toxic than predicted from the toxicity of the individual components of the mixture. This phenomenon of increased toxicity of a mixture is known as *synergism,* and it results from an interaction of one component with the pharmacokinetics or pharmacodynamics of the second component. For example, the first component might interfere with the elimination of the second component, so that a given exposure of the second component produces higher concentrations in the body when applied in the mixture. *Antagonism* is the observation of less than predicted toxicity from a mixture, e.g., when one component induces a higher rate of inactivation of the second component, resulting in a higher concentration of a less toxic metabolite.

Drugs, pesticides, industrial chemicals, etc., when used in mixtures or when given in simultaneous exposure, should be evaluated empirically for interactions to determine whether there is synergism or antagonism. When one component is nontoxic, this test is relatively simple: the nontoxic component can be administered at a high concentration with the complete range of doses of the toxic component. If the dosage–mortality line of the mixture differs significantly from that of the toxic component alone, then an interaction is indicated. If both components are toxic, then the test for an interaction is more complex. One approach is to prepare a mixture containing each component at its median lethal dose. Dilutions are made and administered; then the observed dosage–mortality line is compared with a line predicted by adding the expected mortalities for the individual components at each dilution.

TOXICITY, HAZARD, AND RISK

TOXICITY AND HAZARD

For any substance, the term *hazard* can be used to describe the actual risk of poisoning. Thus, an estimate of toxicity is not a direct estimate of hazard. In fact, toxicity is only one variable to be considered in predicting how hazardous a substance will be during practical use. Another significant variable that must be considered is the potential level of human exposure to the substance. This must be predicted on the basis of factors such as the concentration and circumstances of use of the substance. Although the intrinsic toxicity of a substance cannot be altered because it is a basic property of that substance, it is nonetheless possible to reduce the hazard of a toxic substance by reducing the practical risk of exposure. A simple example is the invention of childproof packaging of nonprescription drugs, which reduces the hazard associated with some drugs by making access to the drug more difficult. In another example, hazards posed by pesticides to the pesticide applicator have been reduced by preparing the pesticide in dissolvable polymer bags containing premeasured quantities designed to be dropped into the sprayer tank without opening. This innovation greatly reduced the risk of exposure to formulated pesticide concentrates by eliminating measuring and mixing by the applicator.

THE ROLE OF LABORATORY TESTING IN ESTIMATION OF HAZARD

Toxicological data from laboratory studies such as those described here are often used by regulatory agencies in an attempt to estimate the hazard to human health posed by a particular toxicant. Even though there may be issues in extrapolating from animal data to humans, and from higher to lower exposure levels, these studies are still extremely useful in estimating human hazard.

To help with both experimental design and interpretation of toxicological data, the mathematical tools of *statistics* are used. In terms of experimental design, statistics can help with issues such as randomization of subjects and choice of sample size. Then, because toxicological data sets are often quite large, *descriptive statistics* are useful to help summarize the data. Examples of descriptive statistics are the *mean, standard deviation* (SD), or *standard error of the mean* (SEM). The SEM is defined as

$$ \mathrm{SEM} = \frac{\mathrm{SD}}{\sqrt{N}} $$

where N is the number of data points in the data set.

Statistics can also be used to help identify *differences* and *trends* in data sets. Normally distributed data that are continuous (such as weight and volume) may be analyzed using *parametric statistics. Nonparametric statistics* are used to analyze data sets that are not normally distributed or that are made up of *discrete data* (data that occur only as integer values).

Tests such as *Student's t-test* (a parametric test) or the *Mann–Whitney U-test* (a nonparametric test) test the hypothesis that two sets of data are *significantly different.* These tests deliver a *p*-value that tells you the probability that the differences between the two groups are simply due to random chance. General scientific consensus states that if there is a 5% or less chance that the differences between the two sets of data are due to chance (in other words, a *p*-value less than or equal to 0.05), then the difference can be termed significant. Multiple groups can be compared using *analysis of variance* (ANOVA) tests, of which there are both parametric and nonparametric forms. If the results of an ANOVA indicate a significant difference, a variety of tests known as *post hoc tests* may be used to further analyze where the differences lie.

One final use of statistics is in the analysis of trends. Tools such as *linear regression analysis* can help determine the relationship between two variables such as, for example, dose and response. A statistic called the *correlation coefficient* (also known as r^2) measures the accuracy with which the data fit the hypothesized linear relationship.

It can be quite difficult to extrapolate from laboratory studies to real-world situations, which is one reason why the processes of risk assessment and management are often fraught with controversy. Although laboratory animals can serve as models for humans in toxicological testing, species differences do exist. Also, many scientists have criticized the practice of using very high doses of toxicants during laboratory testing and then attempting to apply the results to a situation in which human exposure levels are actually very low. Therefore, data must be interpreted with caution.

Epidemiological Data

Along with laboratory data, data from *epidemiological* studies are also used in risk estimation. These studies examine the relationship between exposure to a toxicant (usually accidental or voluntary exposure) and either disease *incidence* (the rate at which new cases of the disease appear in a human population) or disease *prevalence* (the number of existing cases of the disease at a particular point in time).

Epidemiological studies do have some drawbacks. Because of the variability in genetic and environmental factors between individual humans, it can be extremely difficult to be sure that differences in disease incidence or prevalence between exposed and control groups are really due to the factor being tested and not to some other *confounding factor* (a factor that can cause a difference between the groups, but is not the one being tested). Also, exposure levels may be difficult to estimate (particularly if exposure to the toxicant occurred sometime in the past). To maximize the reliability of results, exposed and control groups are often matched as closely as possible for potential confounding factors such as age, sex, lifestyle factors, working conditions, or living conditions. Also, the larger the number of individuals participating in the study, the easier it is to detect small differences between the exposed and control groups.

Recently, the technique of *meta-analysis* has been added to the tools of epidemiologists. This technique involves combining results from different studies in order to acquire the statistical power necessary to determine whether two groups (generally exposed and control) differ with respect to development of disease. A meta-analysis, however, is only as good as the data that it is based on, and there are many disagreements as to how to select which studies to include. Even in published papers, data are not always complete, and in fact, there may be some bias inherent in the pool of available published papers, as negative results may be less publishable than positive results.

Finally, one additional caveat must be kept in mind in terms of epidemiological studies (and, of course, of laboratory studies as well). This is the important concept that *correlation is not causation*. Even if a so-called risk factor is shown to be positively associated with an increased risk of disease, it does not necessarily mean that the risk factor causes the disease.

RISK ASSESSMENT AND RISK MANAGEMENT

In the process of *risk assessment*, hazard is weighed against benefit as regulatory decisions are made concerning potentially toxic substances. The National Academy of Sciences/National Research Council published a report in 1983 outlining the steps involved in risk assessment and risk management. They identify four main components of the process of risk assessment: (1) *hazard identification*, in which it is determined whether a substance is a potential health hazard; (2) *dose–response evaluation*, in which the dose–response relationship is quantified; (3) *exposure assessment*, in which potential exposure levels are estimated; and (4) finally, this information is merged in the process of *risk characterization*, in which effects on the exposed population are estimated. Descriptions of risk are often phrased in terms of the chances of contracting a particular disease during a lifetime of exposure to a particular toxicant at a given level of exposure.

Risk assessment is then followed by *risk management*, which is the process by which regulatory decisions are made concerning health risks. Risk management takes not only risk assessment results but also other political, social, and economic factors into account when making decisions about regulating potential toxicants. Government agencies involved in risk management include the *Occupational Safety and Health Administration* (OSHA), the *Food and Drug Administration* (FDA), and the *Environmental Projection Agency* (EPA).

CASE STUDY   Risk, Perception, and Vaccination

Risk assessment, the systematic scientific evaluation of risk, is not the only thing that must be considered in risk management. Another factor is *risk perception*, which is what people perceive risk to be and which may or may not align with actual risk. There are quite a number of factors that have been shown to influence people's perception of risk. Many people tend to underestimate risks involving technologies with which they are familiar (lawnmowers and bicycles) and to overestimate risks associated with new or unfamiliar technologies (nuclear power). This is also tied to the fact that many people tend to overestimate risks in situations in which strong emotions such as a feeling of *dread* are involved. These emotions, of course, are much more likely to be associated with unfamiliar than with familiar situations.

Two other major factors in risk perception are *choice* and *control*. Many people tend to underestimate risks under circumstances which are voluntary (smoking) or in which they have a high degree of control (driving a car) versus circumstances which are involuntary (pesticide residues in food) or in which they do not have a high degree of control (riding in a commercial airliner).

Finally, another factor influencing risk perception is *availability*. If we have heard fairly frequently about an event happening, we judge it as more likely to happen than other events that we hear about less often. However, availability as a measure of probability is often skewed by whether or not a given event is "newsworthy" (generally meaning catastrophic). Thus, death by plane crash may be judged as a much greater risk than death by automobile accident, because news about plane crashes is featured on the first page of newspapers and websites worldwide (generally for days or weeks following a single incident), whereas news of highway fatalities is given much less prominence. The internet can play a particularly significant role with respect to the availability factor, as it allows a platform for the proliferation of websites which promote a particular viewpoint with no guarantee of scientific validity.

To see how this applies to a real example, we can take the case of vaccination. Although overwhelming medical evidence supports the efficacy of vaccination in disease prevention, as well as confirming the low risk of vaccine-related complications, an anti-vaccine community continues to exist. Medical professionals have debated the reasons for this, but one of the most insidious is likely to be the availability factor.

First of all, as disease rates drop due to successful vaccination campaigns, health issues which follow vaccination (whether or not they are actually related to the vaccination), may seem on the basis of availability to be a greater risk than complications from the disease itself. Additionally, anti-vaccine websites abound on the internet. One study (Kata, 2010) found that out of internet sites available in the United States appearing in Google searches on vaccination, 24% of the top 10 results for "vaccine" were anti-vaccination, whereas an astounding 75% of the top 10 results for "vaccination" were anti-vaccination. Many of these sites were found to feature not only misinformation and misrepresentation of fact, but also emotional appeals and testimonials of parents who blamed vaccines for children's health problems (typically of a frightening nature, thus invoking the dread factor). Another theme found was the objection to the mandatory nature of immunization, claiming an infringement on parental rights (the choice/control factor at play). Thus, education alone is unlikely to solve this issue, unless the other factors contributing to risk perception are addressed as well.

BIBLIOGRAPHY

Beck, B.D., Slayton, T.M., Calabrese, E.J., Baldwin, L., and Rudel, R., The use of toxicology in the regulatory process, in *Principles and Methods of Toxicology*, Hayes, A.W., Ed., Taylor & Francis Group, Philadelphia, 2001, chap. 2.

Bolon, B., Genetically engineered animals in drug discovery and development: A maturing resource for toxicologic research, *Basic Clin. Pharmacol. Toxicol.*, 95, 154, 2004.

Braun, L.A., Risky business—or is it? Risk perception and integrative medicine, *Adv. Integr. Med.*, 1, 48, 2014.

Bruce, R.D., An up-and-down procedure for acute toxicity testing, *Fundam. Appl. Toxicol.*, 5, 151, 1985.

Faustman, E.M. and Omenn, G.S., Risk assessment, in *Casarett and Doull's Toxicology*, Klaassen, C.D., Ed., McGraw-Hill, New York, 2013, chap. 4.

Gad, S.C., Statistics for toxicologists, in *Principles and Methods of Toxicology*, Hayes, A.W., Ed., Taylor & Francis Group, Philadelphia, 2001, chap. 7.

Joly, J.M. and Brown, T.M., Metabolism of aspirin and procaine in mice pretreated with O-4-nitrophenyl methyl(phenyl)phosphinate or O-4-nitrophenyl diphenylphosphinate, *Toxicol. Appl. Pharmacol.*, 84, 523, 1986.

Kata, A., A postmodern Pandora's box: Anti-vaccination misinformation on the Internet, *Vaccine*, 28, 1709, 2010.

MacDonald, J., French, J.E., Gerson, R.J., Goodman, J., Inoue, T., Jacobs, A., Kasper, P. et al., The utility of genetically modified mouse assays for identifying human carcinogens: A basic understanding and path forward, The Alternatives to Carcinogenicity Testing Committee ILSI HESI, *Toxicol. Sci.*, 77, 188, 2004.

MacDonald, N.E., Smith, J., and Appleton, M., Risk perception, risk management and safety assessment: What can governments do to increase public confidence in their vaccine system?, *Biologicals*, 40, 384, 2012.

McGonigle, P. and Ruggeri, B., Animal models of human disease: Challenges in enabling translation, *Biochem. Pharmacol.*, 87, 162, 2014.

Morton, M.G., Risk analysis and management, *Sci. Am.*, July 1993, p. 32.

Rider, C.V., Boekelheide, K., Catlin, N., Gordon, C.J., Morata, T., Selgrade, M.K., Sexton, K., and Simmons, J.E., Cumulative rise: Toxicity and interactions of physical and chemical stressors, *Toxicol. Sci.*, 137, 3, 2014.

Robertson, C., Idris, N.R.N., and Boyle, P., Beyond classical meta-analysis: Can inadequately reported studies be included?, *Drug Discov. Today*, 9, 924, 2004.

Scala, R.A., Risk assessment, in *Casarett and Doull's Toxicology*, Amdur, M.O., Doull, J., and Klaassen, C.D., Eds, Pergamon Press, New York, 1991, chap. 31.

Sim, M., Case studies in the use of toxicological measures in epidemiological studies, *Toxicology*, 181/182, 405, 2002.

Simon, T., *Environmental Risk Assessment. A Toxicological Approach*, CRC Press/Taylor & Francis Group, Boca Raton, Florida, 2014.

Slovic, P. and Peters, E., Risk perception and affect, *Curr. Direct. Psychol. Sci.*, 15, 322, 2006.

Smith, G.D., Reflections on the limitations to epidemiology, *J. Clin. Epidemiol.*, 54, 325, 2001.

2 Toxicokinetics

INTRODUCTION

Interactions of a poison with an organism can be considered in three phases: exposure, toxicokinetics, and toxicodynamics. During the *exposure phase*, contact is established between the poison and the body via one or more routes, for example, a volatile air pollutant is inhaled into the body. Then, during the *toxicokinetic phase*, the poison undergoes movement (Greek: *kinesis*) within the body. This movement includes absorption into the circulatory system, distribution among tissues (including those that will serve as sites of action), and then elimination from the body. The *toxicodynamic phase* is the exertion of power (Greek: *dynamos*) of the poison through its actions on affected target molecules and tissues. These phases can be overlapping, so that once exposure occurs, all phases of action can be in effect simultaneously in the body.

PHARMACOKINETICS AND TOXICOKINETICS

The principles of *toxicokinetics*, like most principles of toxicology, are derived from the science of pharmacology and *pharmacokinetics*, which is the study of how drugs enter, move through, and exit the body. Toxicology and pharmacology are naturally related because although most drugs are therapeutic (medicinally effective) over a narrow range of doses, they are also toxic at higher doses. In fact, many clinically important drugs are moderately toxic; therefore, they must be administered very carefully to avoid reaching toxic concentrations in the body. Conversely, some well-known poisons are therapeutic at lower doses, as seen in the recent development of *tetrodotoxin* as an analgesic, *botulinum toxin* as a cosmetic, and *ivermectin*, a veterinary anthelmintic (a killer of parasitic filarial worms), as a larvicide for human onchocerciasis (blinding filarial disease, discussed subsequently).

Although the basic principles of pharmacokinetics and toxicokinetics are fundamentally the same, there are some differences that can be seen in their application. A very practical problem in pharmacology, for example, is encountered in determining how to administer repeated doses of a drug in order

Tetrodotoxin

See also:
   Effects of Toxicants on Voltage-Activated Ion Channels
   Chapter 4

   Effects of Toxicants on Electrical Conduction
   Chapter 10

   TTX, STX
   Appendix

Botulinum Toxin

See also:
   Acetylcholine
   Case Study: Botulinum Toxin
   Chapter 10

   Bacterial Toxins
   Chapter 20

15

to maintain the proper therapeutic concentration in the bloodstream. To avoid toxicity, the physician must understand how the concentration changes to predict the amount of the dose and the interval between doses.

In toxicokinetics, a more typical problem might be to estimate how concentrations of a toxicant may change over time. For example, one might want to see whether a toxicant is quickly excreted or whether repeated exposures will lead to accumulation in the body.

There are also differences in chemical properties between drugs and other types of toxicants. There are several environmentally important toxic substances (such as some pesticides, for example) that have both high levels of *lipophilicity* (solubility in lipid) and high levels of *persistence* (due to slow degradation rates). These properties result in very slow toxicokinetics compared with most pharmaceuticals. Many common drugs may be absorbed, exert an effect, and be eliminated in hours or even minutes. By contrast, some lipophilic substances found in the environment may be very slowly accumulated, and once exposure is terminated, they may be very slowly eliminated over days or months.

This chapter examines the principles of absorption, distribution, and excretion of toxicants and looks at some simple mathematical models used to study the way toxicants enter, move through, and exit the body.

ABSORPTION

There are several possible routes of *absorption* for toxicants. What all routes have in common is that they present a cellular barrier that toxicants must cross to enter into the bloodstream. Thus, toxicants that can easily cross cell membranes (through *simple diffusion* or other transport processes) can be more easily absorbed than those that cannot. The primary chemical property that enhances the ability of a toxicant to diffuse across biological membranes is lipophilicity (the ability to dissolve in lipids such as the phospholipids, which make up the bulk of the membrane). A measure of this is the *octanol/water partition coefficient or partition ratio*, which is calculated by measuring the ratio of the concentration of a compound in each phase in a mixture of those two liquids at equilibrium. Often this is expressed as the logarithm of the measured ratio and abbreviated "log P." Lipid soluble compounds have a high positive value of log P, whereas the log P value of water soluble compounds is negative.

Other properties that aid in diffusion include small size (which allows the toxicant to fit more easily between membrane molecules and through aqueous pores) and neutral charge (which allows the toxicant to avoid interactions with charged groups on membrane molecules). The pH at which a weak acid or base is 50% ionized is termed the pK_a or pK_b, respectively. Strong acids have a low pK_a, and strong bases have a high pK_b. The pK_a or pK_b can be used to calculate the percentage of ionization of any compound at a given pH. pH varies in different regions of the GI tract, as well as between species. This may contribute to species differences in the absorption of compounds.

There are other mechanisms through which molecules can pass the membrane barrier. These include *facilitated diffusion* and *active transport*. Active transport and facilitated diffusion both require the participation of protein carriers within the membrane and are generally quite limited in the molecules that they can carry. Thus, toxicants that can enter through these mechanisms are generally those that are structurally quite similar to the molecule normally moved by the carrier. For example, heavy metals such as lead may enter through the transporter that normally carries calcium. Facilitated diffusion does not require energy and moves molecules down the concentration gradient. Active transport, however, can move molecules against a gradient and requires energy.

Examples of active transport systems important in toxicokinetics include the *P-glycoprotein transporter*, one of a class of molecules collectively known as *ATP-binding cassette (ABC) transporters*. Another group of transporters is the *solute carriers (SLCs)*, which move organic ions and play important roles in the kidney and liver.

The three major physiological routes of absorption for toxicants are *oral, respiratory*, and *dermal*. Other routes include various sites for injection such as *intramuscular, intraperitoneal* (into the peritoneal cavity, the space that contains the intestines), *intravenous*, or *subcutaneous*.

P-Glycoprotein

See also:
The Blood–Brain Barrier
Chapter 10

Cyanobacterial Toxins
Chapter 20

THE ORAL ROUTE OF ABSORPTION

One common route of absorption for toxicants is the oral route. Absorption can occur all along the GI tract from the mouth to the large intestine. The opportunity for absorption in any region of the GI tract increases with the time a toxicant spends in that region. Thus, little absorption usually occurs in the mouth because of the limited time a toxicant spends there, although the much longer time it takes a toxicant to move through the small intestine gives plenty of opportunity for absorption to take place. Also, the surface area of the region is a factor, as is pH. Some toxicants tend to *ionize*, which is to gain or lose electrons and thus becomes negatively or positively charged ions. Weak bases ionize in low-pH (acidic) environments, whereas weak acids ionize in high-pH (alkaline) environments. Because ionization decreases the likelihood of absorption, weak bases are more likely to be absorbed in the higher pH found in the small intestine, whereas weak acids are more likely to be absorbed in the lower pH found in the stomach. When all factors are considered, most absorption of toxicants occurs in the small intestine (as is the case with absorption of nutrients from food).

The small intestine has a high surface area, a neutral pH of approximately 6, and is highly vascularized (contains many blood vessels). Most toxicants cross the epithelial cells lining the small intestine and enter the bloodstream by diffusion, although a few enter through specific transport mechanisms such as active transport or facilitated diffusion, where they may compete for absorption with the nutrients that are the normal substrates for those transporters.

Typically, a portion of an orally administered drug or toxicant will be unavailable for absorption into the general circulation. Some of the substances will simply continue through the alimentary tract. Factors that can influence absorption include the presence or absence of other materials in the GI tract, the presence or absence of disease, and the age of the individual. Recently, more and more attention is being paid to the role of normal gut flora in GI processes. Gut microbes can metabolize both nutrients and toxicants, potentially influencing both absorption and toxicity. Because microflora can vary between species, this is another potential contributor toward species differences in toxicant absorption.

The presence of chemicals in the GI tract can also influence absorption. For example, EDTA (a chelator) can increase general permeability of the intestine, most likely through binding of calcium (which plays an important role in the maintenance of tight junctions). The presence of EDTA may also enhance absorption of metals by binding to them and increasing their relative lipid solubility. Interestingly, even such common foods as grapefruit juice may impact drug and toxicant absorption. One of the components of the juice (*naringin*) inhibits the transporters which pump drug substrates out of GI cells, thus leading to increased absorption of the drugs. This has, in some cases, led to unexpected toxicity in patients who are unaware of this interaction.

Also, as blood from the lower GI tract travels through the portal vein to the liver prior to traveling to the rest of the body, a portion of the substance may be metabolized and deactivated in the liver prior to reaching other tissues. This phenomenon is known as the *first-pass effect*, and it means that some drugs or toxicants may have less of an effect when administered orally than when administered through another route.

RESPIRATORY ROUTE OF ABSORPTION

Absorption of Gases and Particulates

See also:
   Exposure to Respiratory Toxicants
                              Chapter 8

Another important route of absorption is through the respiratory system. Inhaled gases pass through the nose, pharynx, larynx, trachea, and bronchi prior to entering the lungs. Water soluble gases tend to be absorbed in the watery mucus that lines the upper part of the respiratory tract, while less water soluble gases continue into the lungs. With particles, size is the determining factor. Larger particles are screened out by cilia and mucus in the upper part of the tract, whereas smaller particles continue into the lungs for absorption. In the lungs, barriers to diffusion are few, because the cells that line the lungs are thin and located in very close proximity to blood vessels (facilitating the transfer of oxygen).

DERMAL ROUTE OF ABSORPTION

Of the three major routes, the dermal route provides the greatest cellular barrier to absorption of toxicants. The skin consists of an epidermal layer made up of many layers of epithelial cells and a dermal layer made of connective tissues. Because blood vessels are located in the dermal layer, a toxicant must pass through the epidermis first in order to be absorbed. The top layer of epidermal cells provides the greatest barrier to diffusion, because these cells not only have thickened cell membranes but are also filled with a protein called *keratin*. These modifications effectively block all but the most lipid soluble substances from penetrating farther into the epidermis or dermis. Factors that influence dermal absorption include differences in thickness and degree of keratinization between skin in different body regions as well as the condition of the skin. Injuries (e.g., scrapes, burns, and cuts) that remove the keratinized layer of epithelial cells can significantly increase the potential for dermal absorption. There are, of course, also species differences in epidermal and dermal structures that can impact absorption; on average, the dermis of laboratory animals is thinner than that in humans, but is commonly covered with hair. This, along with differences in dermal blood flow (human dermis is more highly vascularized than that of most laboratory animals), complicates the use of laboratory animals as good models for dermal absorption.

DISTRIBUTION

When a drug is administered intravenously (i.v.) at a known dose, it is assumed to be instantaneously dissolved in the blood or plasma. (The *plasma* is the fluid portion of the blood, whereas the *serum* is the more clear supernatant portion remaining after removal of the fibrin clot and coagulated cells by centrifugation.) Plasma can then be sampled immediately upon administering the dose and at intervals after dosing. The concentrations of the toxicant in the plasma are measured, and the concentration at time zero can be found by extrapolation. The *apparent volume of distribution* of the toxicant is found by dividing the amount of drug in the body (the total amount of drug administered, in milligrams) by the concentration in plasma at time zero. If all of a drug remained in the plasma, this volume of distribution would be equal to the volume of plasma (which for a 70 kg human is approximately 3.0 L). For example, if 3 mg was administered to a 70 kg human and the concentration at time zero was 1.0 mg/mL, then the apparent volume of distribution would be 3.0 L (or 0.043 mL/kg).

Frequently, however, the measured value for apparent volume of distribution of a drug or toxicant exceeds the volume of plasma in the body. This indicates that some of the drug or toxicant has left the plasma and has entered the fluid between and within the cells. Thus, volume of distribution gives an indication of the ability of a drug or toxicant to leave the bloodstream and distribute into the tissues.

The rate at which a drug or toxicant is distributed to the various tissues of the body depends on several factors. Of course, the ability of the drug or toxicant to cross cell membranes is one primary factor. Another important factor is the rate at which blood is supplied to a given tissue. Tissues with a high rate of *perfusion* (rate of blood flow), such as the heart, liver, and kidney, will thus also have a high rate of delivery of toxicants.

There are some tissues, however, that have anatomical and physiological modifications that act to limit the delivery of toxicants. The tissues of the central nervous system, for example, are protected by a system known as the *blood–brain barrier*. This barrier is a result of tighter junctions between cells

Blood–Brain Barrier

See also:
   The Blood–Brain Barrier
                                        Chapter 10

that make up the capillaries in the central nervous system, as well as the wrapping of capillaries in a cellular blanket that increases the width of the cellular barrier to diffusion that toxicants must cross. There are also transporters (such as ABC transporters) that act to limit accumulation of drugs and toxicants in the brain tissue. As a result, only highly lipid soluble toxicants have easy access to the central nervous system.

There are also some toxicants with high affinity for certain tissues. For example, many toxicants bind to plasma proteins such as *albumins*. This tends to hold the toxicant in the bloodstream and to delay release to other tissues. Liver and kidney tissues contain proteins called *metallothioneins*, which bind and hold heavy metals such as cadmium and zinc. Heavy metals may also accumulate in bone, and highly lipid soluble compounds (such as the insecticide DDT) can accumulate in fat tissues. The tissues where toxicants accumulate are sometimes referred to as storage depots.

ELIMINATION

Biotransformation

See also:
 Biotransformation
                            Chapter 3

Elimination is the loss of the parent drug or toxicant from the body due either to *biotransformation* of the *parent drug* to *metabolites* or to excretion of the parent drug in the urine or feces. Metabolites are products of chemical changes in the drug that are catalyzed by various enzymes in the body. These products may vary in toxicity and therapeutic effect from the parent drug. When biotransformation results in a less toxic product, the process is detoxification; however, some reactions form more toxic products from the parent and are known as intoxication or bioactivation reactions.

Urinary Excretion

See also:
 Function of the Kidneys
                            Chapter 12

The major route of excretion of most drugs and toxicants is through the kidneys. Drugs or toxicants and their metabolites are filtered from the blood and excreted in the urine. Other drugs and toxicants and their metabolites may be removed from the blood by the liver, excreted into the bile, and eliminated through the GI tract. Other routes of elimination include through the respiratory system or through secretions such as saliva, sweat, or milk.

TOXICOKINETIC MODELS

MATHEMATICAL MODELS OF ELIMINATION

Many common drugs and toxicants are water soluble and are carried through the body in the blood. The blood-borne concentration can be measured by *chromatographic analysis* of a small sample of blood. Administration of a drug or toxicant is followed by very rapid absorption into the blood (which results in increasing concentration) and the slower elimination of drug from the body (which causes the concentration in the blood to decline). These processes can be mathematically modeled, and the *elimination rate* can be used to predict drug or toxicant concentrations.

Absorption and elimination are opposite processes, and as such, they cannot be estimated when both are in operation. To measure elimination from the blood without needing to take into account absorption, a drug or toxicant can be injected intravenously. In a typical experiment, the drug or toxicant is introduced into a large vein of a rat by injection in a small volume of carrier. Assuming that the drug or toxicant has a high solubility in water, it will dissolve throughout the blood almost instantaneously and thereafter will decline in concentration. Repeated sampling of the blood and measurement of the declining concentrations with time will then allow the elimination kinetics to be determined.

The rate of elimination of a drug, where C is the concentration of the drug in the body, can be described mathematically as

$$\frac{\Delta C}{\Delta t}.$$

The rate of elimination bears units such as mg $*$ kg^{-1} $*$ min^{-1} or mg $*$ kg^{-1} $*$ h^{-1}.

For some drugs or toxicants, the rate of loss from the body is constant over time, independent of the concentration of the drug in the body. In this case, the elimination is described as a *zero-order kinetic process* (Figure 2.1). In practice, few drugs are eliminated in a zero-order process. One that, however, is ethanol. Although unusual, ethanol pharmacokinetics are very important due to the common problem of excess alcohol drinking. Unfortunately, a high dose of ethanol will not be eliminated at a higher rate than a normal dose because the elimination rate is independent of the concentration in the body. For the zero-order process, since the drug or toxicant will be eliminated at the same rate continuously, even as the concentration declines, the equation becomes

Ethanol

See also:
Examples of Teratogens
                                        Chapter 7

Cardiomyopathies and Other
    Effects on Cardiac Muscle
                                        Chapter 9

Other Neurotoxicants
                                        Chapter 10

Fatty Liver
Liver Cell Death: Necrosis and
    Apoptosis
Fibrosis and Cirrhosis
                                        Chapter 11

Forensic Toxicology and Alcohol
    Use
                                        Chapter 16

Ethanol
                                        Appendix

$$\frac{\Delta C}{\Delta t} = -K_0,$$

where K_0 is the *zero-order elimination rate constant*. Knowing the concentration of the drug in the body at any given time, C_o, and the rate of elimination (in the case of zero-order kinetics, $-K_0 * t$), the concentration at a later time, t can be predicted by the equation:

$$C_t = -(K_0 * t) + C_o.$$

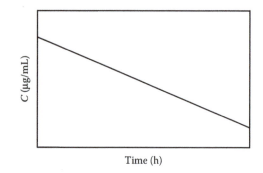

FIGURE 2.1   Zero-order process of elimination.

The time required to eliminate a given portion of the drug via a zero-order process will vary and will lengthen as the initial concentration is increased.

If the rate of elimination of a drug or toxicant is first very rapid and then becomes less rapid with subsequent sampling, it might follow a *first-order process of elimination*. The first-order process is observed as a logarithmic decay in the concentration of the drug in the body, in which the rate of loss is dependent on the concentration of the drug in the body (Figure 2.2). The mathematical relationship describing the first-order elimination kinetics is

$$\frac{\Delta C}{\Delta t} = -KC.$$

In this case, the proportion (not the amount) of the drug lost per unit time is constant. The exponential change in the concentration in the body is a constant value, the *first-order elimination rate constant, K*. Units of K are per time, such as min^{-1} or h^{-1}.

To predict the concentration of the drug after a period of time, the following equation is used:

$$\ln C = -K*t + \ln C_o,$$

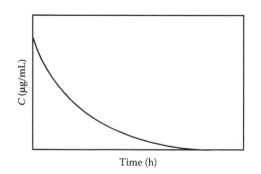

FIGURE 2.2   First-order process of elimination, arithmetic plot.

where C_0 is the initial concentration of the drug or toxicant in the plasma. This equation may be more useful in the following form, in which the natural log (ln) has been converted to the base 10 log (log):

$$\log C = \frac{-K*t}{2.303} + \log C_0.$$

This equation takes the form of a straight line if log C is plotted against time (Figure 2.3). (An easy way to do this is to use semi-logarithmic paper, which makes the logarithmic conversion automatically.) The slope of this line is $-K/2.303$.

The *elimination half-life* of a drug is the amount of time it takes for the original concentration C_0 to fall by one half and can be calculated by using the following equation:

$$t_{1/2} = \frac{0.693}{K}.$$

The time it takes for a drug or toxicant to be nearly totally eliminated (in other words, for C to fall to nearly zero) can be estimated to be approximately $7 \times t_{1/2}$.

COMPLICATING FACTORS

Most drugs are eliminated in a process that resembles first-order pharmacokinetics; however, in practice, the process is usually more complicated. It is not unusual, for example, for drug or toxicant kinetics to vary depending on dose. Some, for example, are eliminated following first-order elimination kinetics at lower doses, but show zero-order kinetics at higher doses. Dose-dependent kinetics are often seen when the mechanism of elimination is saturable. As long as the concentration of the compound of interest remains below the saturation level, an increase in the concentration will result in an increase in the elimination rate (first order). However, once the elimination mechanism is saturated, raising the concentration cannot produce any

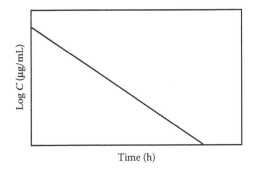

FIGURE 2.3 First-order process of elimination, logarithmic plot.

additional increase in the rate of elimination, resulting in a flat maximum rate (zero order). This, of course, complicates the modeling process.

Also, rather than behaving as if the body were only one single *compartment*, most drugs and toxicants actually move between multiple body compartments. Typical experimental results suggest that there may be two, three, or even more first-order processes by which the drug is eliminated. This would correspond to a body containing several compartments, each eliminating the drug by a slower first-order process.

As described earlier, the first-order elimination constant from a single compartment can be determined simply from the log concentration versus time plot of the observed data. When a *multicompartment model* is necessary, then only the final rate constant, representing loss from the slowest compartment, can be determined directly from a plot. It should be noted that the experimental detection limits for the compound of interest in the plasma might obscure the lower range of this experiment, so that a slow compartment might be observable for some drugs and not observable, but still present, for other drugs. Prior to final elimination, the observed data represent simultaneous loss from multiple compartments, and faster processes must be estimated by working back from the final compartment by a more complex mathematical modeling process.

Finally, exposure to some drugs and toxicants occurs in a repeated or chronic fashion. This leads to a pattern of rising and falling concentrations culminating in a steady state where plasma concentrations vary between an C_{max} and a C_{min}. The magnitude of this range depends on the elimination rate, as well as the timing and dosage of the repeated exposures.

CLEARANCE

Clinical administration of drugs in a multiple-dose regimen requires the estimation of *drug clearance*, which is the volume of blood or plasma that would be cleared (purged) of the drug per unit time to account for the elimination of the drug. Total *systemic clearance* is simply the dose (e.g., in mg/kg) divided by the area under the plasma concentration versus time elimination curve (AUC) for that dose (generally expressed as mg $* mL^{-1} *$ min)

$$\text{Clearance} = \frac{\text{Dose}}{\text{AUC}}.$$

Thus, the typical value for clearance is given in units of mL $* min^{-1} * kg^{-1}$. Considering that AUC is inversely related to clearance, we see that for two drugs administered at identical doses, the one with the smaller AUC will have a larger value for clearance.

ABSORPTION AND BIOAVAILABILITY

However, because not all drugs or toxicants are administered intravenously, methods have been developed to quantify rates of absorption as well as elimination. For

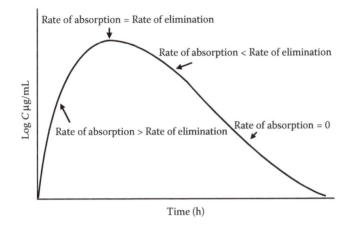

FIGURE 2.4 Plot of absorption and elimination of an oral dose.

typical drugs or toxicants, absorption from the gut is a faster process than elimination. Thus, the concentration of the drug in the blood will initially depend on both the absorption and elimination rates and will rise rapidly to some peak concentration at which the rate of absorption is equal to the rate of elimination. Concentrations will then fall again, as elimination becomes the dominating process (Figure 2.4). Because absorption and elimination are simultaneous processes from an oral dose, it is more accurate and far simpler to estimate an elimination rate from the intravenous experiment described earlier rather than attempting to derive it from an oral dose experiment. From the elimination rate and the data obtained in the oral dose experiment, the absorption rate can then be determined also.

The AUC when the blood concentration of a drug or toxicant is plotted versus time can also be used to estimate drug bioavailability. *Bioavailability* describes the extent to which a drug or toxicant is absorbed orally compared with the intravenous administration and is defined as

$$\frac{AUC_{oral}/Dose_{oral}}{AUC_{i.v.}/Dose_{i.v.}}.$$

The closer this number is to 1, the better the absorption of the compound through the oral route and the greater the bioavailability.

CONTRASTING KINETICS OF LIPOPHILIC SUBSTANCES

Many xenobiotics are of interest to toxicologists because they are slowly eliminated and accumulated in the body. Such compounds, when administered in the experiments typical of pharmacokinetics as described earlier, sometimes have very large volumes of distribution and very low clearance values.

An example of a drug with apparently slow pharmacokinetics is *ivermectin*, a veterinary anthelmintic, which is also used as a long-lasting larvicide for human onchocerciasis (blinding filarial disease). Popularly used as Heartguard® for dogs, ivermectin pharmacokinetics have been studied primarily by oral administration; however, intravenous administration in cattle demonstrated at least two phases of elimination with a rapid distribution phase followed by one or more slower phases of decay. The estimated $T_{1/2}$ was 2.8 days in cattle and 1.8 days in dogs. It should be noted that ivermectin is fluorescent and detectable at low levels (pg/mL) in plasma; therefore, the very slow terminal phase of elimination is detectable, but the quantity of drug is very minute. The slow-release effect of this terminal compartment combined with very high toxicity to the target filarial worms produces the high efficacy of this anthelmintic.

Heartguard® needs to be administered only monthly to dogs, compared with a much more frequent dosing for alternatives. Similarly, ivermectin for human filarial blindness (river blindness) needs to be given only weekly, compared with daily for alternatives.

With some very lipophilic xenobiotics, the simple one-compartment model is not realistic because there is poor solubility in plasma and high solubility in body lipids; therefore, the compound is highly dissolved in fatty compartments, from which it is slowly eliminated over days or weeks, as has been observed for the toxicants DDT and TCDD (both of which will be discussed more in detail later). These compounds would exhibit very low bioavailability in terms of the proportion of the dose actually available in the plasma.

Typical clinical drugs are eliminated in just a few hours; however, there is a recent trend for some categories of pharmaceuticals to have slower kinetics. This is due to the targeting of drugs toward membrane-bound receptors and the need to increase lipophilicity for effectiveness. For example, compared with the earlier nonsteroidal analgesic drug acetaminophen, the drugs rofecoxib and celecoxib are 100–1000 times more lipophilic (fat soluble) and more likely to associate with lipid membranes.

TABLE 2.1

Calculated Hydrophobicity Estimates for Nonsteroidal Analgesics Including COX-2 Inhibitors when Compared with Other Hydrophobic Toxicants

Compound	X log P^a	Mol. wt. (g/mol)
Acetaminophen	0.917	151.163
Aspirin	1.426	180.157
Rofecoxib	3.019	314.357
Ibuprofen	3.481	206.281
Parecoxib	3.822	370.423
Celecoxib	4.157	381.373
TCDD	6.202	321.970
DDE	6.862	318.024
Ivermectin	9.950	1736.160

[a] Calculated log P value as listed at http://pubchem.ncbi.nlm.nih.gov/; estimated hydrophobicity is directly proportional to log P.

This is predicted by the calculated log P values (Table 2.1). This characteristic also would predict slower elimination from the body, giving greater opportunity for side effects, contingent, of course, on relative biotransformation rates.

For comparison, DDE, the dehydrochlorination product of DDT, is one of the most lipophilic of xenobiotics and was found to accumulate in the lipids of fish and other wildlife. DDE is 100,000 times more lipophilic than acetaminophen.

BIBLIOGRAPHY

Breton, J., Daniel, C., Dewulf, J., Pothion, S., Froux, N., Sauty, M., Thomas, P., Pot, B., and Foligne, B., Gut microbiota limits heavy metals burden caused by chronic oral exposure, *Toxicol. Lett.*, 222, 132, 2013.

Fink, D.W. and Porras, A.G., Pharmacokinetics of ivermectin in animals and humans, in *Ivermectin and Abamectin*, Campbell, W.C., Ed., Springer-Verlag, Berlin, 1989.

Gilman, A.G., Rall, T.W., Nies, A.S., and Taylor, P., *The Pharmacological Basis of Therapeutics*, 8th edn, Pergamon Press, New York, 1990.

Joly, J.M. and Brown, T.M., Metabolism of aspirin and procaine in mice pretreated with O-4-nitrophenyl methyl(phenyl)phosphinate or *O*-4-nitrophenyl diphenylphosphinate, *Toxicol. Appl. Pharmacol.*, 84, 523, 1989.

Klaassen, C.D. and Rozman, K., Absorption, distribution, and excretion of toxicants, in *Casarett and Doull's Toxicology*, Amdur, M.O., Doull, J., and Klaassen, C.D., Eds, Pergamon Press, New York, 1991, chap. 3.

Lehman-McKeeman, L.D., Absorption, distribution, and excretion of toxicants, in *Casarett and Doull's Toxicology*, Klaassen, C.D., Ed., McGraw-Hill, New York, 2013, chap. 5.

Mekapati, S.B., Kurup, A., Verma, R.P., and Hansch, C., The role of hydrophobic properties of chemicals in promoting allosteric reactions, *Bioorg. Med. Chem.*, 13, 3737, 2005.

Prasanna, S., Manivannan, E., and Chaturvedi, S.C., QSAR studies on structurally similar 2-(4-methanesulfonylphenyl)pyran-4-ones as selective COX-2 inhibitors: A Hansch approach, *Bioorg. Med. Chem. Lett.*, 15, 313, 2005.

Shargel, L. and Yu, A.B.C., *Applied Biopharmaceutics and Pharmacokinetics*, 2nd edn, Appleton-Century-Crofts, Norwalk, CT, 1985.

Shen, D., Toxicokinetics, in *Casarett and Doull's Toxicology*, Klaassen, C.D., Ed., McGraw-Hill, New York, 2013, chap. 7.

3 Biotransformation

INTRODUCTION

The previous chapter began a discussion of the *toxicokinetic phase*. This phase consists of movement (Greek: *kinesis*) of a poison in the body, including absorption into the circulatory system, distribution among tissues including sites of action, and elimination from the body. This chapter focuses in more detail on the aspect of elimination that involves the loss of the parent drug from the body due to *biotransformation* of the *parent drug* to *metabolites*. This biotransformation generally aids in the excretion of the parent drug in the urine or feces.

Metabolites are the products of enzyme-catalyzed chemical changes in a drug or toxicant. These products may vary in toxicity or therapeutic effect from the parent drug or the toxicant. When biotransformation results in a less toxic product, the process is generally known as *detoxification*. Some reactions, however, lead to the formation of products that are more toxic than the parent. These reactions are then known as *intoxication* or *bioactivation* reactions. In cases in which the parent drug or toxicant is reversibly bound to a protein, or otherwise temporarily sequestered, only to be released later into the blood, the metabolite may, in fact, be the parent compound itself.

In general, blood-borne drugs and toxicants are most capable of crossing cell membranes and binding to protein targets if they are *lipophilic*, small, and neutrally charged. Detoxification, then, is most efficient when the parent compound can be altered to a metabolite that is hydrophilic, large, and carries a charge. These changes typically occur in two stages. First, in what are often referred to as *phase I reactions*, the parent compound is typically hydrolyzed or oxidized (or in some cases, reduced). This leads to the formation of a metabolite that is then conjugated (bound) to a larger, much more hydrophilic molecule, which often carries a charge. This second conjugation step is what is often referred to as a *phase II reaction*. Usually, this stepwise biotransformation is necessary as many parent compounds lack a functional group to which a larger molecule can be attached. Therefore, phase I adds the necessary functional group and phase II attaches the larger molecule. However, it should be noted that for a given substrate, phase II conjugation does not necessarily follow a phase I reaction; nor does phase I metabolism always precede conjugation.

As an example, carbohydrates are often involved in conjugation reactions. Many carbohydrates are very soluble in water (consider the teaspoons of sucrose that dissolve readily to sweeten your coffee or tea), and when a toxicant is conjugated to a carbohydrate, it becomes very water soluble as well. The

Toxicogenomics

See also:
  Toxicogenomics

functional result is that this product molecule, or conjugate, shows little tendency to partition into lipids or membranes and is instead easily excreted in the urine.

A revolution is occurring in the study of biotransformation with methods developed in the field of molecular genetics. Enzymes involved in xenobiotic biotransformation are being studied by proteinless sequencing. In this process, the gene of interest is cloned and sequenced, and then the protein sequence is inferred by translating the codons present in the gene sequence. This method is much faster and easier than enzyme purification and protein sequencing methods. Much of the information in this chapter was derived through application of such techniques. Another important contribution of molecular biology is the development of diagnostic tests for genetic deficiencies. Patients with a low level of detoxicative capacity can be identified, and in some cases, the deficiency in these poor metabolizers has been traced to a specific change in the sequence of a gene.

Advances in urinalysis with high-resolution nuclear magnetic resonance (NMR) have also introduced the potential to relate metabolism to genetics using urinary profiles as a phenotype. This topic is known as *metabolomics* or *metabonomics*. Generally, these methods consider the alteration of the normal profile of bodily metabolites, but could include the profile of xenobiotics if detected.

In the following chapter, we will look at the basic types of xenobiotic biotransformation, as well as the enzyme systems involved. Interestingly, there are relatively few families of enzymes, each with minor variants, which deal with a vast plethora of chemical compounds (both intrinsic and extrinsic) to which the body can be exposed. Thus, these systems generally have (of necessity) very broad substrate specificity (although it is also true that stereoisomers, for example, may be metabolized rather differently). It should also be emphasized that many of these systems metabolize not only xenobiotics (our focus here) but also endogenous compounds such as steroid hormones, to name one example. The highest levels of the enzymes that we will discuss are generally found in the liver; however, other tissues (lung, kidney, skin, to name a few) are also capable of significant xenobiotic metabolism.

PRIMARY BIOTRANSFORMATION (PHASE I REACTIONS): HYDROLYSIS

A drug ingested into the body is subject to the same type of biotransformation as a molecule of food. In both cases, the type of reaction depends on the chemistry of the substance and on whether or not the substance is a substrate for various enzymes that enhance the efficiency of biotransformation. *Hydrolysis* and *oxidation* are two of the most important primary or phase I reactions of biotransformation.

The enzymes that catalyze the hydrolysis reaction are called *hydrolases*. Hydrolases include *amidases* and *peptidases* that are important in the digestion of protein in the diet, as well as the *lipases* that cleave fatty acid esters and glycerides. In addition, *cholinester hydrolases* are active in the hydrolysis of choline esters. Also, among the hydrolases of the liver are several that detoxify important endogenous and xenobiotic carboxylesters. These *carboxylester hydrolases*, although primarily known for their detoxification of xenobiotics, are likely also to function as lipases in lipid digestion.

As an example of a compound that undergoes hydrolysis, consider *methoprene*, an insecticide used for mosquito and fly control. Methoprene possesses chemistry similar to that of a fatty acid alkyl ester (Figure 3.1). During hydrolysis, esters are split through the addition of a molecule of water to yield an acid and an alcohol. (This is the reverse of *esterification*, in which an acid and an alcohol react to produce an ester accompanied by water.) Hydrolysis proceeds more rapidly in alkaline conditions because it is actually the hydroxyl ion that attacks an electrophilic carbon in the reaction. When fed to cattle, methoprene does not accumulate because it is hydrolyzed to produce isopropanol and an aliphatic acid metabolite. The acid product is then oxidized to carbon dioxide and water. Drugs such as *aspirin*, *propanidid*, and *procaine*, as well as pesticides such as *malathion*, its biotransformation product, *malaoxon* (Figure 3.1), and *pyrethrins*, also contain an ester linkage (a carboxylester) in the molecule and can also be hydrolyzed.

At the opposite extreme is the insecticide and fire retardant *mirex*—a carcinogen that is no longer being used. The unusual structure of this insecticide (a cage consisting of only carbon and chlorine; see the appendix) renders it practically impervious to biotransformation. Mirex has no ester present; therefore, there is no opportunity for ester hydrolysis. Mirex was found to accumulate in human adipose and in wildlife tissues due to the combination of lipophilicity and very slow biotransformation.

SERINE HYDROLASES

Peptidases, carboxylester hydrolases, and cholinester hydrolases together form a group known as *serine hydrolases*. These enzymes have, as a catalytic site, a serine residue that reacts with the substrate to form a transiently alkylated enzyme as the ester bond of the substrate is cleaved. Serine hydrolase genes have been cloned, and the enzymes have been found to contain highly conserved regions of amino acid sequences, especially a Phe-Gly-Glu-Ser-Ala-Glu sequence that includes the reactive serine of the catalytic site.

In the peptidases chymotrypsin and trypsin, the three-dimensional structure of the enzyme is known from *x-ray crystallography*. Serine[195] (the serine located at amino acid site number 195 in the enzyme) of the catalytic site is part of a catalytic triad of amino acid residues that also includes histidine[57] and aspartic acid[102]. The nucleophilicity of serine[195] is enhanced by charge transfer from aspartic acid[102] to histidine[57], which accepts the proton of the serine hydroxy[195] group as it attacks the substrate. Site-directed mutagenesis (the technique of inducing a specific mutation in a gene) consisting of the substitution of asparagine[102] for the aspartic acid[102] destroyed the activity of the enzyme without disturbing the configuration of the triad, demonstrating the importance of the charge transfer phenomenon and the role of the aspartic acid[102] in the activity of serine[195].

Shape is also important in the active sites of hydrolases. Carboxylesters and amides form a tetrahedral transition state when attacked by serine[195], prior to the cleavage of the leaving alcohol or amine. The three-dimensional shape of the enzyme forces the substrate into the active site in an orientation that favors formation of this transition state, which is the conformation from which the reaction can proceed most readily. By favoring this orientation, the enzyme lowers the activation energy required for hydrolysis.

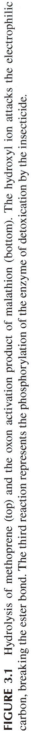

FIGURE 3.1 Hydrolysis of methoprene (top) and the oxon activation product of malathion (bottom). The hydroxyl ion attacks the electrophilic carbon, breaking the ester bond. The third reaction represents the phosphorylation of the enzyme of detoxication by the insecticide.

The carboxylester hydrolases are a very diverse group, with more than 20 genes in mouse and multiple genes in rat and humans. In mouse, there are two clusters of carboxylester hydrolase genes on chromosome 8 and several additional genes on other chromosomes. This characteristic suggests a multigene family. The evolution of multigene families may have involved the duplication of genes on a chromosome followed by the divergence of the duplicated genes, leading to a cluster of different, but related, genes encoding enzymes of similar function whose activity taken together can catalyze a broad spectrum of reactions. In humans, the two major forms of carboxylester hydrolases are CES1 and CES2. Genetic variations in these can affect metabolism of a variety of drugs and toxicants, including methylphenidate (Ritalin®).

Multigene Families

See also:
   Model Organisms and
      Comparative Genomics
                  Chapter 5

Cholinester hydrolases are active against choline esters, which are carboxylesters in which the leaving group alcohol is the quaternary amine, choline. *Acetylcholinesterase* (acetylcholine hydrolase) is a cholinester hydrolase that serves as a critical enzyme for clearing the neuromuscular synapse of the neurotransmitter, *acetylcholine.* Acetylcholinesterase is the target of poisoning by many organophosphorus and carbamate pesticides. It is present in many tissues, including the erythrocytes, from which its activity can be monitored conveniently. The activity against carboxylester substrates declines as the acyl group is lengthened from acetate. In the serum, *butyrylcholine hydrolase* is a similar enzyme with activity against longer-chained acyl esters. There is only one gene each for acetylcholinesterase and butyrylcholinesterase in humans. There have been no reports of genetic variations in human acetylcholinesterase, but many variants of butyrylcholinesterase are known.

Acetylcholinesterase

See also:
   Effects of Toxicants on
      Enzymes
                  Chapter 4

   Acetylcholine
                  Chapter 10

   Organophosphates
                  Appendix

Recent x-ray crystallography of acetylcholinesterase from *Torpedo californica* has revealed a gorge extending very deep into the enzyme. At the bottom of the gorge is the active site, with serine as part of a catalytic triad composed of glutamic acid[327], histidine[440], and serine[200] (similar to the aspartic acid[102]–histidine[57]–serine[195] triad of chymotrypsin). A pocket at the bottom of the gorge is lined with phenylalanine residues that restrict the acyl group of the substrate to acetate. Sequence comparison indicated that a more spacious pocket is present naturally in butyrylcholinesterase and in lipases that must accept substrates bearing large acyl substituents.

Organophosphates

See also:
   Effects of Toxicants on
      Enzymes
                  Chapter 4

   Acetylcholine
   Distal Axonopathies
                  Chapter 10

   Criminal Poisonings
                  Chapter 16

   Pesticides
                  Chapter 17

   Organophosphates
                  Appendix

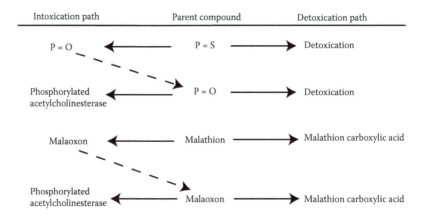

FIGURE 3.2 Pathways of metabolism of organophosphates leading to detoxication and intoxication. Both a general case and a specific case (malathion) are shown.

It is possible by site-directed mutagenesis to create a version of the acetylcholinesterase enzyme with a much more open active site acyl pocket. The effect of this change is to introduce very efficient butyrylcholinesterase activity while retaining acetylcholinesterase activity. This type of mutation appears to have happened spontaneously in the insecticide-resistant acetylcholinesterases of several agricultural pest insects (which formerly lacked butyrylcholinesterase activity).

One group of toxicants that inhibit serine hydrolases is the organophosphate insecticides. These compounds are esters that react rapidly and irreversibly with the active serine of carboxylester hydrolases and peptidases, inhibiting these enzymes. Although this reaction with serine hydrolases is a factor in the detoxification of these potent poisons, the reaction leaves the serine phosphorylated and the enzyme activity lost. Thus, one mole of serine hydrolase is sacrificed for each mole of organophosphate that is broken down. Malathion, as mentioned before, is a substrate for carboxylester hydrolases and in fact is primarily metabolized through the hydrolysis of its carboxylester group. The metabolic product of malathion is malaoxon, which is an even more reactive acetylcholinesterase inhibitor than malathion. Malaoxon possesses both a substrate carboxylester group and an inhibitor phosphorothionate group. Malathion hydrolysis proceeds linearly; however, malaoxon hydrolysis is progressively inhibited, resulting in a progressive decline of the rate of hydrolysis. Although malaoxon is unusual in serving as both substrate and inhibitor for carboxylester hydrases, the general pathways of intoxicative and detoxicative biotransformation illustrated by malathion apply to most organophosphate pesticides (Figure 3.2). Toxicity from these compounds depends on the rates of bioactivation versus detoxification—both of which are affected by factors that alter metabolism.

PARAOXONASES

In mammalian liver and serum, there are enzymes known as *paraoxonases* (PON1, PON2, and PON3 in humans) that can also catalyze organophosphate hydrolysis (but by an unknown mechanism). The structure of a paraoxonase is quite different from that of the serine hydrolases, in that it contains nothing resembling the serine-containing conserved sequence of the serine hydrolases. Instead of a serine active site, this enzyme perhaps uses a cysteine residue (as do *thiol hydrolases* such as the enzyme papain). Also, when a number of *chiral* organophosphate inhibitors (in other words, those that possess four unlike substituents of phosphorus and therefore exist as either (+) or (–) enantiomers) were evaluated as substrates for several types of enzymes, acetylcholinesterase and the peptidase chymotrypsin usually preferred the opposite enantiomer compared with paraoxonases. This suggests that the shape of the active site in a paraoxonase is the mirror image of the active sites of acetylcholinesterase and chymotrypsin. Still, acetylcholinesterase and paraoxonase likely have active sites of similar size (but smaller than either chymotrypsin or carboxylester hydrolase).

Paraoxonases are common in mammals, but very rare in birds and insects. In humans, there is a polymorphism in the activity of paraoxonase in serum (PON3), with a low-activity allele and a high-activity allele. Caucasians have a low-activity allele with a frequency of 50%–70%, whereas Africans and Asians have primarily the high-activity allele and South Pacific aboriginal tribes have only the high-activity form.

EPOXIDE HYDROLASE

Another hydrolase that plays a role in phase I metabolism is *epoxide hydrolase*. This enzyme acts on *epoxides*, compounds in which oxygen shares a single bond with each of two carbon atoms that also share a single bond with each other. Epoxide hydrolase converts epoxides into diols, a reaction that can be important in the detoxification of epoxides (which are quite reactive compounds). The active site of epoxide hydrolase consists of a nucleophilic amino acid (usually aspartic acid), a basic amino acid (often histidine), and an acidic amino acid (either aspartic acid or glutamic acid).

PRIMARY BIOTRANSFORMATION (PHASE I REACTIONS): OXIDATION

THE ROLE OF CYTOCHROME P450

Oxidation is a second mechanism of primary metabolism by which xenobiotics are either detoxified or bioactivated to a more toxic product. A wide variety of oxidation reactions are catalyzed by *cytochrome P450 enzymes* of various forms. These enzymes are found in *microsomes* (smooth endoplasmic reticulum) primarily in the liver, but in other tissues as well. Also known as *monooxygenases*, P450 enzymes include a heme (iron-containing) group, in which molecular oxygen is bound to iron. P450 is named for the characteristic peak of light absorbance at 450 nm, when carbon monoxide, a potent inhibitor, is bound to the reduced enzyme and scanned versus the reduced enzyme as a reference. This is called the *ferrous–CO 450 nm Soret band*.

P450 (*CYP*) genes are members of an anciently evolved multigene superfamily in bacteria, plants,

Cytochrome P450

See also:
  Metabolomics
  Personalized Susceptibility
    and Tailored
    Therapeutics
              Chapter 5

and animals. In humans, to date, there are 57 known P450 genes and 58 *pseudogenes* (probable former genes that no longer produce functional proteins). The functional human enzymes belong to 18 families, based on protein sequence homology (similarities between protein sequences).

In these P450 families, there is a conserved Phe-X-X-Gly-X-X-X-Cys-X-Gly sequence near the cysteine residue that holds the heme molecule in position. Apart from this heme-binding site, however, there is a great diversity of sequences, with each P450 family consisting of proteins with >40% sequence homology. Some families have a relatively broad range of substrates that overlap with other families. P450 families are numbered using a somewhat arbitrary system, in which some numbers were assigned to preserve continuity in the literature. The four major P450 families are CYP1, CYP2, and CYP3 (which play major roles in xenobiotic metabolism) and CYP4 (which metabolizes fatty acids and related endogenous compounds). Subfamilies are designated by adding letters, e.g., subfamily CYP1A; different enzymes within the subfamilies are designated by numbers, e.g., enzyme CYP1A1 or enzyme CYP1A2. There are also individual variations (*polymorphisms*) that occur with these enzymes and that can have significant functional consequences for the individuals that express them. Table 3.1 gives some examples.

Induction

See also:
Stress, Repair, and Recovery
Chapter 4

Many of the P450 enzymes can undergo the process of *induction*, whereby enzyme activities increase following exposure to substrates. Major inducing agents include *phenobarbital*, which induces the synthesis of one group of P450 enzymes (CYP2 and CYP3 families), and *3-methylcholanthrene* (3-MC) and *tetrachlorodibenzodioxin* (TCDD), which induce the synthesis of a separate group of P450 enzymes (the CYP1 family). Induction by 3-MC and related compounds has been explained through a series of experiments. Investigations in mice led to the discovery of a soluble protein receptor, or transcription

TABLE 3.1
P450 Families and Examples of Polymorphisms

CYP Family	Examples of Substrates	Examples of Human Polymorphisms/Effects
CYP1	Polycyclic aromatic hydrocarbons, dioxins, nitrosamines, acetaminophen, caffeine	CYP1A1, CYP1A2—some forms implicated in increased cancer risk
CYP2	Bilirubin, carbon tetrachloride, benzene, ethanol, nicotine, opioids, amphetamines, antihistamines, coumarin, halothane, nonsteroidal antiinflammatory drugs (NSAIDs), tetrahydrocannabinol (THC)	CYP2B6—different forms can either increase or decrease drug clearance rates CYP2C9, CYP2D6—some forms can decrease drug clearance rates
CYP3	Steroids	CYP3A4—some forms show possible increase in risk for prostate cancer progression
CYP4	Fatty acids	CYP4A2—some forms show possible link to increased risk for hypertension

factor, called the *Ah receptor*, which was lacking in homozygous mutant Ah⁻ mice. Mice with the Ah receptor, which has a high affinity for TCDD, were susceptible to poisoning by TCDD, which must undergo bioactivation to produce toxicity. Ah⁻ mice, however, were not responsive to TCDD. Further studies demonstrated that after binding of 3-MC or TCDD to the Ah receptor, the whole complex translocates to the nucleus, where interaction with DNA results in transcription of the appropriate P450 genes. Other inducers act through activation of other nuclear receptors, including the *CAR (constitutively active receptor)* and *PXR (pregnane X receptor)*, which are activated by phenobarbital.

P450 activity can also be *inhibited*. Not only can inhibitors of protein synthesis block the induction process, but also some compounds can act as specific competitive inhibitors for binding to the P450 enzyme. Examples of inhibitors include *carbon monoxide*, which competes with oxygen for binding to the heme site. Compounds such as *piperonyl butoxide* and a compound called *SKF 525-A* compete for binding at the substrate binding site.

Of course, changes in P450 levels and activity brought about by xenobiotics (such as induction and inhibition) can also affect metabolism of endogenous compounds and vice versa.

In P450-catalyzed reactions, molecular oxygen is activated by two electrons, and then one oxygen atom is inserted into the substrate drug and the other oxygen atom is reduced to yield water. During these steps, P450 must be reduced first and then bind substrate and molecular oxygen to complete the oxidation reaction. Reduction is accomplished by electron transport from the enzyme *NADPH cytochrome P450 reductase*. Both enzymes are embedded adjacently in the microsomal membrane, allowing efficient reduction of P450. Reduced P450 then activates molecular oxygen for substrate oxidation. The general P450-catalyzed oxidation reaction is

Carbon Monoxide

See also:
   Effects of Toxicants on
    Transport Proteins
              Chapter 4

   Effects of Toxicants on
    Hemoglobin
              Chapter 9

   Carbon Oxides
              Chapter 17

   Carbon Monoxide
              Appendix

$$\text{Substrate(H)} + O_2 + \text{NADPH} + H^+ \xrightarrow{\text{P450}} \text{substrate(OH)} + H_2O + \text{NADP}^+.$$

Common oxidation reactions catalyzed by P450 include *hydroxylation, dealkylation*, and *epoxidation*. Oxidation reactions change hydrophobic substrates into more polar products, which can then be conjugated and eliminated through excretion in the urine. Oxidative biotransformation via P450 occurs for a great number of these hydrophobic substrates, including endogenous biochemicals such as steroid hormones, fatty acids, and retinoids, as well as many important xenobiotics such as drugs, antibiotics, pesticides, industrial solvents, dyes, and petroleum. Natural toxins from microorganisms and plants can also be substrates of P450.

One major category of oxidation reaction carried out by P450 is *aliphatic hydroxylation*. An example of an aliphatic hydroxylation reaction is shown in Figure 3.3, which shows the P450-mediated hydroxylation of lauric acid. Other substrates for aliphatic hydroxylation include straight-chain *alkanes* from hexane to hexadecane,

FIGURE 3.3 Aliphatic hydroxylation of lauric acid.

cyclohexane, hexobarbital, and *prostaglandins.* This reaction is among those catalyzed by enzymes from CYP4 (cytochrome P450 family 4), a family characterized
by its induction by *clofibrate,* an antihyperlipoproteinemic drug used to decrease
lipid levels in the blood after surgery to reduce the risk of clotting. A gene coding for
one of these proteins has been found in the cockroach and shows greater similarity
to mammalian *CYP4* genes than to several insect genes from other P450 families.
This supports the hypothesis that genes for P450 evolved before divergence of the
vertebrates and the invertebrates. A unique sequence that serves as a fingerprint for
proteins from this family is a sequence of 13 consecutive amino acids that are fully
conserved within CYP4, but found intact in no other families.

Aliphatic oxidation reactions of steroid hormones are also catalyzed by specific
P450 families. The family CYP2A is specific for testosterone 7-α-hydroxylation and
is inducible by 3-methyl cholantrene (3-MC). In the biosynthetic pathway from cholesterol to steroid sex hormones, oxidations in positions 17 and 21 (side chain) are catalyzed by families CYP17 and CYP21, and 11-β-hydroxylation is catalyzed by CYP11
(Figure 3.4).

**Individual Differences
in Metabolism**

See also:
  Metabolomics
  *Individual Variations
    in Metabolism and
    Pharmacogenomics*
                  Chapter 5

The antihypertensive *debrisoquine* is hydroxylated by CYP2D, a family that may be involved in the
biotransformation of many drugs. Metabolic capabilities of this enzyme family can vary from individual
to individual, and patients can be characterized as
extensive metabolizers or poor metabolizers on the
basis of a polymerase chain reaction assay of their
DNA. This was, in fact, the first P450 polymorphism

FIGURE 3.4 Aromatic hydroxylation of testosterone.

discovered. Poor metabolizers may suffer from deficiency in the detoxication of drugs. On the other hand, poor metabolizers among Nigerian cigarette smokers were found to be less susceptible to cancer than extensive metabolizers who smoke, perhaps due to reduced bioactivation of carcinogens in tobacco smoke.

A second major type of oxidation is *aromatic hydroxylation*, a characteristic reaction catalyzed by CYP1 and induced by TCDD or 3-MC. Some aromatic substrates can be hydroxylated by direct insertion of oxygen to produce hydroxylated aromatic rings, such as in the hydroxylation of *chlorobenzene* to form *ortho-*, *meta-*, and *para-*chlorophenols (Figure 3.5). Others are hydroxylated through initial formation of a transient epoxide (arene oxide), which then may undergo hydrolysis to form a more stable metabolite. An example of this is the hydroxylation of *benzo[a]pyrene* to form

FIGURE 3.5 Aromatic hydroxylation of chlorobenzene (top) and benzo[a]pyrene (bottom). A transient epoxide is formed, which may then undergo conversion to a diol by epoxide hydrolase.

FIGURE 3.6   Alkene oxidation of aldrin.

benzo[a]pyrene-7,8-epoxide. This product can then be converted to benzo[a]pyrene-7,8-diol, by addition of water as catalyzed by the enzyme epoxide hydrolase (epoxide hydrase) (Figure 3.5). Aromatic hydroxylation reactions are implicated in bioactivation of carcinogens through the formation of these reactive epoxide intermediates. For example, benzo[a]pyrene-7,8-diol is a proximate carcinogen that can be further epoxidated to form benzo[a]pyrene-7,8-diol-9,10-epoxide, which is even more reactive with DNA than the parent molecule.

Another common reaction is *alkene oxidation*. One example of alkene oxidation is the epoxidation of the cyclodiene insecticide *aldrin* to the 6,7-epoxide product *dieldrin* (Figure 3.6). Another cyclodiene, *chlordane*, is metabolized to heptachlor epoxide. Dieldrin and chlordane were registered for many uses in agriculture, and chlordane was the principal termite-proofing agent used over the past 30 years. However, the registrations of these and most other cyclodiene insecticides have been canceled by the EPA due to persistence and suspected carcinogenicity.

Another type of reaction carried out by P450 is *O-*, *S-*, or *N-dealkylation*. One example of this type of reaction is the dealkylation of *chlordimeform* to form *N*-demethyl chlordimeform, which is again *N*-dealkylated to form *N,N*-didemethyl chlordimeform (products that are successively more toxic; Figure 3.7). These reactions proceed through the formation of reactive oxygen-inserted intermediates that may be carcinogenic.

Dialkylnitrosamines may be activated in this fashion. The final products may also be bioactivated, as seen for several pesticides, including chlordimeform and chlorfenapyr. Oxidative removal of both methyl groups from chlordimeform gives a more potent activity *in situ* in the firefly, and piperonyl butoxide, a P450 inhibitor, blocks the pesticidal activity against cattle ticks. Chlorfenapyr is a new proinsecticide that

FIGURE 3.7   *O-*, *S-*, or *N*-dealkylation of chlordimeform.

FIGURE 3.8 *S*-oxidation of aldicarb.

is 1000-fold less active as an uncoupler of oxidative phosphorylation than is the *N*-dealkylated product. New drugs and pesticides may be designed with these types of bioactivation reactions in mind. A more hydrophobic *prodrug* (an inactive compound that has an active metabolite) could provide efficient transport to the site of action and then be metabolized to an active polar molecule at the site of action.

S- or *N-oxidation*, such as the *S*-oxidation of *aldicarb* to aldicarb sulfoxide, then to aldicarb sulfone (Figure 3.8), often leads to products that are equal to or more toxic than the parent molecules.

Most organophosphorothioate and organophosphorodithioate insecticides are propesticides that are biotransformed to much more toxic and reactive organophosphates through *desulfuration* reactions. For example, malathion-P=S is transformed through the oxidized intermediate (malathion–P–S–O) to form malathion-P=O (Figure 3.9). The lower toxicity parent compounds are used rather than the metabolites because the metabolites are much too toxic for practical handling and application. About 400 organophosphorothioate active ingredients are applied in agriculture and household pest control and in the control of mosquitoes.

OTHER ENZYMES CARRYING OUT OXIDATION

Another class of monooxygenases is the *flavin-containing monooxygenases*, which are known for *N*-oxidation of tertiary amines. These enzymes overlap with P450 in catalysis of certain biotransformation reactions. They are important in *S*-oxidation reactions and in the desulfuration of phosphonates—those organophosphorus pesticides that possess one phosphorus-to-carbon bond. They are stabilized by the presence of

FIGURE 3.9 Desulfuration of malathion.

NADPH, which, along with molecular oxygen, is required for activity. Like P450, activity is found in the liver and kidney; however, flavin-containing monooxygenases lack a heme group and are not inhibited by carbon monoxide, nor by piperonyl butoxide. Also, inducers of P450 do not regulate levels of expression of this enzyme.

Alcohol dehydrogenase (ADH) is another enzyme that can carry out oxidation in a variety of tissues. This enzyme converts small alcohols to aldehydes and is found in a variety of tissues such as liver, kidney, and lung. There are four major classes of isozymes for this enzyme, with class I isozymes primarily responsible for ethanol metabolism. Individuals within the human population differ in which isozymes they express, and differences in the speed of ethanol metabolism between populations depend at least in part on which isozymes are expressed by individuals in that population. Forms of ADH also differ between tissues. Higher levels of class I ADH are located in the liver, which is where most ethanol metabolism takes place. A type of ADH known as class IV ADH is located in the gastrointestinal (GI) tract and may play some role in ethanol metabolism as well. This form of ADH may also be responsible for the increased risk of gastric cancer seen with heavy ethanol consumption, as it leads to the production of the suspected carcinogen *acetaldehyde* in the upper GI tract.

Along with ADH, another enzyme important in ethanol metabolism is *aldehyde dehydrogenase* (ALDH), which converts aldehydes to carboxylic acids (a step that requires NAD+ as a cofactor). There are at least 19 ALDH isozymes occurring in various human tissues, some of which show polymorphisms that may vary between individuals. Some individuals, for example, have a form of ALDH that works more slowly than other forms. If this variation happens to occur in conjunction with a variant form of ADH that produces an increased rate of conversion of alcohol to aldehyde, then aldehyde levels can build and cause unpleasant physiological symptoms such as flushing of the skin. Genetic variations in these enzymes are, in fact, considered to be one of the many potential factors that must be considered in attempting to assess the risk for alcoholism.

Parkinson's Disease

See also:
   Other Cytotoxic Compounds
                    Chapter 10

There are many additional phase I reactions that will not be described in detail here, as few other mechanisms are as important for a wide spectrum of xenobiotics as those discussed already. One final example, *monoamine oxidase* (MAO), is a critical enzyme in the brain (and other tissues) for deamination of neurotransmitters and also for some xenobiotics. It is quite significant in pharmacodynamics because it is the target for many enzyme-inhibiting antipsychotic drugs, including the original MAO inhibitor, *iproniazid*. Also, two forms of MAO, MAO-A and MAO-B, have been identified, each with different substrate specificities. This discovery has opened the door for additional fine-tuning of pharmacological interventions in the MAO system. MAO-B activity increases with age and has been hypothesized to play a role in the development of *Parkinson's disease*, a common neurodegenerative disease of age.

PRIMARY BIOTRANSFORMATION (PHASE I REACTIONS): REDUCTION

Some compounds are not oxidized, but are in fact *reduced* during phase I metabolism. These compounds probably act as electron acceptors, playing the same role as oxygen. In fact, reductions are most likely to occur in environments in which oxygen levels are low. Reductions can occur across nitrogen–nitrogen double bonds (*azo reduction*) or on nitro (NO_2) groups (*nitro reduction*). These reactions are primarily catalyzed by cytochrome P450 and *NADPH–quinone oxidoreductase*. Often, the resulting amino compounds can then be oxidized to form toxic metabolites, making this an activation reaction rather than a detoxification reaction. Nitro reduction may also occur in the GI tract, where it is carried out by intestinal microorganisms. There are also NADPH-dependent reductases that reduce carbonyls.

Another major type of reduction that typically results in activation is *dehalogenation*. Halogens can be removed from compounds such as carbon tetrachloride by replacing them with hydrogen or oxygen or through elimination of two adjacent halogens and replacement with a double bond. These reactions are catalyzed by cytochrome P450 and glutathione *S*-transferase and produce free radicals and other reactive intermediates with the strong potential to bind to cellular macromolecules.

Disulfides, sulfoxides, *N*-oxides, and quinones can also be reduced through a number of different pathways. These reductions often reverse the oxidation carried out by P450 and other enzymes and can lead to cycling between oxidized and reduced forms and ultimately an increase in the elimination half-life of the compound.

SECONDARY METABOLISM (PHASE II REACTIONS)

Products of phase I may enter a secondary phase of biotransformation in which they are rendered highly polar by conjugation to carbohydrates, amino acids, or small peptides. These products, or *conjugates*, are excreted from the body more efficiently than the parent or the phase I products. Enzymes that catalyze phase II reactions are most commonly located in the cytosol and appear to be coordinately regulated along with phase I, so that products do not accumulate when detoxication rates increase. Research in phase II biotransformation is complicated by the need to hydrolyze conjugates with enzymes, such as glucuronidase, peptidase, or sulfatase, in order to recover, identify, and measure the quantity of the phase I product that had been conjugated.

GLUCURONIDATION

Conjugation of phase I products with the activated nucleoside diphosphate sugar *uridine diphosphoglucuronic acid* (UDPGA), a process known as *glucuronidation*, is catalyzed by the enzyme *UDP-glucuronosyltransferase*, which is found in mammals primarily in rough and smooth endoplasmic reticulum from the liver, kidney, alimentary tract, and skin. At least eight forms of UDP-glucuronosyltransferases exist in the rat liver, differing in substrate specificity and inducibility by phenobarbital, 3-MC, and other inducers. Analogous enzymes are encoded by four major families of at least 22 highly homologous genes in humans. This multigene family displays a diversity of genes within one species and includes individual genes that are conserved among species.

Glucuronidation employs UDPGA (Figure 3.10) as a *cofactor*. The first step in the synthesis of this cofactor is the reaction of uridine triphosphate with glucose-1-phosphate as catalyzed by UDP-glucose pyrophosphorylase. Although readily reversible, this reaction is pulled toward synthesis by the rapid hydrolysis of a pyrophosphate to orthophosphate (catalyzed by pyrophosphatase), leaving only one of the products (and thus blocking the back reaction). The UDP-glucose thus formed is then converted to UDPGA by oxidation of the glucosyl C-6 methanol moiety to a carboxyl group.

Bilirubin

See also:
   Function of the Liver
                      Chapter 11

Glucuronidation is naturally important in the conjugation of *bilirubin*, an endogenous compound produced when heme released from the hemoglobin of dead erythrocytes is oxidized in the spleen. Bilirubin possesses two proprionic acid groups, one of which becomes esterified with glucuronyl from UDPGA to yield bilirubin monoglucuronide. This product is then converted to bilirubin diglucuronide for excretion as the major

FIGURE 3.10   Glucuronidation.

pigment in bile (although it appears that the second glucuronidation is accomplished by a different mechanism).

The UDP-glucuronosyltransferase responsible for bilirubin conjugation differs from other, apparently distinct, enzymes that catalyze conjugation of steroids or phenolic xenobiotics. Examples of these reactions include the conjugation of *testosterone* with UDPGA to form testosterone glucuronide and the conjugation of *1-naphthol* with UDPGA to form naphthyl glucuronide.

In the conjugations described earlier, the glucuronide conjugation occurs on an oxygen atom, resulting in carboxylic acid or ether products. However, *N*-, *S*-, and *C-glucuronidation* can also occur. While generally detoxicative, the formation of *N*-glucuronides of arylamines and *N*-hydroxyarylamines may enhance bladder cancer by aiding transport of the carcinogen from the liver to the bladder, where the conjugate may undergo acid hydrolysis, thus releasing the carcinogen.

In some cases, the parent drug rather than a phase I metabolite may be conjugated directly. Examples of direct glucuronidation include the antihistamine *tripelennamine*, which is conjugated to form a quaternary amine *N*-glucuronide, and the antibiotic *sarafloxacin hydrochloride*, for which the most significant biotransformation in chickens and turkeys is direct glucuronidation.

Glucuronidation is a principal conjugation reaction in most mammals; however, a minor amount of *glucosidation* (conjugation with glucose) has also been detected in mammals. The domestic cat, lion, and lynx fail to produce glucuronides of certain small substrates, but do conjugate bilirubin. In contrast, glucosidation is the primary conjugation reaction in insects (which coincidentally lack hemoglobin) and plants. In many insects, phase II reactions are very efficient because certain insecticides are oxidized and excreted primarily as conjugated metabolites.

GLUTATHIONE CONJUGATION

Glutathione is the tripeptide *L*-glutamyl-*L*-cysteinylglycine. This compound has a particularly nucleophilic thiol (sulfhydryl) in the central cysteinyl residue. Glutathione possesses an unusual glutamyl linkage to cysteine; the more common peptide linkage of glutamate is through the carboxyl group bonded to the carbon atom. Many tissues are rich in glutathione.

Glutathione can react spontaneously with peroxides and other potentially damaging electrophilic compounds, including certain phase I products of xenobiotics. Some of these reactions are catalytically enhanced by *glutathione transferases*, enzymes with a high affinity for lipophilic substances. However, all catalyzed reactions can proceed (although at some slower rate) without the enzyme.

Glutathione transferases are *dimers*, consisting of two identical or nonidentical subunits (enzymes with nonidentical subunits are called heterodimers). There are now seven known cytosolic classes of these enzymes: alpha, mu, pi, theta, kappa, sigma, and zeta. Enzymes within a class have subunits that share a much greater homology (around 70%) than the subunits of enzymes in different classes. In human tissues, there are five alpha class subunits (GSTA1 through GSTA5 for glutathione *S*-transferase class alpha subunits 1 through 5), five mu class subunits (GSTM1 through GSTM5), one pi class subunit (GSTP1), two theta class subunits (GSTT1 and GSTT2), one zeta class subunit (GSTZ1), and one mitochondrial form and two microsomal forms (which are significantly different from any of the other categories). These forms differ in isoelectric point, but they are similar in size at approximately 23,000 Da per subunit. Besides in the liver, activity is found in the kidney, small intestine, and some other tissues. Activity is found in microorganisms, plants, and throughout the animal kingdom. Some forms have been found to be inducible.

Site-directed mutagenesis studies have suggested that a conserved tyrosine residue may be necessary for catalytic activity; however, the precise interactions with substrates have not been determined. Because all reactions occur even without catalysis, it appears that the glutathione sulfhydryl group is the active site and that the enzyme holds it in a position that enhances the nucleophilic attack on the substrate (which is typically electrophilic) and thus enhances the rate of transition to products.

A wide variety of biotransformation reactions are catalyzed by glutathione transferases. In these reactions, reduced glutathione is conjugated to a reactive electrophilic carbon, nitrogen, or oxygen atom. One major class of electrophilic compounds that are attacked by glutathione includes products of primary detoxication reactions, e.g., arene oxides produced by P450-catalyzed oxidation of xenobiotics discussed in phase I. The resulting glutathione conjugate is metabolized through several reactions, converting the glutathione portion to *N*-acetylcysteinyl (*mercapturic acid*), the derivative of the conjugate most commonly detected in the urine, and several other products (Figure 3.11). Glutathione transferases may also play a role in the detoxication of endogenous lipid and nucleic acid hydroperoxides produced by superoxide radical attack. Linoleic acid hydroperoxide is a good substrate for several glutathione transferases.

Acetaminophen

See also:
   Liver Cell Death: Necrosis
      and Apoptosis
                     Chapter 11

   Acetaminophen
                     Appendix

The significant role of glutathione transferases in xenobiotic detoxication can be observed in the biotransformation of the common analgesic drug *acetaminophen*, which is transformed to *N*-acetyl-*p*-benzoquinonimine, a potential hepatotoxin, in a phase I *N*-oxidation. This reactive product is rapidly conjugated to glutathione in a reaction that is catalyzed readily by glutathione transferases. Depletion of glutathione by acetaminophen overdose negates this detoxication and can lead to liver toxicity or death in extreme cases.

Chemical weed control with *atrazine* in maize is possible because the maize glutathione transferase efficiently detoxifies the pesticide. Atrazine is used in greater quantity than any other pesticide in the United States primarily for maize. Both reduced glutathione and glutathione *S*-transferase activity are increased in corn and sorghum crops by adding *herbicide safeners* such as dichlormid and flurazole to certain herbicides. These additives increase the margin of safety, or selectivity between killing the competing weed species and the crop plant. The mechanism of increasing glutathione conjugation is unknown, but the enhancement of glutathione *S*-transferase appears to be via induction of transcription from secondary genes.

Glutathione transferases are also important in various aspects of cancer research. *Aflatoxin B1* is a mutagen in the Ames test when activated by liver microsomal

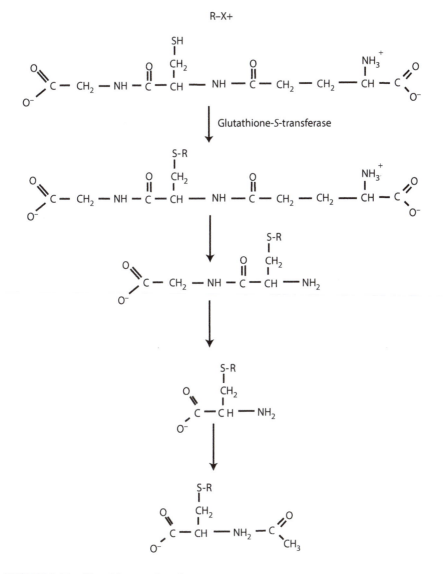

FIGURE 3.11 Glutathione conjugation.

fraction. The epoxide of aflatoxin B1 formed by P450-catalyzed oxidation can be conjugated with glutathione at a slow rate, and induction of higher rates of conjugation decreases the oncogenicity. In contrast, *1,2-dibromoethane (ethylene dibromide)*, a once common agricultural fumigant canceled by the EPA, is biotransformed to a highly carcinogenic glutathione conjugate that acts as a sulfur mustard to alkylate DNA. This activation is probably responsible for the high rate and rapid formation of stomach and nasal carcinoma observed in rats administered 1,2-dibromoethane. In some cases, tumors become resistant to antineoplastic agents due to the high expression of glutathione transferases.

ACETYLATION AND OTHER PHASE II REACTIONS

Acetylation is an important transformation of arylamines and hydrazines (but not phenols), catalyzed by an *acetyl–CoA-dependent N-acetyltransferase*. In humans, there are two NATs: NAT1 and NAT2. Although often detoxicative, acetylation may not lead to a more hydrophilic product and, in some cases, may yield a better substrate for P450 or other phase I enzymes. Common xenobiotic substrates include *isoniazid, benzidine, procainamide*, and *p-aminobenzoic acid*. This mechanism is greatly reduced in dogs and foxes, and there is a common recessive allele of the NAT2 gene that produces slow acetylation in humans.

In addition to glucuronidation, glutathione conjugation, and acetylation, many other phase II reactions are known. These include conjugations of phase I products to sulfate, glucose, glycine, and other molecules catalyzed by the respective transferase enzymes.

FACTORS THAT INFLUENCE METABOLISM

There are a number of factors that can affect both phase I and phase II metabolism. One obvious factor is species. Both qualitative (differences in enzymes expressed) and quantitative differences (differences in the level of expression) in drug-metabolizing enzymes exist between species. Even within a single species, of course, there are differences, as which alleles for a polymorphic enzyme a particular individual carries will most certainly influence metabolic capabilities.

Age is another important factor. The capacity for metabolism is generally low during development, and in humans, it does not reach adult levels until well after birth. Likewise, rates of xenobiotic metabolism seem to decline with age, a factor that must be taken into account when physicians prescribe drugs for the elderly. Gender differences in metabolism also exist in many species, but to this point, they have not been shown to play a significant role in xenobiotic handling in humans. Diet and other environmental factors may also impact metabolism in some cases.

THE ROLE OF METABOLISM BY GUT FLORA

Although it has long been recognized that metabolism of xenobiotics by the gut microbiome could play a significant role in the physiological effects of drugs and toxicants, the importance of this factor may well have been underestimated. This is a complicated question to address, however, as it involves not only the characterization of gut flora (which, of course, varies from individual to individual), but also the characterization of the metabolic activity of those organisms. Also, not only does the microbial community affect the metabolism of xenobiotics, but exposure to xenobiotics in turn affects the metabolic activities of the microbial community. Studies on this subject continue, though, and suggestions have been made that clinical manipulation of the gut microbiome may actually prove to be an effective tool in managing patients' response to medications.

BIBLIOGRAPHY

Brown, T.M. and Bryson, P.K., Selective inhibitors of methyl parathion-resistant acetylcholinesterase from *Heliothis virescens, Pestic. Biochem. Physiol.,* 44, 155, 1992.

Clayton, T.A., Baker, D., Lindon, J.C., Everett, J.R., and Nicholson, J.K., Pharmacometabonomic identification of a significant host-microbiome metabolic interaction affecting human drug metabolism, *Proc. Natl. Acad. Sci. USA,* 106, 14728, 2009.

Daly, A.K., Pharmacogenetics of the major polymorphic metabolizing enzymes, *Fundam. Clin. Pharmacol.,* 17, 27, 2003.

deBethizy, J.D. and Hayes, J.R., Metabolism: A determinant of toxicity, in *Principles and Methods of Toxicology,* 4th edn, Hayes, A.W., Ed., Taylor & Francis, Philadelphia, 2001, chap. 3

Dickinson, D.A. and Forman, H.J., Cellular glutathione and thiols metabolism, *Biochem. Pharmacol.,* 64, 1019, 2002.

Gerhold, D., Lu, M., Xu, J., Austin, C., Caskey, C.T., and Rushmore, T., Monitoring expression of genes involved in drug metabolism and toxicology using DNA microarrays, *Physiol. Genomics,* 5, 161, 2001.

Grothesen, J.R. and Brown, T.M., Stereoselectivity of acetylcholinesterase, arylester hydrolase and chymotrypsin toward 4-nitrophenyl alkyl(phenyl)phosphinates, *Pestic. Biochem. Physiol.,* 26, 100, 1986.

Hodgson, E. and Goldstein, J.A., Metabolism of toxicants: Phase I reactions and pharmacogenetics, in *Introduction to Biochemical Toxicology,* Hodgson, E. and Smart, R.C., Eds, Elsevier, New York, 2001, chap. 5.

LeBlanc, G.A. and Dauterman, W.C., Conjugation and elimination of toxicants, in *Introduction to Biochemical Toxicology,* Hodgson, E. and Smart, R.C., Eds, Elsevier, New York, 2001, chap. 6.

Matsumura, F., *Toxicology of Insecticides,* Plenum Press, New York, 1984.

Maurice, C.F., Haiser, H.J., and Turnbaugh, P.J., Xenobiotics shape the physiology and gene expression of the active human gut microbiome, *Cell,* 152, 39, 2013.

Nebert, D.W. and Gonzalez, F.J., P450 genes: Structure, evolution and regulation, *Annu. Rev. Biochem.,* 56, 945, 1987.

Nebert, D.W. and Nelson, D.R., p450 gene nomenclature based on evolution, in *Cytochrome P450,* Waterman, M.R. and Johnson, E.F., Eds, Academic Press, San Diego, CA, 1991, p. 3.

Parkinson, A., Biotransformation of xenobiotics, in *Casarett and Doull's Toxicology,* Klaassen, C.D., Ed., McGraw-Hill, New York, 2001, chap. 6.

Parkinson, A., Ogilvie, B.W., Buckley, D.B., Kazmi, F., Czerwinski, M., and Parkinson, O., Biotransformation of xenobiotics, in *Casarett and Doull's Toxicology,* Klaassen, C.D., Ed., McGraw-Hill, New York, 2013, chap. 6.

Ravichandran, K.G., Boddupalli, S.S., Hasemann, C.A., Peterson, J.A., and Deisenhofer, J., Crystal structure of hemoprotein domain of P450BM-3, a prototype for microsomal P450s, *Science,* 261, 731, 1993.

Stine, K.E., Environmental and endogenous factors determining the expression of toxicity, in *General and Applied Toxicology,* 3rd edn, Ballantyne, B., Marrs, T., and Syversen, T., Eds, John Wiley and Sons, UK, 2009, chap. 23.

Sussman, J.L., Harel, M., and Frolow, F., Atomic structure of acetylcholinesterase from *Torpedo californica*: A prototypic acetylcholine-binding protein, *Science,* 253, 872, 1991.

Wislocki, P.G., Miwa, G.T., and Lu, A.Y.H., Reactions catalyzed by P450, in *Enzymatic Basis of Detoxication,* Jacoby, Ed., Academic Press, New York, 1981, pp. 135–182.

Zanger, U.M. and Schwab, M., Cytochrome P450 enzymes in drug metabolism: Regulation of gene expression, enzyme activities, and impact of genetic variation, *Pharmacol. Ther.,* 138, 103, 2013.

4 Cellular Sites of Action

INTRODUCTION

Many toxic substances are known to cause poisoning through a chain of events that begins with action at a very specific target. Often, this target is a biological molecule with which the toxicant binds or reacts, such as one or more of the various types of proteins, lipids, and nucleic acids within the cell. Symptoms resulting from exposure to a toxicant may relate directly to this molecular event, or may be complicated by secondary effects, just as symptoms of a disease may be due to physiological imbalances that are secondary to the initial infection. Therefore, identification of the primary site of action requires careful collection and interpretation of biochemical and physiological evidence. For some toxicants, this initial event in poisoning, or molecular lesion, has been characterized. For many other toxicants, the precise interaction of the toxicant with one or more specific biological molecules has yet to be demonstrated. With the increasingly powerful experimental techniques available, however, it is likely that precise molecular sites of action will be described for many more toxicants in the near future.

The nature of the interaction between the toxicant and the binding site is also important. Toxicants may bind covalently to cellular macromolecules, leading typically to long-lived or virtually permanent changes within the cell. This is a typical pattern for toxicants which are *electrophilic*, behaving as electron acceptors and interacting readily with *nucleophilic* electron donors (including nucleophilic sulfurs, nitrogens, and oxygens in proteins and nucleic acids). Noncovalent binding (such as formation of ionic bonds or hydrogen bonds) tends to be much more easily reversible. In some cases, the toxicant may even affect a target molecule indirectly, through alteration of the cellular environment in which the target molecule resides and functions. One straightforward example of this would be effects mediated through toxicant-induced alterations in pH; another might be the disruption of the membrane structure produced by some lipophilic compounds, which may indirectly impact membrane-associated proteins.

Of course, not all binding of toxicants to molecules within a cell results in toxicity. Binding to "nontarget" molecules either may be physiologically inconsequential or may even divert binding of toxicants from other targets (examples of which will arise in later chapters).

This chapter discusses some of the ways in which toxicants are known to interact with biological molecules, as well as some of the techniques used to study these interactions.

INTERACTION OF TOXICANTS WITH PROTEINS

Proteins are composed of a linear chain of amino acids linked together by a type of covalent bond known as a peptide bond. The order of the amino acids in a particular protein (known as the protein's *primary structure*) is encoded in a molecule called DNA, which is located in the nucleus of the eukaryotic cell. Specifically, the sequence of nucleotides in a segment of DNA known as a *gene* directs the order in which amino acids will be assembled to build a particular protein.

DNA contains many genes, each of which contains the instructions for producing specific amino acid sequences. When proteins are manufactured by a cell, the proper DNA nucleotide sequence is first *transcribed* (copied) into messenger RNA (mRNA). This mRNA is then spliced, with segments called *introns* being cut out of the final product, while segments called *exons* remain. This processing allows one gene to actually code for several different proteins through a mechanism called *alternative splicing*, in which *exons* are combined together in different ways to produce different final mRNAs. The final mRNA copy then leaves the nucleus for the cytoplasm, where the encoded instructions are then *translated* on structures called ribosomes and an amino acid chain is constructed. The order of the amino acids that appear in a protein is known as its *primary structure.*

There are around 22 different amino acids that commonly appear in proteins, each of which carries an amino group (NH_2) and a carboxyl group (COOH), which tend to ionize at a physiological pH to form NH_3^+ and COO^-. These charged groups interact together to fold regions of many proteins into the form of helices or pleated sheets, in a pattern that is referred to as the *secondary structure* of the protein. An additional layer of folds and twists is referred to as the *tertiary structure* of the protein. This tertiary structure is produced by interactions of the side chains (R groups) of the amino acids (which differ from one amino acid to the next) with each other and with the aqueous environment of the cell. One such interaction is *hydrogen bonding*, an attraction between a positively charged region on one R group and a negatively charged region on another. *Hydrophobic interactions* (the folding of hydrophobic R groups to the interior of the protein and hydrophilic region to the exterior) can also impact tertiary structure. Finally, some protein molecules are an aggregation of two or more subunits that may be identical or may be coded for by different genes. The way in which these subunits fit together is called the *quaternary structure* of the protein.

Proteins can play a variety of structural and functional roles within the cell. Tubulin, actin, and other structural proteins comprise the *cytoskeleton* of the cell, providing physical support and also figuring prominently in cell motility (movement of a cell or structures within a cell). Some proteins function as *hormones*, carrying messages between cells; others function as the *receptors* on cell surfaces to which hormones and other messengers bind. Proteins also make up the *ion channels* that regulate the flow of ions across cell membranes. *Transport proteins* such as hemoglobin move substances through the bloodstream, and other proteins called *antibodies* defend the body as part of the immune system. Finally, protein catalysts called *enzymes* regulate biochemical reactions. Although toxicants can and do interact with all of these functional types of proteins, this chapter focusses on the effects of toxicants on enzymes, receptors, ion channels, and transport proteins. And, although gene expression will be examined in more detail in a later chapter on carcinogenesis, we will also briefly discuss here the impact of toxicants on proteins that play a role in genetic regulation.

EFFECTS OF TOXICANTS ON ENZYMES

Enzymes are catalysts, meaning that they enhance the rate of various biochemical reactions in the cell. Usually, an enzyme only catalyzes one specific type of reaction. In a catalyzed reaction, *substrates* (molecules that participate in the reaction) interact with the *active site* of the enzyme and are converted to *product*, leaving the enzyme chemically unchanged.

The rate of an enzyme-catalyzed reaction depends primarily on the concentration of the substrate. The greater the substrate concentration, the more enzyme becomes bound to the substrate and the faster the reaction proceeds. As substrate concentration increases, however, the enzyme eventually becomes *saturated* with substrate, and the rate (velocity) of the reaction approaches a maximum. This maximum rate is called the V_{max} or *maximum velocity*. The concentration of substrate at which the rate of reaction is half of the V_{max} also has a specific designation. It is called the *Michaelis constant* (K_m) for that substrate. The V_{max} and K_m for a reaction can be determined experimentally by measuring the reaction rate in a series of samples with differing substrate concentrations. This will produce a curve such as the one shown in Figure 4.1.

The rate of an enzyme-catalyzed reaction may be modified through other means as well. Some enzymes have sites known as *allosteric* sites, which are located in a different region of the molecule than the active site. Binding of molecules to these sites can change the shape or conformation of the enzyme molecule, thus affecting its catalytic ability. Allosteric sites commonly play a role in negative feedback mechanisms. For example, the product of an enzymatic reaction may bind to an allosteric site on that enzyme, thus preventing overactivity of the enzyme and the resulting accumulation of too much of the product.

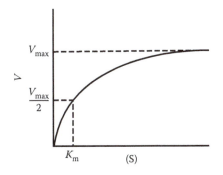

FIGURE 4.1 Enzyme kinetics. This graph of velocity versus substrate concentration shows the relationship between K_m (the substrate concentration when velocity is one half V_{max}) and V_{max}.

Enzymes are a common target for toxicants within the cell, and enzyme inhibition is a common molecular mechanism of poisoning. First of all, inhibition may be characterized as either *reversible* or *irreversible*, depending on the strength of the bond formed between the enzyme and the inhibitor (noncovalent binding generally produces reversible inhibition; covalent binding may produce irreversible inhibition). We will look first at the characteristics of reversible inhibition and then look at an example in which irreversible inhibition plays a role.

Acetylcholinesterase

See also:
  Serine Hydrolases
         Chapter 3

  Acetylcholine
        Chapter 10

  Organophosphates
        Appendix

There are two basic types of reversible inhibition, and they can be distinguished experimentally on the basis of their different effects on the K_m and V_{max} of the reaction. Inhibitors that compete with the substrate for binding to the active site of an enzyme are called *competitive inhibitors*. Because of their direct competition with the substrate, competitive inhibitors increase the K_m. In other words, a higher concentration of the substrate is required to achieve a given reaction rate. However, the effects of competitive inhibitors can be overcome with large excesses of substrate, and at high enough substrate concentrations, the reaction can achieve the same V_{max} as would be expected in the absence of the inhibitor.

Organophosphates

See also:
  Serine Hydrolases
         Chapter 3

  Acetylcholine
  Axonopathies
        Chapter 10

  Criminal Poisonings
        Chapter 16

  Pesticides
        Chapter 17

  Organophosphates
        Appendix

In contrast, *noncompetitive* inhibitors bind to and act on an allosteric site and not on the active site of the enzyme. As a result, the ability of these inhibitors to alter the enzyme's catalytic ability is independent of the concentration of substrate. In terms of kinetics, these inhibitors reduce the V_{max}; however, they do not alter the K_m.

One example of an enzyme that serves as a target for toxicants is the enzyme *acetylcholinesterase*, which is inhibited by a category of compounds known as *organophosphate insecticides*. Acetylcholinesterase is an enzyme that is important in the passage of impulses between neurons. Communication between neurons is carried out by a group of molecules called *neurotransmitters*, which are released from one neuron and then diffuse across the space between the neurons (called a *synaptic gap*), where they bind to *receptors* on the membrane of the adjoining neuron. Once this signaling process is complete, the synaptic gap must be cleared of neurotransmitters in order to be ready for the next signal. Although some neurotransmitters are reabsorbed by the releasing neuron, others are broken down by enzymes. Acetylcholinesterase is one such enzyme, clearing the synaptic gap by breaking down the neurotransmitter acetylcholine at synapses where it is in use.

Functionally, acetylcholinesterase catalyzes the hydrolysis (splitting through the addition of water) of acetylcholine to form choline and acetate. During the catalytic process, an acetyl group (COCH₃) from acetylcholine becomes covalently bound to a serine (an amino acid) in the active site of the enzyme. (This is a process called acetylation.) Next, a choline molecule is released through hydrolysis. Finally, because the covalent bond between the enzyme and the acetyl group is not very stable, that bond is also hydrolyzed and the acetate released. This process is shown in Figure 4.2. Inhibition of acetylcholinesterase by organophosphates involves a similar reaction between the enzyme and the organophosphate inhibitor at the active site. First, a covalent bond forms between the enzyme and the inhibitor, with the serine residue (which normally becomes acetylated) becoming phosphorylated instead (Figure 4.3). Next, hydrolysis occurs and part of the inhibitor molecule is released. The final step, however, is a little different. Whereas the covalent bond between the acetate and the enzyme is weak and easily hydrolyzed, the covalent bond formed between the phosphate group and the enzyme is quite stable and may last for at least several hours.

Although several hours is an extremely long period of inhibition (based on the timescale of a cell), recovery of the enzyme can and does occur. The rate of recovery of the phosphorylated enzyme depends on the chemical characteristics of the specific organophosphate inhibitor. Although most commercial organophosphate insecticides produce a phosphorylated enzyme that takes several hours to recover half its activity, inhibition by other organophosphate insecticides results in a phosphorylated enzyme with a half-life of several days. And with some organophosphate

FIGURE 4.2 The hydrolysis of acetylcholine by the enzyme acetylcholinesterase. The acetyl group of acetylcholine becomes bound to a serine residue at the active site of the enzyme, and the molecule undergoes hydrolysis to release choline. Further hydrolysis releases the acetate.

FIGURE 4.3 The phosphorylation of the enzyme acetylcholinesterase by the organophosphate insecticide methyl paraoxon. Instead of becoming acetylated, the serine residue becomes phosphorylated.

insecticides, the phosphorylated enzyme may undergo a process called *aging*, during which a chemical change to the phosphoryl group (generally a dealkylation) occurs (as shown in Figure 4.4). Aged, phosphorylated enzyme does not reactivate at all; so, this event, in effect, converts a reversible inhibition event into an *irreversible* inhibition event. Irreversible inhibition is characterized by the formation of a relatively permanent covalent bond with an amino acid at the active site of the enzyme. Because of the permanence of the reaction, this effect is not overcome by excess substrate.

FIGURE 4.4 The recovery or "aging" of phosphorylated acetylcholinesterase.

Due to the long recovery times and tendencies of inhibited enzyme to undergo aging, exposure to organophosphate insecticides leads to the accumulation of phosphorylated enzyme and thus a decline in active enzyme levels. The potency of inhibitors can be compared by measuring the rate at which enzyme activity declines. Symptoms of inhibition result when about 60%–70% of the enzyme is phosphorylated and are related to an excess of acetylcholine which results in overstimulation at synapses (1) between nerve and skeletal muscle, (2) in the central nervous system, and (3) in the parasympathetic branch of the autonomic nervous system. These symptoms in humans include constriction of the pupil, slowing of heart rate, bronchoconstriction, excessive salivation, and muscle contraction. The cause of death in poisoning by organophosphate insecticides is usually asphyxiation, caused by either malfunctioning of the diaphragm and the related muscles involved in breathing or failure of the respiratory center in the brain, which controls this process. Additionally, long-term exposure to organophosphates can contribute to secondary damage to neurons, neuronal cell death, and resulting persistent neurological deficits.

Antidotes for organophosphate toxicants include the compound pralidoxime (2-PAM) and *atropine*. 2-PAM is a base that reacts with the phosphorylated acetylcholinesterase to remove the aged phosphoryl group and regenerate the enzyme. Atropine is a compound that acts as an antagonist (blocker) at one type of acetylcholine receptor (the receptor found in the parasympathetic system) and primarily counteracts the symptoms of parasympathetic overstimulation. Organophosphate poisoning is also treated by providing artificial respiration and general life support.

Another potential set of targets for toxicants are the various mitochondrial enzymes. Mitochondria (Figure 4.5) are cellular organelles that extract usable energy from glucose and other molecules and transfer it to be stored in the high-energy molecule ATP. This process begins with the metabolic pathways of glycolysis and the citric acid cycle, where glucose is broken down, leading to the reduction of (addition of high-energy electrons to) electron acceptor molecules such

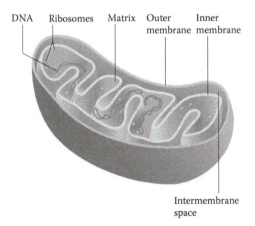

DNA   Ribosomes   Matrix   Outer   Inner
                           membrane   membrane

Intermembrane
space

FIGURE 4.5   A mitochondrion. (Modified with permission from http://www.shutterstock.com/pic-142194019/stock-vector-easy-to-edit-vector-illustration-of-mitochondria-diagram.html?src=&ws=1. Accessed on June 21, 2014.)

as NAD^+ or FAD^+. The electrons gained by these molecules are then transferred to a series of enzyme complexes and electron carriers called the *respiratory chain*, located within the mitochondrial inner membrane. Electrons are then passed down from one member of the chain to the next (releasing energy as they go), until they are eventually donated to oxygen at the end of the line.

According to the *chemiosmotic hypothesis*, as electrons are transported along the chain, the energy released is used to pick protons up from inside the mitochondrion and move them across the inner membrane to the space between the inner and outer membranes. Because the inner mitochondrial membrane is not very permeable to protons (due in part to the presence of a phospholipid called cardiolipin), a concentration gradient develops. Although this gradient cannot push protons back across the impermeable membrane itself, it can push protons back across the membrane through hydrophilic channels that are part of an enzyme called an *Mg^{2+} ATPase*. The movement of a pair of protons through the ATPase channel causes a part of the ATPase (a sort of a molecular rotor) to actually turn, much the way that water flowing over a dam can turn a water wheel. This turning then supplies the energy necessary to drive the synthesis of a molecule of ATP. This entire process is illustrated in Figure 4.6.

The effects of toxicants on several of the enzymes in this mitochondrial system are well documented. For example, one of the most notorious toxic chemicals, *cyanide*, blocks electron transport by inhibiting the actions of the enzyme complex cytochrome oxidase, a member of the electron transport chain. Other toxicants, including the pesticide *rotenone*, the barbiturate *amytal*, and the antibiotic *antimycin A*, inhibit the activity of other enzyme complexes in the chain. Because electrons cannot be passed on to an enzyme complex that cannot accept them, the effect of any of these toxicants is to completely shut down the flow of electrons in the blocked chain. This blockage can be detected in the laboratory by measuring the resulting decrease in mitochondrial oxygen uptake (which can be done with a tool called an *oxygen electrode*), as blockage of the flow of electrons prevents the final transfer of electrons to an oxygen molecule. In terms of functional consequences for the cell, because the proton gradient does not develop properly, there is a reduction in the ATP production (which is also measurable in the laboratory).

Another mitochondrial enzyme that has been shown to be affected by toxicants is the Mg^{2+} ATPase. ATPase inhibitors such as *oligomycin* inhibit this enzyme, thus blocking the formation of ATP. Inhibition of the ATPase also prevents the discharge of the proton gradient, and because the resulting pressure opposes further proton pumping, electron transport, and thus oxygen uptake, is also reduced.

A third group of toxicants that act on mitochondria are known as *uncouplers*. As the name suggests, these toxicants uncouple the two processes of electron transport and ATPase production. The result is a stimulation of electron transport (again, measured by oxygen uptake), along with a reduction in ATP production. There are many different uncouplers, and the mechanism of action of many is not completely clear. Most uncouplers probably interact with membrane proteins or lipids to alter permeability of the membrane to protons, thus interfering with the maintenance of the proton gradient. The oldest synthetic organic insecticide, dinitro orthocresol (DNOC), is an uncoupler and was once used as a drug to enhance weight loss, because uncoupling allows food to be oxidized without producing ATP. This

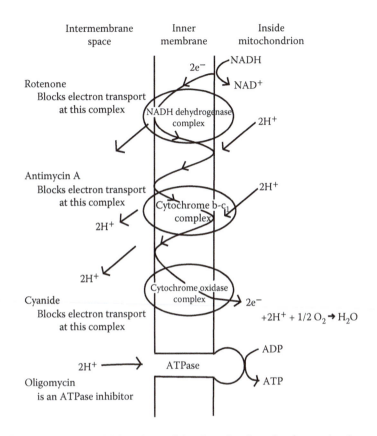

FIGURE 4.6 Mitochondrial function and the sites of action of toxicants. As electrons flow down the electron transport chain, protons are pumped across the inner membrane, establishing a gradient. The proton motive force developed moves protons back through the channel of the ATPase, catalyzing the formation of ATP. Rotenone, antimycin A, and cyanide block this process by blocking the electron transport chain, and oligomycin blocks the action of the ATPase. Other compounds such as DNOC act as uncouplers and dispel the gradient.

target has been rediscovered recently with the development of a new insecticide, chlorfenapyr (American Cyanamid; then BASF). Another well-known uncoupler is 2,4-dinitrophenol.

Finally, in an interesting twist, some toxicants themselves may actually act as enzymes, producing their effects through enzymatic action on cellular substrates. The toxin ricin, for example, has hydrolase activity and blocks protein synthesis by breaking down a component of eukaryotic ribosomal RNA. Several bacterial toxins have enzymatic activity, as do some components of animal venoms.

EFFECTS OF TOXICANTS ON RECEPTORS AND ION CHANNELS

Receptors are proteins that respond to the binding of a signal molecule generally referred to as a *ligand*. Receptors may be found embedded in membranes, as transmembrane channels, or in the cytoplasm of cells. Ligands may be hormones, neurotransmitters, other internal signaling molecules, or even xenobiotics (compounds foreign to the body). Generally, binding of ligands to receptors is quite specific, in that only a limited number of ligands (which are generally closely related structurally) will bind to a given receptor. Upon binding to the receptor (typically in a noncovalent bond), ligands form a ligand–receptor complex with kinetics that are similar to those in the formation of an enzyme–substrate complex. Receptor systems are part of the cell's *signal transduction system*, facilitating communication both between and within cells.

Experimentally, a specific receptor can be detected and identified using a ligand that has been labeled with a radioactive element such as tritium. First, the amount of labeled ligand bound to all proteins (including both specific binding to the receptor and nonspecific binding to other molecules in the cell) is measured. Then, the binding experiment is repeated, but this time with a 100-fold excess of nonlabeled ligand also added. The idea is that the nonlabeled ligand will bind to the specific receptor, a saturable process, leaving only the nonspecific binding sites for the labeled ligand. When this difference between total and nonspecific binding is measured over a series of labeled ligand concentrations, the binding of ligand to the specific receptor can be estimated.

The protein structures of several receptors have been inferred from the nucleotide sequences of the genes that encode them. Such new techniques in molecular genetics have the potential to revolutionize the understanding of toxicology on the cellular level by allowing for a detailed analysis of many proteins that occur at concentrations far too low for conventional purification. This technique involves extraction of mRNA from an organism known to produce the receptor protein of interest in relative abundance. The mRNA is then partially purified by chromatography and used as a template for preparing complementary DNA by reverse transcription (making a copy of DNA from RNA), a process that is catalyzed by the enzyme reverse transcriptase, as found in RNA viruses. The resulting complementary DNA can then be cut into pieces by enzymes called restriction endonucleases, and the pieces inserted into host bacteria using vectors such as bacterial plasmids or viruses. The transformed bacterial cells then are grown into colonies, with the bacteria in each colony containing a fragment of the DNA of interest.

If a part of the gene sequence is known (or can be inferred from a known sequence of amino acids in the protein), then a radioactively labeled complementary probe can be synthesized and used to identify the colonies containing the gene of interest. If the gene sequence is not known, then identification of the proper colonies must be made by immunoassay for the protein (if an antibody for that protein is available) or by testing for the characteristic activity of the protein of interest. Once the proper piece of DNA is identified, its nucleotide sequence can be determined and then confirmed by matching it with mRNA or DNA from the original organism through a technique called hybridization. Once several sequences are known, polymerase chain reaction (PCR) techniques can be applied to obtain sequences from additional species without further gene cloning.

In contrast to enzymes, receptor proteins do not catalyze chemical reactions; however, binding of a ligand to a receptor may ultimately result in profound changes such as triggering the transcription of a gene, opening of an ion channel through a membrane, or activation of enzymes through the actions of second messengers. The interaction of toxicants with a receptor can be categorized on the basis of the response of the receptor to that toxicant. Toxicants that mimic the action of the natural neurotransmitter are known as *agonists*; they bind to the receptor in its active state and elicit the same response as the endogenous ligand normally does. *Antagonists* are toxicants that bind to the receptor site but do not produce a response. They can negate the action of a neurotransmitter or agonist. *Partial agonists* produce a response; however, that response is not as strong as the response produced by the endogenous ligand.

Neurotransmitter Receptors

See also:
Effects of Toxicants on
Synaptic Function
Chapter 10

Several receptors are actually ligand-activated transmembrane ion channels. Many of these are found in the nervous system, where communication between neurons is mediated by chemicals called neurotransmitters. Neurotransmitters are synthesized and released by *presynaptic* neurons, migrate across the *synapse* (the space in between neurons), and ultimately bind to receptors on neighboring *postsynaptic* neurons. The presence of a particular receptor determines the nature of the synapse. When acetylcholine receptors are present in the postsynaptic membrane, the synapse is termed cholinergic because it will respond to acetylcholine. Adrenergic synapses have receptors that respond to norepinephrine, GABAergic synapses contain receptors that respond to *gamma-aminobutyric acid* (GABA), and so on for the many different neurotransmitter substances that exist in the nervous system.

Two of these neurotransmitter/receptor systems in the nervous system are the *GABA receptor* system and the *nicotinic acetylcholine receptor* system (Figure 4.7). These receptors incorporate ion channels that are opened in response to binding by neurotransmitters. GABA receptors incorporate a chloride ion channel and are composed of primary alpha, beta, and gamma subunits. Acetylcholine receptors are more complex, with four different genes encoding protein subunits. The alpha subunit, which binds the neurotransmitter, occurs twice in the molecule, along with one of each of the beta, gamma, and delta subunits, arranged in a rosette that has

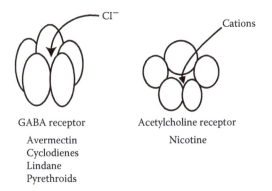

GABA receptor	Acetylcholine receptor
Avermectin	Nicotine
Cyclodienes	
Lindane	
Pyrethroids	

FIGURE 4.7 The GABA and the acetylcholine receptors and toxicants that bind to them.

been observed by electron microscopy. Agonists for the GABA receptor include drugs such as barbiturates and benzodiazepines; antagonists include lindane, an insecticide. *Nicotine* is an agonist of the nicotinic cholinergic receptor, whereas *curare*, the muscle relaxant and Amazonian hunter's arrow-tip poison, acts as a blocker at that receptor.

Toxicants may also interact with receptors that are coupled to intracellular effectors through intermediaries such as *G proteins* or by enzymes such as tyrosine kinases. G proteins are comprised of three subunits: an alpha subunit, a beta subunit, and a gamma subunit. In their resting state, these three subunits can be found to be associated together in the membrane near their affiliated receptor. In this resting state, the alpha subunit of the G protein is bound to a molecule called guanosine diphosphate (GDP). When a ligand binds to the receptor, the structure of the receptor is altered and it develops a high binding affinity for the G protein complex. Binding of the complex to the receptor triggers the replacement of GDP with the molecule guanosine triphosphate (GTP), which in turn leads to the dissociation of the alpha subunit from the rest of the protein. The alpha subunit is now activated and can bind to various intracellular targets and either activate or inactivate them. Binding of the alpha subunit to a target stimulates hydrolysis of GTP to form GDP, which inactivates the alpha subunit. The inactivated subunit returns to the other two subunits, and the G protein returns to its resting state, ready to be activated again. This process is shown in Figure 4.8. One example of a receptor that is linked to G proteins is the *muscarinic acetylcholine receptor. Muscarine toxin* from the fly agaric mushroom, *Amanita muscaria*, is an agonist for this receptor; *atropine*, another toxin of botanical origin, is an antagonist.

FIGURE 4.8 The steps involved in the activation of a target protein, following the binding of a ligand to a G-protein-linked receptor.

Many ligand–receptor systems (such as G protein-linked receptors) ultimately produce their effects through activation of *second messenger systems*, which link the signal produced by binding of the ligand to the receptor with functional changes within the cell. Often, binding activates the enzyme *adenylate cyclase*, which catalyzes the conversion of ATP into *cyclic AMP*. Cyclic AMP is then involved in the activation of a group of enzymes called protein kinases, which catalyze the phosphorylation of other proteins, leading to changes in the cell function. In an alternative system, activation of an enzyme, *phospholipase C*, leads to the production of the molecules diacylglycerol and inositol (1,4,5)-triphosphate, which can themselves activate protein kinases or produce increases in intracellular calcium levels.

Second messenger systems can also be targets for toxicants. There is some evidence that interference with calcium metabolism, for example, may be a factor in the mechanisms by which some chemicals produce cell death. Other studies, however, point to abnormal increases in calcium levels in dying cells as a result of rather than as the cause of cellular injury (more about this when we discuss cell death later in the chapter).

EFFECTS OF TOXICANTS ON VOLTAGE-ACTIVATED ION CHANNELS

Action Potential

See also:
  Effects of Toxicants on
  Electrical Conduction
                Chapter 10

Another category of transmembrane ion channels includes those that respond to local voltage change in charged membranes such as the nerve axon. The passage of an impulse through the nerve axon is an electrical phenomenon, in that the impulse is actually a wave of ionic depolarization. Because neurons are polarized with a negative internal charge, the opening of sodium ion channels has a depolarizing effect as positive ions flow inward with the sodium gradient. Following the opening of sodium channels, a second wave of channels, in this case potassium channels, also opens. The opening of these channels allows the potassium gradient to be dispelled through the movement of potassium ions out of the neuron. Once initiated, the wave of depolarization initiated by the sodium channel opening passes completely down the axon in a wave known as the *action potential*. The action potential requires no energy; however, once completed, ATP is used to drive ion pumps to restore the resting sodium and potassium gradients.

Upon reaching the presynaptic membrane at the end of a neuron, the electrical impulse must be converted into a chemical signal for transmission of the signal to neighboring neurons. First, voltage-activated calcium channels open and calcium flows into the cells due to a very strong concentration gradient. The change in the intracellular calcium concentration then leads to the release of a neurotransmitter, which can then cross the synapse to bind to receptors on the postsynaptic neuron.

Both sodium and potassium ion channels have been described by cloning and sequencing of the genes that encode them. In the process of cloning genes for channels, the expression of activity can be observed by injecting mRNA into *Xenopus* oocytes, which then produce the protein and incorporate the channels into the oocyte membrane. If the protein is produced and incorporated, voltage stimulus results in channel opening.

Voltage-activated sodium channels are the molecular sites of action of *tetrodotoxin* (TTX), an alkaloid found in skin and gonads of the globe fish, *Spheroides rubripes*, and in certain newts and frogs. *Saxitoxin* (STX), from the dinoflagellates *Gonyaulax catenella* and *G. tamatensis* (of poisonous red tide), also acts specifically on the sodium ion channel. Both toxins block a tetrodotoxin-sensitive sodium channel (binding in a noncovalent manner) and are among the most toxic naturally occurring chemicals known, with median lethal doses in mice of approximately 10 mg/kg by intraperitoneal injection. *Batrachotoxin*, on the other hand, increases permeability to sodium. It appears that poisoning of only a small percentage of sodium ion channels can be lethal. This is due to the necessity of maintaining a resting potential that is more negative than the threshold for the action potential. Leakage of sodium ions through a few poisoned channels results in a resting potential that is too close to the threshold, causing hyperexcitability of the neuron.

Tetrodotoxin, Saxitoxin

See also:
  Effects of Toxicants on
    Electrical Conduction
                        Chapter 10

  Cyanobacterial Toxins
  Dinoflagellates and
    Paralytic Shellfish Toxins
  Mollusks
  Fish, Birds, and Mammals
                        Chapter 20

  TTX, STX
                        Appendix

Effects of Toxicants on Transport Proteins

Carbon Monoxide

See also:
  Effects of Toxicants on
    Hemoglobin
                        Chapter 9

  Carbon Oxides
                        Chapter 17

  Carbon Monoxide
                        Appendix

Although there are several different proteins that function in transport, probably the best known is *hemoglobin*, an oxygen-carrying molecule found in red blood cells. In vertebrates, the hemoglobin molecule is made up of four amino acid chains, with each chain possessing an iron-containing structure called a heme group, which is capable of carrying a molecule of oxygen. The binding of oxygen to the heme group of one of the amino acid chains alters the conformation of the remaining chains to make oxygen binding easier. Likewise, the release of oxygen by one heme group produces a conformational change that encourages oxygen release by the other heme groups as well. The net result is that hemoglobin loads oxygen easily in areas rich in oxygen and unloads it promptly in areas low in oxygen. Around 98% of the oxygen in the bloodstream is carried by hemoglobin and the remaining 2% is dissolved in the blood plasma.

Carbon monoxide (CO) is a toxicant that interferes with the functioning of hemoglobin by competing with oxygen for binding to the heme groups. In humans, hemoglobin has a much higher affinity for carbon monoxide than for oxygen (by a factor of 200), so even very small amounts of carbon monoxide can effectively block oxygen binding. Carbon monoxide poisoning is particularly insidious because the potential compensatory responses to oxygen deprivation are not triggered by lowering of oxygen binding by hemoglobin, but only by changes in dissolved oxygen levels (which are affected little, if at all, by moderate levels of carbon monoxide). Thus, in carbon monoxide poisoning, there is no sensation of struggling for breath, but just gradual loss of consciousness. Poisoning is treated with oxygen, often with a little carbon dioxide added to stimulate breathing. Under normal circumstances, less than 1% of a person's circulating hemoglobin carries carbon monoxide; for smokers, however, the figure is closer to 5%–10% because of exposure to the carbon monoxide given off in cigarette smoke.

Nitrates and Nitrites

See also:
Vascular Spasms and Blood
    Pressure
Effects of Toxicants on
    Hemoglobin
                        Chapter 9

Phosphorus and Nitrogen
                        Chapter 17

Preservatives and
    Antioxidants
                        Chapter 19

Nitrates, Nitrites
                        Appendix

Another alteration that can prevent hemoglobin from carrying oxygen is the oxidation of the heme iron from the ferrous (Fe^{+2}) to the ferric (Fe^{+3}) state. A hemoglobin molecule with these oxidized heme irons is called *methemoglobin*. A small percentage of circulating hemoglobin exists normally in this state; however, exposure to certain chemicals can dramatically increase methemoglobin levels, leading to a condition called *methemoglobinemia*. One such group of chemicals is the nitrates. These highly water-soluble compounds found in sewage wastes and fertilizers are frequent groundwater pollutants. They are also used in the processing and preservation of meats. Nitrates are converted in the gastrointestinal system to nitrites, which then oxidize the heme iron to produce methemoglobin.

Methemoglobinemia is relatively rare in adults, because of the existence of a biochemical system that can reduce the oxidized heme iron, converting methemoglobin back to hemoglobin. Infants, however, are deficient in this enzyme, a factor that puts them at special risk. Most cases of methemoglobinemia occur in infants in rural areas where nitrate-contaminated well water is used to prepare formula. Affected babies literally turn blue, but termination of exposure and prompt medical attention usually lead to recovery. One possible treatment for methemoglobinemia is administration of the compound *methylene blue*, which stimulates reduction of the oxidized heme iron.

Interestingly enough, there is one case in which formation of methemoglobin is deliberately induced and that is the treatment of cyanide poisoning. Nitrites are administered in order to produce moderate levels of methemoglobin, to which cyanide binds with an even higher affinity than it does to cytochrome oxidase. Administered along with the nitrites is a compound called sodium thiosulfate, which converts the cyanide in the bloodstream to thiocyanate (which can then be excreted).

EFFECTS OF TOXICANTS ON PROTEINS INVOLVED IN THE REGULATION OF GENE EXPRESSION

The information necessary for a cell to build the RNA and protein molecules that carry out its functions is, of course, encoded in its DNA. Because the DNA in each somatic cell of a eukaryotic organism is identical, whereas cellular functions are not, there must clearly be mechanisms in place to regulate the way in which the information in DNA is utilized by each individual cell. Cells are also obviously able to alter their function in response to both short- and long-term changes in environmental conditions, a capability that would be difficult, if not impossible, if DNA expression was static and unchanging.

The mechanisms by which DNA expression is controlled have been gradually elucidated in recent years and have been found to involve the interaction between proteins known as *transcription factors* and sites on DNA known as *promoter regions* where transcription is typically initiated. Transcription factors bind to these regions (as well as other regulatory regions on the DNA molecule) and then interact with polymerases and other molecules to either promote or block transcription. They may be constitutively present and active or require activation through interaction with other molecules. Examples include CREB, NF-κB, PPARα, and c-Myc, to name just a few of the approximately 2000 known human transcription factors.

Cytochrome p450

See also:
  The Role of Cytochrome P450
                                    Chapter 3

  Metabolomics
                                    Chapter 5

Because transcription factors typically contain not only a DNA-binding domain, but also *trans*-activating domains for binding with other endogenous ligands, they are also potential targets for toxicant binding. This binding may either increase or decrease activity of the transcription factor.

One example of how interaction of a toxicant with a transcription factor can trigger transcription is the induction of the enzyme cytochrome P450 by *xenobiotics* (foreign substances). The *cytochrome P450-dependent monooxygenases* are a large group of enzymes important in the metabolism and detoxification or activation of many xenobiotic compounds. Levels of the various forms of P450 are responsive to the concentrations of xenobiotics in the body. In other words, the presence of certain xenobiotics can trigger an increase in the synthesis of P450 enzymes, a process called *induction*. For at least some forms of P450, induction appears to be mediated by effects on a transcription factor. For example, the xenobiotic chemical dioxin (also known as TCDD) has the ability to induce the synthesis of one form of P450 (CYP1A) through interaction with a specific transcription factor known as the *Ah receptor*. TCDD enters liver cells (where most xenobiotic metabolism occurs) and binds to the Ah receptor in the cytoplasm. The *receptor–ligand complex* then moves into the nucleus, where it acts as a transcription factor, interacting with DNA to initiate the transcription of several genes, including the gene that encodes the form of P450 that metabolizes TCDD. Many other xenobiotics with chemical structures similar to that of TCDD also act as agonists for this same cytosolic receptor and can also initiate synthesis of the enzyme. Other forms of P450 are induced by other chemicals (including phenobarbital, ethanol, and other drugs and toxicants). Inhibitors of P450 include carbon monoxide and piperonyl butoxide.

Of course, toxicants can also impact transcription through direct interaction with the regulatory regions of DNA, a mechanism that will be explored further in the chapters on carcinogenesis and teratogenesis.

EFFECTS OF TOXICANTS ON LIPIDS

Organic Solvents

See also:
  Other Neurotoxicants
                                   Chapter 10

Lipids, of course, play a major role in the structure and function of cell membranes. Cell membranes are composed of a *phospholipid bilayer* (Figure 4.9), with an internal hydrophobic region consisting of the hydrocarbon tails and an external hydrophilic region made up by the phosphate heads of the phospholipids. Certain proteins, including receptors and ion channels, have a tertiary structure with strongly hydrophobic regions, allowing them to be imbedded in the membrane. Membranes are dynamic, having fluidity as well as constant turnover with the incorporation of newly synthesized components.

Highly lipophilic substances may dissolve readily into membranes because they reach much higher equilibrium concentrations in lipid than in the blood. Organic solvents and anesthetic gases are among the substances with narcotic activity that appears to be correlated with lipid solubility. These compounds may dissolve in membranes of the central nervous system, the heart, and other organs, most likely exerting their toxic effects through alteration of the membrane structure and function. This creates a situation in which a compound can affect cellular function without specific binding to the molecule involved. For example, alterations of membrane fluidity can affect the function of membrane-bound proteins by altering the physical and chemical characteristics of their surroundings. There is, in fact, evidence that high concentrations of anesthetics increase lipid fluidity of membranes; however, this effect is not observed at normal anesthetic concentrations. Membranes can also be disrupted by detergents, as well as by strong acids and bases.

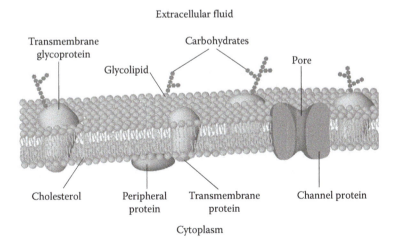

FIGURE 4.9 The phospholipid bilayer and proteins as typically found in membranes. (Modified with permission from http://www.shutterstock.com/pic-97190063/stock-vector-structure-of-plasma-membrane.html?src=&ws=1. Accessed on June 21, 2014.)

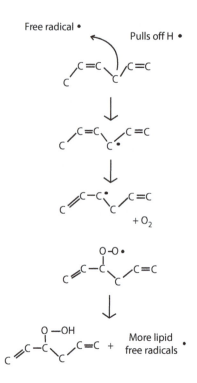

FIGURE 4.10   The steps involved in lipid peroxidation. A hydrogen atom (1P+1E) is pulled from a polyunsaturated fatty acid (which has weak C—H bonds). The resulting radical reacts with oxygen to form a peroxyl radical, which can pull a hydrogen off another fatty acid, thus propagating the cycle.

The fatty acid chains of many membrane phospholipids are unsaturated (contain double bonds). These unsaturated fatty acids are susceptible to damage through a process called *lipid peroxidation*. In lipid peroxidation, *free radicals* (molecules with unpaired electrons) formed from halogenated hydrocarbons and other xenobiotics attack fatty acids, removing hydrogen atoms and converting the fatty acids into free radicals themselves. These fatty acid free radicals may then react with oxygen to form additional free radicals and unstable peroxides (which can also break down, yielding even more free radicals). Thus, the process spreads, which can lead to structural and functional damage not only to the plasma membrane, but also to the membranes of cellular organelles, such as the endoplasmic reticulum (Figure 4.10). Damage done by free radicals can be moderated through the presence of glutathione or other molecules, which can scavenge free radicals.

EFFECTS OF TOXICANTS ON NUCLEIC ACIDS

The nucleic acid DNA is built of units called *nucleotides*, each of which consists of a base (adenine, guanine, cytosine, or thymine), a phosphate, and a deoxyribose sugar. Nucleotides can be linked together through covalent bonds between the phosphate of one nucleotide and the sugar of the next. The DNA molecule itself consists of two *complementary* chains of nucleotides, meaning that each nucleotide in one chain is paired opposite a specific partner in the other chain. A nucleotide containing the base cytosine will always pair with a nucleotide containing the base guanine. Likewise, a nucleotide containing adenine will always pair with a nucleotide containing thymine. The two chains are held together by hydrogen bonds that form between the base pairs and twist the chains together in a form called a double helix (Figure 4.11).

Mutagenesis

See also:
   The Mutational Theory of
   Carcinogenesis
                                    Chapter 6

The toxicants that interact with and produce changes in the cellular DNA are described by the general term *mutagens*. Most mutagens interact with the base portions of nucleotides. Some mutagens, such as nitrous acid, may delete portions of bases, removing, for example, an amine group from adenine or cytosine (a process called *deamination*). Other compounds (mustard gas, for example) act as *alkylating agents*, adding alkyl groups to bases, while still others replace nucleotides that they closely resemble. Also, exposure to ultraviolet light and ionizing radiation can cause cross-linking between DNA bases. Enzyme systems exist that can deal with such damage, repairing or removing and replacing damaged nucleotides. Unrepaired damage, however, can lead to mutagenicity or carcinogenicity due to misreading or incorrect replication of DNA during cell division. These topics will be discussed in more detail in Chapter 5.

FIGURE 4.11 The double helix structure of DNA. (Modified with permission from http://www.shutterstock.com/pic-158504498/stock-vector-vector-drawing-of-a-double-helix-double-helix-science-image-of-a-dna-double-helix-straight-and.html?src=&ws=1. Accessed on June 21, 2014.)

A relatively recently discovered class of nucleic acids which play an important role in the cell are the *microRNAs* (*miRNAs*). Hundreds of genes have been discovered, which code for these small RNAs which, after processing, are little more than 21 or 22 nucleotides in length. The genes for these highly conserved molecules (found in multiple eukaryotic species) are not infrequently found within introns of other genes. miRNAs play a role in regulating protein synthesis, by binding to and inhibiting translation of complementary messenger RNA (mRNA). The function of miRNAs has been studied through the use of anti-miRNAs, which are complementary oligo-nucleotides designed to bind to and block miRNA function. Perhaps surprisingly, detecting developmental effects produced by knocking out individual miRNAs has proven to be difficult, a phenomenon that may be due to redundancy (miRNAs with overlapping function). However, there is some evidence that effects can be seen in responses to environmental stressors, and patterns of miRNA expression have been seen to differ in disease states (including cancer and cardiovascular disease). Thus, miRNAs may well turn out to play an important role in response to toxicants.

MECHANISMS OF CELL DEATH

Cell death is a complicated phenomenon. At times, of course, cell death is a neces-sary event in the life of a multicellular organism. During the process of development, for example, structures are formed that will be removed before the process is com-plete (flaps of tissues between the fingers, for example). At other times, however, cell death may be a response to overwhelming damage inflicted upon a cell by external agents. Such a mechanism could also, of course, still be advantageous to the organ-ism by removing cells that are damaged and potentially dysfunctional.

Observations of cells have shown that there are at least two distinct patterns of cell death, which can be differentiated on the basis of the steps that the cells undergo. These have been termed *apoptosis*, or programmed cell death and *necrosis*, typically triggered by injury to the cell. However, a third cellular process, *autophagy*, or self-digestion, may also play an important role in cell death. Beginning with apoptosis, we will discuss the characteristics of each, as well as the circumstances under which each comes into play.

Apoptosis

Apoptosis, or programmed cell death, involves a number of readily observable cellu-lar changes. In cells that are undergoing apoptosis, the cellular chromatin condenses near the nuclear membrane, and the nucleus and the cytoplasm shrink and then dis-sociate into fragments (*apoptotic bodies*) that are then phagocytized by neighboring cells. In fact, rearrangement of cell surface molecules can serve to mark apoptotic cells for macrophage binding.

Research in recent years has been focused on understanding the molecular steps involved in apoptosis. It is clear that in most cells, proteins known as *caspases* are the ultimate "executioners." These cysteine-dependent aspartate-specific proteases attack various structural and functional protein targets within the cell, leading to cellular death. They also activate some of the DNases that produce the fragmentation of DNA that is characteristic of apoptosis. But what triggers caspase activation? Evidence indicates that in many cases, a mitochondrial protein called *cytochrome c* may play a significant role. Cytochrome c is coded for by a nuclear gene. Following its synthesis, the molecule is transported into the mitochondria, where further modifications (including the addition of a heme group) take place. Cytochrome c is released from the mitochondria during the process of apoptosis, after which it binds to a protein called *Apaf-1* and initiates a cascade of caspase activation.

Release of cytochrome c from the mitochondria appears to occur as part of what has been called the *mitochondrial permeability transition* (MPT), where the opening of a channel termed the *permeability transition pore* (PTP) leads to a dramatic increase in permeability that allows the release of cytochrome c and other molecules. During the MPT, mitochondria become permeable to anything smaller than 1500 kDa, and the opening of even a single pore may be sufficient to produce the loss of membrane potential that accompanies the MPT. Loss of cytochrome c itself, of course, will shut down the electron transport chain, and this along with MPT-related depolarization effectively derails oxidative phosphorylation. In fact, mitochondria that have undergone the MPT hydrolyze, rather than synthesize, ATP, potentially leading to significant cellular ATP depletion. Some ATP, however, appears to be required in order for cells to complete the apoptotic pathway in what has become known as the *intrinsic* or *mitochondrial pathway* of apoptosis.

There are a number of molecules that play a role in either promoting or inhibiting apoptosis. Evidence has indicated that the influence of many pro-apoptotic regulator molecules, including the proteins *Bax*, *Bid*, and *Bad*, as well as anti-apoptotic regulator proteins such as *Bcl-2* and *Bcl-xL*, is mediated through their ability to facilitate or block the MPT and thus block the mitochondrial pathway. Other molecules influencing this pathway include *inhibitor of apoptosis proteins (IAPs)*, which bind to and inhibit caspases. IAPs are themselves regulated by the proteins *second mitochondrial activator of caspases (SMAC)* and *HtrA2*, both of which block IAPs from inhibition of caspases.

Some studies, however, have pointed to alternative mechanisms for apoptosis beyond the mitochondrial pathway, some of which may bypass the mitochondria. The *extrinsic* or *receptor-mediated pathway*, for example, leads to activation of caspases through formation of a *death-inducible signaling complex* on the cell surface. Finally, the discovery that cell death can and does occur in mice that are genetically engineered to be deficient in certain caspases has indicated the existence of other pathways that may involve mitochondria, and yet, not caspases. A protein called *apoptosis inducing factor* (AIF), which can induce apoptotic changes, has been identified and may play a role in these pathways, as might the proteins *endonuclease G* and/or *granzyme A*. An additional mechanism for some apoptosis inhibitors may to bind to and block active caspases.

Multiple molecules that have been shown to play a role in the regulation of apoptosis include trophic or growth factors and tumor *necrosis* factor.

NECROSIS

Apoptosis stands in contrast to another type of cell death called *necrosis*, which is characterized by swelling of the cell and nucleus, swelling of mitochondria, and membrane disruption. Necrosis is typically triggered by injury to the cell and can release factors that are harmful to surrounding cells. At one time, the hypothesis was that toxic injury inevitably led to necrosis rather than to apoptosis, but that has proven to be incorrect. Actually, several chemicals are capable of inducing either necrosis or apoptosis, depending on the dose (with higher doses producing necrosis and lower doses producing apoptosis) or on other factors such as the energy state of the cell, i.e., the amount of ATP present. There has even been a description of a type of cell death termed "necroptosis," which resembles necrosis morphologically but involves receptors typically associated with apoptosis.

The process of necrosis, like that of apoptosis, most likely involves initiation of the MPT, and one hypothesis holds that the number of mitochondria undergoing MPT is a major determinant in the apoptosis/necrosis decision. Cells that are less severely damaged have fewer mitochondria involved and may either survive the damage or, at a certain threshold of involvement, trigger their apoptotic program. Cells that are more severely damaged, however, may have virtually all of their mitochondria involved and may not have sufficient cellular resources to carry out the apoptotic program. These cells would then undergo necrosis. Lack of appropriate caspase activation (by reactive oxygen species, for example) might also tend to divert a cell from an apoptotic pathway onto a necrotic one.

One factor that has been associated with necrosis (and also, in some cases, apoptosis) is impairment of calcium regulation. Toxicants may directly affect calcium levels (which are normally quite low in the cytoplasm of the cell) through interference with calcium transporters or through interference with energy metabolism (preventing the cell from carrying out active transport of calcium or other ions). Elevated calcium levels in the cytoplasm can stress mitochondria (which play an important role in cellular calcium homeostasis and signaling), as they work to pick up and sequester the excess calcium. High levels of calcium have been shown to stimulate the respiratory chain and thus can lead to the production of reactive oxygen species. Excess calcium can also interact with the cytoskeletal protein actin (leading to blebbing of membranes) and can activate damaging hydrolytic enzymes. This can lead to the total breakdown of the cellular mechanisms that have been associated with progression of a cell to necrosis.

AUTOPHAGY

Finally, a mechanism which is closely tied to cell death is *autophagy*, a pathway for degradation of cellular components. In autophagy, membrane surrounds an area of cytoplasm to form an *autophagosome*, which then merges with a lysosome, resulting in the digestion of the material contained within. An increase in autophagy can be induced by a variety of stressors (including starvation), and amino acids released by digestion can be used to produce new proteins or to fuel cellular metabolism.

Autophagy undoubtedly plays an important role in normal cellular processes, removing damaged components and recycling cellular materials. The extensive tissue remodeling that takes place in many organs during growth and development is also undoubtedly dependent on autophagy. It may also act as a survival mechanism (for example, in cells that are undergoing starvation) and, in some cases, may even protect against apoptosis, reducing release of cytochrome c. Or it may ultimately lead to apoptosis, necrosis, or to a type of cell death, sometimes termed "autophagic cell death," which is characterized by autophagy without the condensation of chromatin seen in classic apoptosis.

STRESS, REPAIR, AND RECOVERY

Even damaged cells, of course, may undergo repair if the damage is not so severe as to trigger apoptosis or necrosis. On the molecular level, damaged proteins, lipids, and nucleic acids may be repaired or resynthesized. For example, oxidation of sulfhydryl groups on proteins may be enzymatically reduced by thioredoxins and glutaredoxins; also, endogenous DNA repair systems are a major factor in protection against carcinogenesis.

DNA Repair

See also:
Protection Against the
Development of Cancer
Chapter 6

Cells that have been stressed by chemical insult may also alter their basic metabolic processes, dramatically reducing protein synthesis. This response to stress has been documented in many different biological systems and typically features the induction (increased synthesis) of a specific group of proteins, which were first termed heat shock proteins due to their initial discovery in cells exposed to hyperthermia (elevated temperatures). An entire family of these proteins, now more generally known as *stress proteins*, has since been identified. Ranging in size from approximately 15 to 110 kDa in molecular weight, some of these proteins are constitutive (are found in the cell under normal conditions), whereas others have been found to be induced in response to a variety of cellular stresses, including heavy metals, oxidative stress, and ischemia.

The manufacture of the inducible stress proteins seems to be an adaptive response, conferring resistance to further stresses. The mechanism of how induction occurs in eukaryotes is relatively well understood and involves binding of an activated *heat shock factor protein* (HSF) to a responsive *heat shock element* (HSE) in the genome, which initiates the processes of transcription and translation for these proteins.

The existence of stress proteins in many species certainly argues for a central role for these proteins in fundamental cell processes, and heat shock protein induction has, in fact, been explored in relationship to a number of basic cellular phenomena. Many stress proteins seem to function as molecular chaperones by regulating protein folding, whereas others play a role in regulating the function of receptors such as the glucocorticoid receptor. Stress proteins may also play a role in cell death and possibly in oncogenic transformation. Interestingly, recent evidence indicates that stress proteins may play an important role in evolution as well, as overloading stress proteins appears to increase genomic instability and thus increase the occurrence of mutations. This, in turn, would lead to greater phenotypic diversity in offspring, which might enhance survival.

Proteins which are too far degraded to be repaired and/or refolded may be targeted for destruction by the *proteasome*, a complex of proteins with protease activity. Proteins are targeted for degradation through the addition of a small protein called *ubiquitin*.

On a larger scale, cells that have undergone apoptosis or necrosis may be replaced by stimulation of cell division in the cells that remain behind. Regeneration of damaged liver tissue through mitotic division of the remaining hepatoctyes, for example, begins within a few hours after damage occurs. Although mitosis can occur across a variety of cell types, dedicated populations of cells may play an important role in cell replacement. Examples would include Clara and type II cells in the lung and *oval cells* in the liver. Stimulation of this replacement activity is typically mediated by the release of chemical signaling molecules from the damaged cells and may involve macrophages and other cells of the immune system. These would include molecules such as *tumor necrosis factor* (TNF), as well as various interleukins, which activate pathways involved in growth and proliferation.

Inflammation

See also:
　The Inflammatory Response
　　　　　Chapter 13

Fibrosis

See also:
　Chronic Obstructive
　　Pulmonary Disease:
　　Bronchitis and
　　Emphysema
　　　　　Chapter 8

　Fibrosis and Cirrhosis
　　　　　Chapter 11

In general, tissue response to injury is termed *inflammation* and typically involves increase in the blood flow to the injured area, increased capillary permeability, and recruitment of immune system cells such as macrophages, white blood cells, and fibroblasts to the area. Although inflammation is normally a beneficial response, excessive activity of cells such as fibroblasts can lead to the accumulation of extracellular materials like collagen to a degree that hampers the functionality of the tissue. This overproduction is termed *fibrosis* and is a factor in chemical-induced tissue damage in many organ systems.

CASE STUDY　Cyclooxygenase Inhibitors

One example of a group of drugs with a mechanism of action that produces both therapeutic effects and side effects is the *cyclooxygenase* (COX) inhibitors. COXs are enzymes that catalyze the conversion of arachidonic acid (which is released from cell membranes by another class of enzymes called phospholipases) into a compound called prostaglandin H_2 (PGH_2), which serves as the starting point for the synthesis of a variety of other *prostaglandins* (PGE_2 and PGI_2), as well as a compound called *thromboxane* (TXA).

Prostaglandins and thromboxane are found in almost all tissues and mediate a variety of processes, including vasodilation, vasoconstriction, platelet aggregation, chemotaxis, smooth muscle contraction, smooth muscle relaxation, kidney function, and pain sensation. These effects are mediated by the binding of various prostaglandins and thromboxane to as many as nine different categories of G protein-linked receptors. The response of a tissue to prostaglandins depends upon which prostaglandins are produced in that tissue as well as which receptors are present.

It has long been known that prostaglandin-mediated physiological effects could be blocked by compounds that act as inhibitors of COXs. One example of a drug that interferes with the prostaglandin system is *aspirin* (acetyl-salicylic acid, a derivative of the naturally occurring compound salicylate). Aspirin irreversibly inhibits COX through acetylation of a hydroxyl group of a serine residue on the enzyme, which then prevents binding of arachidonic acid. The resulting blockade of prostaglandin synthesis is then responsible for the therapeutic effects of aspirin:

- Antipyresis (lowering of elevated body temperature)
- Analgesia (pain relief)
- Anti-inflammatory effects
- Reduction in platelet aggregation
- Reduction in cancer risk

as well as the side effects of aspirin therapy:

- Gastric ulceration (due to effects on acid secretion)
- Acid–base disturbances (due to effects on respiration, metabolism, and kidney function)
- Hepatotoxicity, including association with Reye's syndrome (see Chapter 11), a rare condition seen most often in children who have been treated with aspirin for viral illnesses
- Prolongation of bleeding time
- Central nervous system effects, ranging from tinnitus (ringing in the ears) to respiratory depression and coma

Recently, it was discovered that the COX enzyme is actually a family of enzymes. The COX-1 isoform is expressed constitutively (in other words, is present all the time) and seems to play a role in the maintenance of normal tissue homeostasis as well as protection of the gastric mucosa. In contrast, the COX-2 isoform is induced during inflammation (the tissue's response to injury; see Chapter 13) and seems to play a major role in that process. Interestingly, the active site of COX-2 was discovered to differ somewhat from the active site of COX-1, raising the possibility of selective inhibition of the two isoforms. In particular, there was a great deal of interest in the possibility of blocking the inflammatory effects mediated by COX-2 without impacting the homeostatic functions mediated by COX-1.

This new opportunity was taken advantage of by a number of pharmaceutical companies, who moved quickly to develop selective COX-2 inhibitors. These drugs included Celebrex® (celecoxib), Vioxx® (rofecoxib), and Mobic® (meloxicam), and they were quickly approved by the FDA (in 1998, 1999, and 2000, respectively) and put to use treating patients with rheumatoid arthritis, osteoarthritis, and other inflammatory conditions. However, in 2004, a long-term study of the effect of rofecoxib on development of recurring colon polyps was terminated early due to data that indicated an increased risk of heart attack and stroke in patients on rofecoxib, compared with patients on a placebo. Merck and Co., Inc., the maker of Vioxx®, then voluntarily withdrew the drug from the market. Shortly thereafter, a second clinical trial, involving the effect of celecoxib on development of recurring colon polyps, was also terminated early following similar results. A study with lower doses of celecoxib did not show increased risk of cardiovascular problems and was not terminated.

The link between mechanism of action and physiological effects of inhibition of the COX enzymes is not yet completely understood, but hypotheses have been suggested. It appears that the decrease in the production of PGI_2, which promotes vasodilation, unaccompanied by a decrease in TXA2 (which is produced by COX-1 and promotes vasoconstriction), may be responsible for the increased risk. Also, questions have been raised as to whether the potential cardiovascular risk apparently associated with these drugs should have been identified earlier, rather than many years following drug approval. The one thing that is certain is that work remains to be done in fully understanding the mechanisms of action of this clinically important category of drugs.

BIBLIOGRAPHY

Antonsson, B. and Martinou, J.-C., The Bcl-2 protein family, *Exp. Cell Res.*, 256, 50, 2000.

Bloom, J.C. and Brandt, J.T., Toxic responses of the blood, in *Casarett and Doull's Toxicology*, Klaassen, C.D., Ed., McGraw-Hill, New York, 2001, chap. 11.

Cannon, C.P. and Cannon, P.J., COX-2 inhibitors and cardiovascular risk, *Science*, 336, 1386, 2012.

Chandrasekharan, N.V., Dai, H., Lamar Turepu Roos, K., Evanson, N.K., Tomsik, J., Elton, T.S., and Simmons, D.L., COX-3, a cyclooxygenase-1 variant inhibited by acetaminophen and other analgesic/antipyretic drugs: Cloning, structure, and expression, *Proc. Natl Acad. Sci. USA*, 99, 13926, 2002.

Chen, Y., Organophosphate-induced brain damage: Mechanisms, neuropsychiatric and neurological consequences, and potential therapeutic strategies, *NeuroToxicology*, 33, 391, 2012.

Danial, N.N. and Korsmeyer, S.J., Cell death: Critical control points, *Cell*, 116, 205, 2004.

Ecobichon, D.J., Toxic effects of pesticides, in *Casarett and Doull's Toxicology*, Klaassen, C.D., Ed., McGraw-Hill, New York, 2001, chap. 22.

FitzGerald, G.A., Coxibs and cardiovascular disease, *N. Engl. J. Med.*, 351, 1709, 2004.

Green, D.R. and Kroemer, G., The pathophysiology of mitochondrial cell death, *Science*, 305, 626, 2004.

Green, D.R. and Victor, B., The pantheon of the fallen: Why are there so many forms of cell death?, *Trends Cell Biol.*, 22(11), 555, 2012.

Gregus, Z., Mechanisms of toxicity, in *Casarett and Doull's Toxicology*, Klaassen, C.D., Ed., McGraw-Hill, New York, 2013, chap. 3.

Hata, A.N. and Breyer, R.M., Pharmacology and signaling of prostaglandin receptors: Multiple roles in inflammation and immune modulation, *Pharmacol. Ther.*, 103, 147, 2004.

Hendrick, J.P. and Hartl, F.-U., Molecular chaperone functions of heat-shock proteins, *Annu. Rev. Biochem.*, 62, 349, 1993.

Herman, B., Gores, G.J., Nieminen, A.-L., Kawanishi, T., Harman, A., and Lemasters, J.J., Calcium and pH in anoxic and toxic injury, *CRC Crit. Rev. Toxicol.*, 21, 127, 1990.

Juni, P., Nartey, L., Reichenbach, S., Sterchi, R., Dieppe, P.A., and Egger, M., Risk of cardiovascular events and rofecoxib: Cumulative meta-analysis, *Lancet*, 364, 2021, 2004.

Kaplan, K.B. and Li, R., A prescription for "stress"—The role of Hsp90 in genome stability and cellular adaptation, *Trends Cell Biol.*, 22, 576, 2012.

Lemasters, J.J., Nieminen, A.-L., Qian, T., Trost, L.C., Elmore, S.P., Nishimura, Y., Crowe, R.A., Cascio, W.E., Bradham, C.A., Brenner, D.A., and Herman, B., The mitochondrial permeability transition in cell death: A common mechanism in necrosis, apoptosis and autophagy, *Biochim. Biophys. Acta*, 1366(1-2), 177, 1998.

Lemasters, J.J., Qian, T., Bradham, C.A., Brenner, D.A., Cascio, W.E., Trost, L.C., Nishimura, Y., Nieminen, A.-L., and Herman, B., Mitochondrial dysfunction in the pathogenesis of necrotic and apoptotic cell death, *J. Bioenerg. Biomembr.*, 31(4), 305, 1999.

Mendell, J.T. and Olson, E.N., MicroRNAs in stress signaling and human disease, *Cell*, 148, 1172, 2012.

Mizushima, N. and Komatsu, M., Autophagy: Renovation of cells and tissues, *Cell*, 147, 728, 2011.

Moreland, D.E., Effects of toxicants on electron transport and oxidative phosphorylation, in *Introduction to Biochemical Toxicology*, Hodgson, E. and Smart, R.C., Eds, Elsevier, New York, 2001, chap. 13.

Morimoto, R.I., Kline, M.P., Bimston, D.N., and Cotto, J.J., The heat shock response: Regulation and function of heat-shock proteins and molecular chaperones, *Essays Biochem.*, 32, 17, 1997.

Moseley, P.L., Heat shock proteins and heat adaptation of the whole organism, *J. Appl. Physiol.*, 83(5), 1413, 1997.

Orrenius, S., Nicotra, P., and Zhivotosky, B., Cell death mechanisms and their implications in toxicology, *Toxicol. Sci.*, 119, 3, 2011.

Penefsky, H.S., Pullman, M.E., Datta, A., and Racker, E. Partial resolution of the enzymes catalyzing oxidative phosphorylation. II. Participation of a soluble adenosine triphosphatase in oxidative phosphorylation, *J. Biol. Chem.* 235, 3330, 1960.

Pullman, M.E., Penefsky, H.S., Datta, A., and Racker, E. Partial resolution of the enzymes catalyzing oxidative phosphorylation. I. Purification and properties of soluble, dinitrophenol-stimulated adenosine triphosphatase, *J. Biol. Chem.* 235, 3322, 1960.

Rabinowitz, J.D. and White, E., Autophagy and metabolism, *Science*, 330, 1344, 2010.

Robertson, J.D. and Orrenius, S., Role of mitochondria in toxic cell death, *Toxicology*, 181/182, 491, 2002.

Vane, J.R. and Botting, R.M., The mechanism of action of aspirin, *Thromb. Res.*, 110, 255, 2003.

Wu, W., Liu, P., and Li, J., Necroptosis: An emerging form of programmed cell death, *Crit. Rev. Oncol./Hematol.*, 82, 249, 2012.

Yang, Y., Xing, D., Zhou, F., and Chen, Q. Mitochondrial autophagy protects against heat shock-induced apoptosis through reducing cytosolic cytochrome c release and downstream caspase-3 activation, *Biochem. Biophys. Res. Commun.* 395, 190, 2010.

Zoratti, M. and Szabo, I., The mitochondrial permeability transition, *Biochim. Biophys. Acta*, 1241(2), 139, 1995.

Zucchini-Pascal, N., de Sousa, G., and Rahmani, R., Lindane and cell death: At the crossroads between apoptosis, necrosis and autophagy, *Toxicology*, 256, 32, 2009.

5 Genomics and New Genetics in Toxicology

INTRODUCTION

The study of *genomics* is based upon information first gathered in an organized way in an undertaking known as the *Human Genome Project*. This project was executed by an international consortium and resulted in a data bank of the linear sequence of DNA over not only all human chromosomes, but also those of a number of model species. These data (GenBank) are freely available to international science through the National Center for Biotechnology Information (NCBI) of the National Library of Medicine, U.S. Building on this foundation, the information in this data bank accumulates as the genomes of additional species are analyzed. The availability of these linear sequence data has led to the growing disciplines of genomics, comparative genomics, bioinformatics, systems biology, and toxicogenomics.

THE HUMAN GENOME PROJECT

Semi-automation of methods for determining the sequence of DNA, combined with the development of very large insert libraries, allowed the contemplation of defining the entire genome of humans and several model species. The Human Genome Project, authored by the U.S. National Institutes of Health, was organized internationally with primary laboratories in the United States, United Kingdom, and Japan, with one laboratory designated to analyze each chromosome. The race was joined competitively by Celera Genomics, a commercial laboratory that adopted a shotgun approach to analysis versus the hierarchical approach of the government project. (Shotgun sequencing is faster but more prone to errors.) This challenge seemed to speed the process; completion of the analysis, with a few gaps outstanding, was announced in a joint bulletin in 2000 and published in February 2002, 2 years ahead of schedule.

The human haploid genome appears to have approximately 3×10^9 base pairs (including both nuclear and mitochondrial DNA), with approximately 21,000 known protein-encoding genes. This is considerably lower than most earlier estimates predicted. When genes were sorted according to known function of their encoded proteins, one initial surprise was that a high percentage of genes appeared to encode factors with regulatory roles over other genes, including the group of proteins known as *transcription factors*. Estimates for the number of transcription factors are around 2000, with the implication that each transcription factor regulates an average of 10 genes. Of course, this is only an average; some transcription factors may regulate only a handful of genes, whereas others regulate hundreds. Perhaps, the most important finding was that approximately half of the apparent genes could not be neatly pigeonholed with other genes of known function. Thus, the project had revealed a large deposit of valuable information to be mined by the government or industry, opening up a vast potential for new drug discovery.

The remainder of the genome was often termed "junk DNA" because it did not possess the attributes of typical genes. However, the intervening years have demonstrated that much of this DNA does, in fact, play a critical role in normal functioning. This part of the genome would include sequences that code for a variety of RNAs, regulatory sequences that control the expression of genes, short and long interspersed nuclear elements (SINEs and LINEs, which are repeated DNA sequences that can move [*transpose*] within the genome), telomeres, and many other functional elements.

SORTING OUT GENES

The grouping together of genes for proteins with known functions is one aspect of the overall study of *gene ontology*, the attempt to sort out the relationships between gene products. For an orphan gene that falls into the no known category, various approaches can be applied to determine its function. One approach is to express the gene in an artificial system such as a cell culture, collect the protein that is manufactured, and analyze its function. Another approach is to *knock out* (inactivate) the gene in a model species of interest and observe the functional impact of the loss of the protein. The approach of knocking out, or negating, genes was accelerated by the serendipitous discovery *of RNA interference* (RNAi), a type of defensive mechanism in eukaryotes in which foreign RNA is degraded. It was discovered that if foreign RNA with a sequence similar to that of a specific messenger RNA (mRNA) produced by a cellular gene was introduced into a cell, the cell would destroy not only the foreign RNA, but the similar endogenous mRNA as well (Figure 5.1). This area of research can yield many knockout mutants whose physiological changes can give clues to the functions of the genes and thus advance the realm of gene ontology (Figure 5.2).

MODEL ORGANISMS AND COMPARATIVE GENOMICS

The Human Genome Project included five model organisms. The relatively small genomes of the bacterium *Escherichia coli*, the yeast *Saccharomyces cerevisiae*, and the nematode *Caenorhabditis elegans* were useful for testing various strategies for acquiring the entire sequence of an organism, and indeed, a complete analysis of those models was accomplished rapidly. The larger genome of the fruit fly, *Drosophila melanogaster*, required more time, but the vast array of genetic manipulations available in fruit flies and the various and detailed genetics maps of this species made this a very important model to include. It was eventually completed, 2 years prior to the completion of the human genome.

The house mouse, *Mus musculus*, serves as the best-known genetic model among mammals, and the completion of the mouse genome 2 years after the human genome provided many insights when the sequences were finally aligned. For example, it became apparent that the "junk DNA" of the mouse resembled that of humans in the

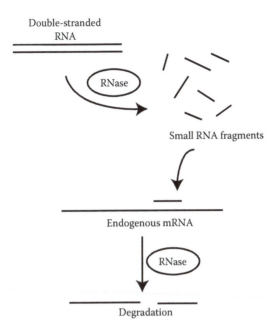

FIGURE 5.1 RNAi. Foreign double-stranded RNA is degraded by RNase. Fragments of the foreign RNA bind to complementary regions of endogenous mRNA and trigger its degradation also.

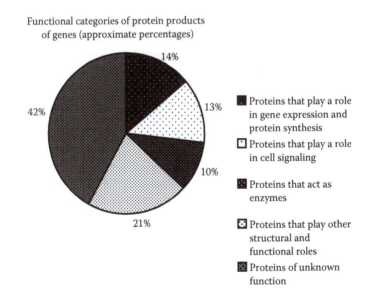

FIGURE 5.2 Functional categories of protein products of genes (approximate percentages). (Based on data from Venter, J.C. et al., *Science*, 291, 1304, 2001.)

large proportion of SINEs and LINEs. These are repeated DNA sequences that can move (*transpose*) within the genome, with LINEs coding for enzymes that can then make copies of the original sequence and insert it elsewhere in the genome. These sequences may then play a role in indirectly regulating nearby genes when they are transcribed or undergo transposition.

Another model species was *Arabidopsis thalia*, a small plant for which much genetic information had been previously accumulated. The complementary nature of a plant model offers many advantages when seeking to understand vital aspects in the structure of the human genome.

Model organisms have proven to be extremely useful in discovering the functions of newly discovered human genes. *Orthologous genes* are those having a highly similar primary sequence in another species; these are revealed from GenBank in minutes, or sometimes seconds, by a computerized search that can be executed over the Internet by simply entering any nucleotide or amino acid sequence of interest. The computerized search employs *hidden Markov models* to align the queried sequence of letters to the database of over 173 million and nearly 162 billion nucleotide bases (accumulated as of June 2014 and, of course, still increasing). GenBank returns a statistically ordered list of potential matching orthologs with links to original entries and *annotations* describing ontology and other information. When a gene has appeared to be conserved among related species in a putative evolutionary lineage, it can be declared a *homologous gene*.

Many genes appear to have been duplicated and diverged over time into *multigene families*. Members of these families are sometimes referred to as *paralogous genes*.

Cytochrome P450

See also:
  The Role of Cytochrome
  P450
                              Chapter 3

  Metabolomics
  Individual Variations
    in Metabolism and
    Pharmacogenomics
                              Chapter 5

If the same family appeared in multiple diverse species, the members would be homologous to each other; such is the case with the large superfamily of *CYP* genes, which are very important in toxicology. Another incredible example of paralogous genes is the set of *homeotic genes* discovered to control aspects of embryonic development. These were discovered first in *Drosophila* (by Edward B. Lewis, Christiane Nusslein-Vollard, and Eric Wieschaus, who were eventually rewarded with the Nobel Prize in Medicine). Work was extended in the beetle, *Tribolium casteneum*, in which it was demonstrated that the homeotic genes are present along chromosome in the same order as the segments of the beetle that each controls in embryonic development. This was the erudite observation of an insecticide toxicologist turned geneticist and now genomicist, R.W. Beeman of the U.S. Department of Agriculture. Homeotic genes are also now recognized as having homologous genes in humans.

Thus, in many cases, the function of the homologous gene in a model species can be determined experimentally and the information then extrapolated to an understanding of the human gene. For example, the technique of RNAi, as described earlier, was developed and exploited in *C. elegans* with astonishing speed and was employed by a consortium to knock out many of the genes of the nematode. The mutants produced by this technique provided clues to the function of many protein products of genes. Similarly, in *Drosophila*, the impact of insertions of known transposable elements resulting in mutant phenotypes can be traced by mapping and analysis of the DNA flanking the new insertion in a technique known as *transposon tagging*.

Comparative genomics has expanded from the original five model species as the techniques honed during the project are applied more efficiently. The next tier of mammalian genomes has included many domesticated species such as cows, pigs, dogs, cats, and horses, etc. (Table 5.1). The rat genome also will be completed and scrupulously compared with the mouse. Also notable among vertebrate species are the zebra fish, important for its amenability to embryological experimentation, and the salmon, for which commercial interests are among the incentives for understanding and manipulating its genome. A major initiative on microorganisms is also adding many bacteria and fungi to the list of genomes being analyzed. As an example of an application with practical utility in combating infectious diseases, the genomes of the malaria organism, *Trypanosoma brucei*, and its important mosquito vector, *Anopheles gambiae*, have been analyzed. This accomplishment had the additional benefit of providing close comparative genomics between dipteran insects when the mosquito and fruit fly sequences were aligned. Finally, the National Science Foundation has established a Tree of Life project, with the goal of using genomic information to further elucidate phylogenetic relationships between the major taxa.

Building on the Human Genome Project, another multidisciplinary project that has been launched is the *Human Proteome Project*, with a goal of identifying and characterizing proteins from the known protein-producing genes. Another offshoot has been the *Human Encyclopedia of DNA Elements (ENCODE)* project. The goal of the consortium involved in this undertaking is to identify not only genes, but all the other functional elements in the human genome as well. Finally, taking the genome idea in an altogether different direction, the *Human Microbiome Project* is attempting to characterize the genomes of the microbes that colonize the human organism.

TOXICOGENOMICS

Knowledge of the genetic code provides new insights into the function of the cell. Expression of each gene can be monitored with time to observe how genes are turned off and on during development. The normal and diseased states can be compared to identify perturbations in gene expression, and this can also be applied to learn the mechanism of poisoning by a chemical. In fact, the *Gene Expression Omnibus* of the NCBI (http://www.ncbi.nlm.nih.gov/geo/) seeks to collect a common database across platforms for experiments measuring gene transcripts.

Monitoring Transcription: Gene Expression and Microarrays

A gene is expressed when its code is transcribed into an RNA molecule, called a primary transcript. The transcript is then converted into a messenger RNA by removing

TABLE 5.1

Examples of Some of the 247 Eukaryotic Species for Which Computerized Maps Linking to Sequences of DNA Are Available through Map Viewer

Protozoans	Fungi	Plants	Animals/Invertebrates	Animals/Vertebrates/Mammals
Babesia bovis	Aspergillus clavatus	Allium cepa (onion)	Aedes aegypti (mosquito)	Ailuropoda melanoleuca (panda)
Cryptosporidium hominis	A. fumigatus	Antirrhinum majus (snapdragon)	Anopheles gambiae (mosquito)	Bos taurus (cow)
C. parvum	A. niger	Arabidopsis thaliana (thale cress)	Apis mellifera (honey bee)	Bubalus bubalis (river buffalo)
Giardia intestinalis	Candida glabrata	Avena sativa (oat)	Drosophila melanogaster (fruit fly)	Canis familiaris (dog)
Leishmania braziliensis	Cryptococcus neoformans	Brassica rapa (mustard)	Caenorhabditis elegans (nematode)	Capra hircus (goat)
L. infantum	Fusarium graminearum	Camellia sinensis (tea)	Hydra magnipapillata	Cavia porcellus (guinea pig)
L. major	Magnaporthe oryzea (rice blast fungus)	Capsicum annuum (pepper)	Nasonia vitripennis (jewel wasp)	Condylura cristata (star-nosed mole)
Paramecium tetraurelia	Neurospora crassa	Cocos nucifera (coconut palm)	Tribolium castaneum (red flour beetle)	Equus caballus (horse)
Plasmodium berghei	Saccharomyces cerevisiae (baker's yeast)	Eucalyptus globulus		Felis catus (cat)
P. chabaudi	Schizosaccharomyces pombe (fission yeast)	Glycine max (soybean)	**Animals/Vertebrates**	Microtus ochrogaster (prairie vole)
P. falciparum		Helianthus annuus (sunflower)	Anas platyrhynchos (mallard)	Monodelphis domestica (opossum)
P. yoelii		Hordeum vulgare (barley)	Anolis carolinensis (green anole)	Mus musculus (mouse)
Tetrahymena thermophile		Manihot esculenta (cassava)	Chrysemys picta (painted turtle)	Mustela putorius furo (ferret)
Trichomonas vaginalis		Nicotiana tabacum (common tobacco)	Columba livia (pigeon)	Orcinus orca (killer whale)
Trypanosoma brucei		Oryza sativa (rice)	Danio rerio (zebra fish)	Ornithorhynchus anatinus (platypus)
		Phaseolus vulgaris (kidney bean)	Falco peregrinus (Peregrine falcon)	Oryctolagus cuniculus (rabbit)
Animals/Vertebrates/Mammals /Primates		Picea abies (spruce)	Gallus gallus (chicken)	Ovis aries (sheep)
Callithrix jacchus (marmoset)		Pinus taeda (loblolly pine)	Meleagris gallopavo (turkey)	Rattus norvegicus (rat)
Gorilla gorilla		Pyrus communis (pear)	Xenopus (Silurana) tropicalis (frog)	Sarcophilus harrisii (Tasmanian devil)
Homo sapiens (human)		Quercus robur (oak)	Zonotrichia albicollis (sparrow)	Sorex araneus (European shrew)
Macaca fascicularis (macaque)		Solanum lycopersicum (tomato)		Sus scrofa (pig)
Nomascus leucogenys (gibbon)		Solanum tuberosum (potato)		Trichechus manatus latirostris (manatee)
Pan troglodytes (chimpanzee)		Theobroma cacao (cocoa)		Tursiops truncates (dolphin)
Papio anubis (baboon)		Triticum aestivum (wheat)		
Pongo abelii (orangutan)		Zea mays (corn)		
Saimiri boliviensis (squirrel monkey)				

Source: Data from National Center for Biotechnology Information, United States in 2014.

Note: Sublists are in the alphabetic and not taxonomic order. Map Viewer can be located at http://www.ncbi.nlm.nih.gov/projects/mapview/.

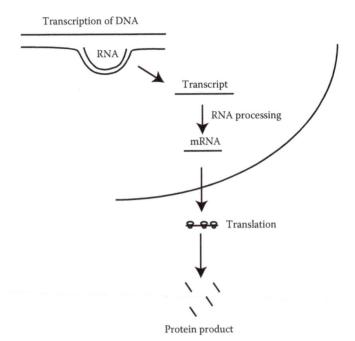

Transcription of DNA

RNA

Transcript

RNA processing

mRNA

Translation

Protein product

FIGURE 5.3 Overview of protein synthesis. During the process of transcription, RNA polymerases build a complementary RNA copy of the base sequence of the gene being transcribed. The transcript is then processed, with introns being removed and exons being spliced together to form mRNA. During translation, which takes place in the cytoplasm on the ribosomes, transfer RNAs bring the correct amino acids (in the order specified by the mRNA) and build the protein product.

introns (noncoding segments) and splicing together *exons* (the coding segments along with leader and trailer segments). The mRNA is then translated into a protein that may have a particular catalytic function in the cell (e.g., it might be among the enzymes in a metabolic pathway to produce energy from food), or it might be a regulator of other genes (a summary of transcription and translation can be found in Figure 5.3).

One emphasis of toxicogenomics is the study of the *proteome*, which is a newly invented term for all the proteins encoded in the DNA. This is often studied by inference, in that it is the quantity of the mRNA that is often measured and not that of the protein per se. The mRNA molecules produced by a cell can be identified, and a complementary DNA (cDNA) molecule can be synthesized to represent each one. These cDNA molecules can then be bound in an array to a glass or silicon surface to form a *microarray chip*. Samples of RNA can then be applied to the chip, and those that are complementary to the DNA on the chip will bind or hybridize to the chip, giving a fluorescent signal. The chip is read by a fluorimeter, and a computerized representation of the intensity of fluorescence is generated. A commercialized example used in medical diagnostics is the GeneChip (Affymetrix).

Another method for analyzing gene expression is to use expressed sequence tags (ESTs), which are a collection of partial cDNA sequences. A third method is serial analysis of gene expression (SAGE). This involves the collection of mRNA, synthesis of the complementary DNA, and cleavage of the cDNA by endonucleases to produce a 10 base pair cDNA tag for every mRNA transcript. The tags are then polymerized, amplified, and sequenced. SAGE is considered the most robust analysis, but it is also very expensive.

Analyzing consecutive samples gathered over the course of time can yield a snapshot of the genes turning on or off during any process of interest, including exposure to toxicants. Also, specialized chips can be made only for the genes involved in a certain process, such as a chip for genes related to immunity. In this way, new perspectives are possible on response of the cell to poisoning, infection, differentiation, hormonal activity, and other processes.

Systems Biology

See also:
  Monitoring Transcription: Gene Expression and Microarrays
                    Chapter 5

  Systems Toxicology
                    Chapter 5

Early experiments with microarrays quickly revealed that genes are expressed dynamically; i.e., the mRNA from a given gene might increase and then decrease with time. Another early revelation from microarrays was that there are approximately as many genes that are turned off or downregulated as there are those that are switched on by various processes. It is now clear that a complex set of pathways of regulation of gene expression exists, involving a host of transcription factors, activators, and repressors. Furthermore, these pathways interact in networks and systems through which parts of pathways can be co-opted into new networks for alternate uses. The complex nature of gene networks has led to the realization that the next level of understanding will require making sense of extremely large blocks of data. The understanding of the complex control of the cell is a goal of a new field called *systems biology*.

Assuming that most mRNA molecules are translated to corresponding proteins, microchip experiments can be considered a view of the universe of proteins in the cell. The dynamics of expression of proteins are due to regulation of the various steps to the protein, including transcription, splicing to mRNA, and translation into protein. It should be noted that the last of these, translation into protein, is not observed in the microarray experiment and must be confirmed by other techniques that determine the presence of the protein itself, such as the use of an antibody to the protein or a measure of the functioning protein apart from its mRNA.

Uranium

See also:
  Case Study: Three Mile Island, Chernobyl, and Fukushima
                    Chapter 18

One example of the application of gene expression technology in toxicogenomics was the analysis by SAGE of uranyl nitrate exposure in mice as a model of how widespread exposure to uranium metal in the environment might affect humans. Approximately 200 genes were significantly altered in expression, most being overexpressed, upon chronic exposure to uranyl nitrate in drinking water. Gene ontology analysis showed that several categories of genes were represented more than others, and those included genes involved in solute transport, oxidative stress response, protein synthesis, and cellular metabolism. Representative genes from each category were chosen, and reverse transcriptase polymerase chain reaction (PCR) was used to confirm the quantities of those genes. (Results were found to be consistent with the SAGE results.) From that study, candidate genes are being studied to develop sensitive biomarkers for the renal disease associated with exposure to uranium.

OTHER ROLES FOR RNA

The role of transcription has been very well defined by studies of DNA and the various RNA polymerases that catalyze transcription; therefore, entering the age of proteomics, it was assumed by many that transcription was the key to the dynamics of mRNA. Recent experiments have demonstrated that although transcriptional regulation is common, the transcript RNA is under the control of a variety of other regulatory molecules on its way to becoming mRNA. The processing of transcripts has been understudied partly due to the experimental difficulties of handling RNA and partly due to ignorance of the nature of the transcripts possible from the genome. It was only recently recognized that conventional genes are floating on a sea of transposable elements in the human genome. As mentioned earlier, this DNA was once considered junk DNA; however, these genes encode not proteins, but RNA for the sake of RNA, a rather radical concept to many biochemists.

miRNA

See also:
 Effects of Toxicants on
  Nucleic Acids
    Chapter 4

The RNA world is now being revealed with a vengeance. It is clear now that there are RNA molecules known as *ribozymes* that can perform catalytic functions similar to those of enzymes. There are also RNA molecules that recognize and bind to other molecules; these are called *aptamers*. Of course, it was clear for many years that the key adapter to translate the genetic code of DNA (the blueprint) into the structure of the cell (the house) happened to be *transfer RNA*, which provides in one small polymer an anticodon to read each codon (code word) in the mRNA and a vehicle to introduce the corresponding amino acid into the polymerizing primary structure of the protein. Finally, miRNAs have been found to play an important role in regulating protein synthesis by binding to and inhibiting translation of complementary mRNAs.

As with proteins, the various functions of RNA molecules depend on the three-dimensional structure, as determined by hydrogen bonding of complementary bases. This is similar to DNA, but with RNA, the structure is produced within a single strand folding on itself, rather than between two different strands. Although the three-dimensional shape of transfer RNA has been recognized for decades, deducing tertiary structure of ribozymes, aptamers, and other RNA molecules has been a fascinating new venture, and many transcripts remain to be analyzed in this regard.

Without RNA, there would be no message and no adapter, so perhaps it should come as no surprise that besides being messenger and adapter, RNA can also be regulator, catalyst, and perhaps transporter of genes between species. An interesting caveat is that as powerful as RNA is likely to be, it does not pass itself to the next generation, but it passes its information to the next generation in the code written in DNA. This is raising the interesting question in evolutionary biology of whether RNA could carry out the entire process of life, including serving as both the blue-print and the house. Given its newly revealed functions, RNA now seems a better prospect than either DNA or protein for the original molecule of life. Thus, the idea that RNA is central to life is likely to shape experimental molecular biology and even evolutionary biology in a new way.

SINGLE NUCLEOTIDE POLYMORPHISMS

With the human and mouse genomes well defined as to the protein coding sequence of nucleotides, several important goals of toxicology can be approached more pow-erfully than imagined a decade ago. Now, not only can mechanisms of poisoning be studied by observing changes in expression of the genes in the cell, but toxi-cologists are also beginning to focus on the possibility of characterizing individual susceptibility to toxicants by cataloging the particular *alleles* (variations on a gene) for genes that code for critical target proteins or for proteins involved in xenobiotic detoxification.

Often, two or more alleles may be found for a single gene that differ by one single nucleotide at one position (for example, an adenine instead of a thymine). This type of variation is known as a *single nucleotide polymorphism* (SNP, pronounced "snip"), and there are probably several hundred thousands of these in the human genome. Over the past decade, many specialized methods have been developed to analyze SNPs, many of which use the polymerase chain reaction (PCR) to amplify (make additional copies of) DNA (Figure 5.4). In this reaction, the DNA of inter-est is incubated with nucleotides, DNA polymerase, and short *primers* (stretches of DNA that are complementary to the 3' ends of the target DNA). This mixture is heated, which causes the double-stranded DNA to separate, and then it is cooled, which allows the primers to bind to the target DNA and the polymerases to build a new complementary copy for each strand. These cycles of heating and cooling can be repeated, with the amount of target DNA doubled in every cycle. Following amplification, the DNA sequence of the product can be determined. A streamlining of this process was recently invented with the use of phage Phi 29 polymerase, which uses a mechanism of rolling circles and needs neither pairs of primers nor thermal cycling to produce enough DNA of any gene of interest, so that analysis of sequence can be executed.

As an alternative to determining the sequence of the amplification prod-uct, allele-specific probes can be used. This allows the SNP to be determined in one tube in a process known as real-time PCR, a technique utilized by the LightCycler™ (Roche Diagnostics Division of F. Hoffmann La-Roche Ltd, Basel, Switzerland) thermal cycler with integrated fluorimeter (originally developed by Idaho Technologies) and similar instruments. TaqMan™ (Roche) fluorescent probes are used for LightCycler, or it is sometimes possible to discriminate between SNPs by melting point analysis, as provided by Idaho Technologies. Other types of probes

FIGURE 5.4 Overview of the polymerase chain reaction (PCR). Double-stranded DNA is heated, causing separation of strands. Cooling then allows specially designed primers to hybridize to the DNA. The primers serve as the starting point for DNA polymerases to build complementary copies of each strand, thus doubling the amount of DNA. This process can then be repeated for multiple cycles.

include hairpin-like molecular beacons with a fluor (light emitter) and a quencher (light absorber) that are initially in contact (thus emitting no fluorescence), but which become separated (and thus begin to fluoresce) upon hybridization to the target.

Faster SNP analyses are possible using oligonucleotide arrays of alternative alleles with amplified sample DNA; however, in both PCR/sequence analysis and hybridization-based techniques for SNPs, it is more difficult to determine the genetic heterozygote (e.g., A/G) than the two corresponding homozygotes (e.g., A/A and G/G) due to the mixture of alleles present in the heterozygotes. Conventional automated sequence analysis with fluorescent dyes, when encountering a genetic SNP, will usually declare an unknown ("N") nucleotide base at the position of the mixture, whereas visual examination of the data via the computer-simulated electrophero-gram will indicate fluorescences from the two nucleotide bases of interest in roughly equal proportions. Analytical software must be modified to search for and to declare two possible nucleotide bases at every position or at least at the position of the anticipated SNP.

All techniques employing amplification of the gene of interest carry the potential for errors introduced in the enzymatic reactions of amplification. Invader™ (developed by Third Wave Technologies, Madison, WI, United States) is a process for detecting SNPs directly, avoiding amplification of the sampled DNA. Invader employs a common sequence on the 5′ end of the interrogating probe, which is cleaved when the allele-specific portion of the probe hybridizes to the target. The common sequence then hybridizes to a reporting beacon, negating quenching and activating fluorescence in a step that can be increased by cycling.

METABOLOMICS

It has been observed that profiling of metabolites can be a sensitive means of discriminating among populations, and this can serve as a point of entry into the toxicogenomics of a poison or other compound of interest. Just as expression of all genes can be compared between the healthy and diseased states using microarrays, *metabolomics* (alternatively known as metabonomics) aims to compare snapshots of thousands of small molecules representing the metabolic pathways of the cell.

Xenobiotic Metabolism

See also:
Biotransformation
Chapter 3

New methods and instrumentation are making it possible to survey this broad landscape of metabolites in a way that will be complementary to genomic analysis. The most important breakthroughs have come in the application of high-resolution 600 MHz hydrogen nuclear magnetic resonance (NMR) spectroscopy, liquid chromatographic separation of metabolites coupled with high-resolution mass spectroscopic detection, and the combining of the techniques in two-dimensional analyses. A chemical analysis of urine or other body fluids is then followed by computerized pattern recognition and principal component analysis to identify ratios of metabolites that are correlated with traits of interest and that represent a kind of chemical phenotype. In the laboratory, for example, metabolomic profiling has provided a phenotype discriminatory of strains of mice and various physiological states in rats and mice.

In some ways, metabolomics can be a more direct view of what is happening in the diseased state than the gene expression analysis; however, this view can become distorted by secondary effects. Also, the fields of metabolomics and gene expression should offer many points of confirmation when both the metabolite of a certain pathway and the expression of the gene encoding the enzyme or other protein responsible for that step in the pathway can be simultaneously monitored. Metabolomics offers the potential for inexpensive and rapid analysis from biological tissues. Although the initial cost of the instrumentation is high in the case of NMR and mass spectroscopy, urine sampling, sample preparation, and analysis can be less expensive compared with gene expression analysis.

Metabolomic profiling revealed significant differences in urinary metabolites in rats exposed to the hepatotoxin hydrazine, an inducer of steatosis, and renal cortical nephrotoxin mercuric chloride ($HgCl_2$) (Figure 5.5). Data revealed significant differences between urinary metabolites produced by Han-Wistar rats exposed to

FIGURE 5.5 Differences between NMR spectra of urine from (a) control HW rats (HW is a strain of rat), (b) HW rats treated with hydrazine, (c) control SD rats (a different strain of rat), and (d) SD rats treated with $HgCl_2$. (From Holmes, E. et al., Chemometric models for toxicity classification based on NMR spectra of biofluids, *Chem. Res. Toxicol.*, 13, 471, Copyright 2000, American Chemical Society.)

hydrazine and by control rats and between Sprague-Dawley rats exposed to mercuric chloride and their controls. Among metabolites considered to be diagnostic were taurine and creatine, increased in hydrazine-treated Han-Wistar rats, and alanine and glucose, increased in mercuric chloride-treated Sprague-Dawley rats. Discrimination of affected individuals using metabolomics can be coupled to gene expression analysis via microarrays to increase the analytical power of toxicogenomics by simultaneously observing the metabolomic phenotype with the transcripts increased or decreased.

Systems Biology

See also:
  Monitoring Transcription:
  Gene Expression and
  Microarrays
  Systems Toxicology
                    Chapter 5

While the place of metabolomics among the new genetics sciences has been challenged as being more biochemical and less genetic in character, it is clear that metabolomics has much to offer in the overarching new discipline of systems biology that is rapidly emerging. One clear advantage is the potential to determine the time course of sublethal poisoning from metabolomic profiles easily collected through time after exposure.

PERSONALIZED SUSCEPTIBILITY AND TAILORED THERAPEUTICS

The draft human genome is the *consensus code* (a record of the nucleotides most commonly found in each position) for approximately 75 individuals of diverse ethnic backgrounds, and as such, it is a generic code. Although the human species is remarkably uniform in that only about 2% of the genome varies across the species, having a consensus sequence alone does not solve the questions of genes and disease or genes and particular traits of interest. Answering those questions requires further analysis of genetic polymorphism and relating of specific sequences with the trait of interest. It requires the organized application of genomics to the study of inheritance. Initial approaches were to analyze comparatively closed populations for the causes of notorious congenital diseases. Although this has resulted in many successful diagnoses, alternative alleles can be associated with the same physiological disease. So ultimately, molecular diagnosis of disease must be applied to each family and individual.

An individual can determine whether or not he/she carries any of the known alleles for a certain disease or trait, but he/she cannot conclude that the trait is not carried in him/her by another allele or polymorphism at a separate locus. Let us consider the case of chestnut coat color in horses. This is a recessive trait controlled at the *mc1r* locus, and chestnut horses are commonly homozygous *e/e*, with carriers having a single *e* allele. The first analysis of horses in the United States resulted in an association of this trait with a point mutation; however, when this analysis was executed in German Black Forest horses, a second allele, *ea*, was identified. If the analysis were applied to unrelated horses, for example, a rare breed in Asia, another (third) allele might be associated with chestnut.

Cytochrome P450

See also:
  The Role of Cytochrome
  P450
                    Chapter 3

  Model Organisms and
  Comparative Genomics
  Metabolomics
                    Chapter 5

There is also the issue of *incomplete or reduced penetrance* to deal with. This is the case in which not all individuals with a particular genotype express the same phenotype. For example, individuals with mutations in the *BRCA1* or *BRCA2* gene, although at increased risk for cancer, do not all develop the disease. One estimate for the penetrance of *BRCA1* mutation for the development of breast cancer by age 70, for example, was 68%. Reduced penetrance is also a major issue in assessing the role of allelic variation in the prediction of teratogenic outcomes. So, until a better understanding of the interaction of genetics and environment in many diseases (including cancer) is reached, knowledge of genotype alone can be useful, but not definitive in predicting individual outcomes.

The study of how factors other than the DNA structure itself influence gene expression is known as epigenetics, and epigenetic phenomena have recently been shown to be of greater importance than once was thought in predicting toxicological outcomes. Epigenetic mechanisms can include DNA methylation, modification of histones, and the action of microRNAs, and evidence indicates that they can be responsive to environmental factors.

INDIVIDUAL VARIATIONS IN METABOLISM AND PHARMACOGENOMICS

As we know from studies of biotransformation, many xenobiotics are chemically altered in the body to either more or less active metabolites. One well-studied mechanism for this is oxidation catalyzed by cytochrome P450 (CYP) enzymes of the liver. Now that all the human *CYP* genes are known, it is possible to examine the expression of these genes under various conditions using a microarray analysis. It is clear that individuals may be poor metabolizers of certain drugs due to genetic deficiencies, usually missense mutations in the *CYP* genes, resulting in lack of the encoded *CYP* enzyme. In this case, the deficiency can be detected either at the level of DNA or at the level of expression in the form of mRNA, such as in the microarray experiment. Less well studied are the few examples in which point mutations affect the quality of the enzyme; these cases might be missed in a microarray analysis.

The biological activity of many common pharmaceuticals is partially dependent on rates of biotransformation. Also, many drugs are found to interact through altering biotransformation in mixtures, and it will be very informative to know the genotype of the patient in this regard. In fact, it might be possible to tailor drug prescriptions on the basis of the genotype of the patient in terms of biotransformation and site of action. Recently, the first kits for this type of analysis were commercialized and involve the cytochrome P450 genes, which encode monooxygenases important in xenobiotic biotransformation.

Xenobiotic Metabolism

See also:
Biotransformation
Chapter 3

A simple example of intoxication is the hepatotoxicity of the analgesic acetaminophen, which is dependent on the pathway of biotransformation in the individual patient. The time course of gene expression was analyzed by Williams et al. (2004) using the GeneChip™ oligonucleotide microarray method for toxic (3.5 mmol/kg) and nontoxic (1 mmol/kg) doses of acetaminophen administered i.p. to CD-1 mice. Inferred expression levels of over 60 genes were significantly altered in a dose-dependent manner. At the nontoxic dose, most genes were

Acetaminophen

See also:
Glutathione Conjugation
Chapter 3

Liver Cell Death: Necrosis and Apoptosis
Chapter 11

Acetaminophen
Appendix

regulated higher at 1 h, and most returned toward the control level of expression by 4 h. At the toxic dose, expression of most genes increased to a much higher level by 4 h, and then expression fell to below the control level by 24 h. Categories of genes studied included antioxidant-related genes such as *heat shock* and *metallothionein*, glutathione-related genes such as *GSH S-transferase mu*, metabolism-related genes such as *CYP2a4, CYP3a11, CYP3a16*, and *S-adenosylmethionine decarboxylase 1*, transcription-related genes such as *Jun* and *c-Fos* oncogenes, immune-related genes such as *cathepsin E*, apoptosis-related genes such as *p21*, and others, including *hemoglobin* alpha *adult chain 1. CYP2a4* under the toxic dose was among only a few genes regulated lower at 4 h and then rebounding to a very high expression at 24 h.

An oligonucleotide chip is being produced that is diagnostic for 100,000 SNPs of humans distributed across the genome and chosen as the more reliable SNPs for assay among over 500,000 tested (Affymetrix). Others are pushing forward with new technologies to produce the "thousand dollar" genome, i.e., new technologies for determining nearly the entire sequence of an individual, the personal genome. A great advantage of defining the personal genome is that it would eliminate the problem of having to analyze again for newly recognized SNPs of interest as they are discovered because the information would already have been collected. Given the new understanding that significant regulation of genes comes from intergenic regions, consideration of the whole genome could be a necessity for rational application of pharmacogenomics.

RACE, ETHICS, AND GENOMICS

An interesting paradox arising in medical genomics is that there is a recent government initiative toward the elimination of ethnic and racial health disparities, which suggests the inclusion of race as a variable to be studied. At the same time, genomic science is discovering that humans of various ancestral and geographical origins are amazingly similar at the level of DNA and the encoded protein.

In fact, human DNA differs less than one might expect from that of other species. For example, the DNA of *Homo sapiens* varies by only about 1% from that of *Pan troglodytes*, chimpanzee. Variation in the DNA is least among the sequences encoding conserved proteins, and it is greater for introns in transcribed genes.

This variation is even less if you look at the primary protein sequence, especially among conserved domains of proteins. A recent search of GenBank entering an inferred amino acid sequence from a putative transcription factor from an insect returned related sequences of 53 amino acids that were identical in humans and chimpanzees, whereas the related mouse sequence differed from humans only by the first amino acid. Another search using 88 amino acids of a putative sugar transporter from an insect returned a human sequence that was conserved in 48 of 88 amino acids compared with the insect, but was identical to both chimpanzee and mouse for all 88 amino acids.

In terms of comparisons between humans, only 0.1% of human DNA varies among individuals, so that racial differences, if they can be defined, will amount to less than one nucleotide base among 1000. The regions of the genome in which the highest level of variation occurs between humans likely correspond to those that contain repeated sequences with few genes (such as those regions encoding

SINEs, LINEs, and other small RNAs). Genomic analysis has suggested that variation among such interspersed nuclear elements is correlated with the striking polymorphism between breeds of dogs, so it is plausible that a closer examination of regions of repeats will be needed for understanding any putative relationship between genetic characteristics and diseases and responses to xenobiotics. Unfortunately, it is those same regions in which results of sequence analysis are least reliable, and more effort will be required to define those areas of the genome more robustly before correlations with human traits can be accurately assessed.

How do these genetic differences between individuals correlate with racial and ethnic groups? Analyses of polymorphic short terminal repeats, restriction fragment length polymorphisms, and SINE insertions between humans in Africa, Asia, and Europe indicated that 86%–90% of the variation was found within continental groups, whereas only 10%–14% was significant between continents. Further analyses have demonstrated that African populations are most diverse genetically and have the greatest genetic distance from Asian and European subpopulations.

However, with very few exceptions, most polymorphisms of medical importance are distributed throughout human populations. In addition, there is much mixing of the human population today, so that skin color, a commonly used qualifier for race, was significantly correlated with self-identified ancestral origin in mixed populations, but the correlations ranged from weak (in Mexicans) to moderately strong (in Puerto Ricans). This points out, in particular, that correlations found for an ethnic group at the geography of origin might not hold for individuals of that origin who have migrated and mixed with those of other origins.

One new tool for better understanding human population genetics is to use *haplotypes* to trace geographical origins. A haplotype is a piece of a chromosome that has not recombined or mutated for several generations and is traceable back through the lineage. HapMap is a program to identify haplotypes using the developing SNP database, microsatellites, and other forms of polymorphisms combined with knowledge of ancestral genetics where available. Studies of lineage and origin can thereby advance from reliance on mitochondrial DNA and Y chromosome sequences to include more of the genome and to include genes of interest for disease and pharmacogenomics. Most identified haplotypes have been found across continental groups (for more information, see Gibson and Muse, 2002); therefore, it is to be expected that most genes of interests in pharmacogenomics will also be distributed among races.

Finally, there are clearly ethical considerations that need to be taken into account when considering individual drug design and susceptibility to disease or toxicant exposure. Primary is the question of the right to privacy for an individual regarding this sensitive genetic information, revelation of which could jeopardize not only the individual but also descendants. Potential misuses of the new genomics include insurance and forensic investigations as well as military applications. For example, weapons could be targeted toward specific alleles or haplotypes of humans or animals, or biological weapons could be redesigned and engineered for more virulence, gene targeting, or even to avoid detection. Ethical considerations would also extend to the use of genomics for well-intentioned research. For example, a recent controversy has arisen regarding whether to destroy all stocks of smallpox or to allow the genetic engineering of the smallpox genome for the purpose of understanding the mode of virulence of this disease.

SYSTEMS TOXICOLOGY

Considering the exponential growth of information in the various subdisciplines of the new genetics introduced here, it is clear that a new approach must be taken in order to apply the flood of data to understanding the process of poisoning more clearly. Increasingly, the genomic disciplines are converging with *systems engineering* as it has been applied in the *metabolic control analysis*, to form the emerging field of *systems biology*. It is becoming apparent that "toxicology is gradually evolving into a *systems toxicology* that will eventually allow us to describe all the toxicological interactions that occur within a living system under stress and use our knowledge of toxicogenomic responses in one species to predict the modes-of-action of similar agents in other species" (Waters and Fostel, 2004). Only the development of systems toxicology can enable the rational combining of information gathered from comparative genomics, proteomics, metabolomics, and related research into predictive models useful in such pursuits as developing tailored pharmaceuticals with lessened risk of side effects or other contraindications or understanding the impact of environmental pollutants on wildlife species. Systems biology, while now at the rudimentary level of providing "snapshots of a currently poorly mapped molecular landscape" (Waters and Fostel, 2004), can be developed toward the lofty goal of understanding the life process in a computational and testable framework as more details are revealed to us.

CASE STUDY   Using GenBank and Online Tools in Genomics

Identifying genes in sequence data (strings of nucleotides) is not a trivial proposition. *Annotation* is the process of assigning candidate genes from the draft sequence of the genome. DNA is of course double stranded, and for a given gene, one strand (the *minus strand*) is transcribed and the other strand (the *plus strand*) is not. However, either strand can serve as the minus strand, so that genes running in opposite directions can overlap like trains passing on two tracks. Genes are generally located within *open reading frames*, which begin with an initiation codon and end with a stop codon. (Not all open reading frames contain functioning genes, however.) Annotation employs various kinds of information, but relies heavily on the recognition of previously known conserved domains (areas that are found to be the same in related proteins) and comparative genomics, in which genes known from other species are found to align to the candidate gene. Annotation of human and various model draft sequences is certainly not complete as of this writing, and many candidate genes cannot be classified at this time.

Alternatively, many genes were cloned from mRNA that was collected from cells and then converted into what is called *cDNA* in a reaction catalyzed by the enzyme *reverse transcriptase*. The cDNA sequence can then be aligned with the genomic sequence found in the Human Genome Project. However, one must keep in mind that the cDNA, like the mRNA, contains only exons (introns were removed by splicing), and the genomic sequence contains the entire gene with introns and exons. One revelation from this analysis was that there are often multiple mRNAs from one gene, a result of alternate splicing of the premessenger RNA transcribed from the gene.

Suppose you have a nucleotide sequence

aaacattgttatgctggaaatgctagaata

and you would like to determine whether it encodes a protein or polypeptide. How could you go about answering that question? Sit down at a computer that is connected to the Internet and follow the instructions below. This exercise will give you some practice in using a few of the many tools and techniques available online to help researchers study gene sequences.

1. First, go to http://au.expasy.org/. This will connect you to the Expert Protein Analysis System (ExPASy) server of the Swiss Institute of Bioinformatics. This tool can help you analyze protein sequences.
   a. The list of "categories" in the left-hand column should offer "proteomics" as a choice. Clicking on this will bring up a screen with two columns: "Databases" and "Tools." Under Tools, scroll down the (very long) alphabetical list of tools until you find "Translate." Click on Translate, which will take you to the tool itself. Enter the nucleotide sequence in the box and select "translate sequence."
   b. Six possible translations should be generated: forward translations in the three possible frames and then reverse translations in the three possible frames. Record the results for the possible translations.
   c. Look these translations over. Look for sites where initiation of transcription might occur (signified by "Met," which is methionine) as well as where termination of transcription might occur (signified by a special "stop" codon). Open reading frames will be highlighted for you in red.
2. Now let us see whether a particular gene possesses this DNA sequence. For this, we will use NCBI GenBank, located at http://www.ncbi.nlm.nih.gov. Click on the link that says "BLAST." Or, alternatively, you can go directly to BLAST at http://blast.ncbi.nlm.nih.gov/Blast.cgi.
   a. On the BLAST page, you are looking to match a nucleotide string, so choose "Nucleotide BLAST," which should take you to a search engine called BLASTn.
   b. Now you are ready to use BLASTn to search for your sequence. Enter the nucleotide string in the box and run BLAST. When the results are returned, examine the matches and answer the following questions:
      i. How many perfect matches were found?

      ii. How many human genes were found among the matches?

      iii. Were there any matches among several species that could represent a set of homologous genes?

3. Return to the NCBI home page. Select "Gene" in the search box and enter the name or abbreviation of the gene you found by BLASTn.

    a. Examine the information returned about the gene of interest. Select the human version of the gene. From the following page, under "Genomic Content," follow the computer link to Map Viewer.

    b. On which human chromosome is this gene located?

    c. What other genes are located in the vicinity of the gene of interest? Scroll up or down, or zoom out to see the region.

    d. A *contig* is an abbreviation for a contiguous sequence of genomic DNA. *Synteny* is the common genetic linkage of genes to the same chromosome. If several genes are found in one contig, then those genes can be assumed to be found on the same chromosome. Synteny can be confirmed genetically by test crosses (in rodents) or by a pedigree analysis (in humans). By going back and choosing the mouse version of the gene, you can compare the synteny of genes near the gene of interest between humans and rodents.

    e. Return to the human gene report and under "Genomic Regions," click on "Reference sequence details." Under "mRNA and Proteins," you will see several isoforms listed. Under the first one listed, choose "Source sequences." Choose the mRNA (not the cDNA) sequence link and look at the source sequence for this mRNA. Can you find your sequence of interest? (Hint: use your browser's "find" tool to search for a match for at least part of your original nucleotide sequence.) Then go back to the gene page and select "UniProtKB/Swiss-Prot" for that isoform. This will take you to a protein database, which shows the sequence of all of the isoforms. Do any contain the amino acid sequence you found in your ExPASy translations (Part 1)?

BIBLIOGRAPHY

Alberts, B., Johnson, A., Lewis, J., Raff, M., Roberts, K., and Walter, P., *Molecular Biology of the Cell*, Garland Science, New York, 2002.

Altschul, S.F., Madden, T.L., Schäffer, A.A., Zhang, J., Zhang, Z., Miller, W., and Lipman, D.J., Gapped BLAST and PSI-BLAST: A new generation of protein database search programs, *Nucleic Acids Res.*, 25, 3389, 1997.

Burczynski, M.E., Ed., *An Introduction to Toxicogenomics*, CRC Press, Boca Raton, FL, 2003.

Chen, M., Zhang, M., Borlak, J., and Tong, W., A decade of toxicogenomic research and its contribution to toxicological science, *Toxicol. Sci.*, 130, 217, 2012.

Evans, D.G., Shenton, A., Woodward, E., Lalloo, F., Howell, A., and Maher, E.R., Penetrance estimates for *BRCA1* and *BRCA2* based on genetic testing in a clinical cancer genetics service setting: Risks of breast/ovarian cancer quoted should reflect the cancer burden in the family, *BMC Cancer*, 8, 15, 2008.

Gasteiger, E., Gattiker, A., Hoogland, C., Ivanyi, I., Appel, R.D., and Bairoch, A., ExPASy: The proteomics server for in-depth protein knowledge and analysis, *Nucleic Acids Res.*, 31, 3784, 2003.

Gibson, G. and Muse, S.V., *A Primer of Genome Science*, Sinauer Associates, Sunderland, MA, 2002.

Holmes, E., Nicholls, A.W., Lindon, J.C., Connor, S.C., Connelly, J.C., Haselden, J.N., Damment, S.J.P., Spraul, M., Neidig, P., and Nicholson, J.K., Chemometric models for toxicity classification based on NMR spectra of biofluids, *Chem. Res. Toxicol.*, 13, 471, 2000.

Ioannidis, J.P., Ntzani, E., and Trikalinos, T.A., "Racial" differences in genetic effects for complex diseases, *Nat. Genet.*, 36, 1312, 2004.

Kappen, C. and Salbaum, J.M., Gene expression in teratogenic exposures: A new approach to understanding individual risk, *Reprod. Toxicol.*, 45, 94, 2014.

Khan, S.R., Baghdasarian, A., Fahlman, R.P., Michail, K., and Siraki, A.G., Current status and future prospects of toxicogenomics in drug discovery, *Drug Discov. Today*, 19, 562, 2013.

Lewis, C.M., Obregon-Tito, A., Tito, R.Y., Foster, M.W., and Spicer, P.G., The human microbiome project: Lessons from human genomics, *Trends Microbiol.*, 20, 1, 2012.

Li, W., On parameters of the human genome, *J. Theor. Biol.*, 288, 92, 2011.

Lorkowski, S. and Cullen, P., *Analyzing Gene Expression. A Handbook of Methods: Possibilities and Pitfalls*, Wiley-VCH Verlag GmbH & Co., Weinheim, Germany, 2003.

Marshall, A., Taroncher-Oldenburg, G., Aschheim, K., DeFrancesco, L., and Cervoni, N., Focus on systems biology, *Nat. Biotechnol.*, 22, 10, 2004.

Mirbahai, L. and Chipman, J.K., Epigenetic memory of environmental organisms: A reflection of lifetime stressor exposures, *Mutat. Res.*, 754, 10, 2014.

Myers, A.J., The age of the "ome": Genome, transcriptome and proteome data set collection and analysis, *Brain Res. Bull.*, 88, 294, 2012.

National Human Genome Center, Genetics for the human race, *Nat. Genet.*, 36 (Suppl.), 60, 2004.

Omenn, G.S., The strategy, organization and progress of the HUPO Human Proteome Project, *J. Proteomics*, 100, 3, 2014.

Paabo, S., The mosaic that is our genome, *Nature*, 421, 409, 2003.

Parra, E.J., Kittles, R.A., and Shriver, M.D., Implications of correlations between skin color and genetic ancestry for biomedical research, *Nat. Genet.*, 36, 11, S54.

Qu, H. and Fang, X., A brief review on the Human Encyclopedia of DNA Elements (ENCODE) project, *Genomics Proteomics Bioinformatics*, 11, 135, 2013.

Siddeek, B., Inoubli, L., Lakhdari, N., Rachel, P.B., Fussell, K.C., Schneider, S., Mauduit, C., and Benahmed, M., MicroRNAs as potential biomarkers in diseases and toxicology, *Mutat. Res.*, 764, 46, 2014.

Taulan, M., Paquet, F., Maubert, C., Delissen, O., Jacques Demaille, J., and Romey, M.C., Renal toxicogenomic response to chronic uranyl nitrate insult in mice, *Environ. Health Perspect.*, 112, 1628, 2004.

The ENCODE Project Consortium, An integrated encyclopedia of DNA elements in the human genome, *Nature*, 489, 57, 2012.

The International HapMap Consortium, A haplotype map of the human genome, *Nature*, 437, 1299, 2005.

Van Hummelen, P. and Sasaki, J., State-of-the-art genomics approaches in toxicology, *Mutat. Res.*, 705, 165, 2010.

Venter, J.C., Adams, M.D., Myers, E.W., Li, P.W., Mural, R.J., Sutton, G.G., Smith, H.O. et al., The sequence of the human genome, *Science*, 291, 1304, 2001.

Waters, M.D. and Fostel, J.M., Toxicogenomics and systems toxicology: Aims and prospects, *Nat. Rev. Genet.*, 5, 936, 2004.

Williams, D.P., Garcia-Allan, C., Hanton, G., LeNet, J.L., Provost, J.P., Brain, P., Walsh, R., Johnston, G.I., Smith, D.A., and Park, B.K., Time course toxicogenomic profiles in CD-1 mice after nontoxic and nonlethal hepatotoxic paracetamol administration, *Chem. Res. Toxicol.*, 17, 1551, 2004.

6 Carcinogenesis

CANCER

Cancer is not a single disease, but rather a general term referring to many kinds of malignant growths that invade adjoining tissues and sometimes spread to distant tissues. *Carcinomas* are cancers of epithelial tissue, whereas *sarcomas* are cancers of supporting tissues (such as connective or muscle tissues).

The presenting symptom of cancer is often a cellular mass or *tumor*. Tumors are often characterized as falling along a continuum between benign and malignant, based on the characteristics of their cells. *Benign* tumors are encapsulated, slowly growing, noninvasive, and can be controlled by excision; *malignant* tumors are nonencapsulated, rapidly growing, invasive, disseminating, and recalcitrant to treatment. Most tumors appear to have originated from a single cell of origin, yet are heterogeneous in nature due to accumulation of different mutations in different cells during subsequent rounds of cellular division.

Staging is a method for describing the status of a cancer for diagnosis and management. The general classification of each case into stages I–IV is based on a scoring system known as *TNM*: development of the tumor (T), involvement of lymph nodes (N) in the region of the tumor, and degree of metastases (M). For example, the tumor is categorized from T0 (no evidence of a tumor) to T4 (a massive lesion with extensive invasion into adjacent tissues). A combination of clinical, radiographic, surgical, and pathological techniques is used to determine TNM scores. Staging is performed in diagnosis and periodically throughout treatment to evaluate management and remediation of the disease.

THE EPIDEMIOLOGY OF CANCER

With approximately a half million deaths annually, cancers are second only to heart diseases among causes of death in the United States, accounting for 20% of all fatalities in the United States. Breast cancer, lung cancer, and colorectal cancer are the most common cancers in women; prostate cancer, lung cancer, and colorectal cancer are the most common cancers in men. Cancer incidence is higher among blacks than whites and among males than females. Cancer incidence is much higher among the middle aged and elderly than among young people; this is most likely due to the multiple steps and latent period that characterize most forms of cancer.

According to the National Cancer Institute, significant increases were seen in incidence rates in cancers of the thyroid, liver, skin (melanoma), kidney, and testis between 1992 and 2002. Incidence rates were significantly lower over that time period for cancers of the brain, colon, ovary, oral cavity, stomach, prostate, lung, cervix, and larynx, as well as for leukemias. Cancer death rates are down slightly overall, perhaps due to earlier and better detection and treatments.

Environmental Factors in Cancer

Although genetic predisposition is recognized for certain populations in some forms of cancer, most cases appear to be due to environmental factors. The first suggestion of this was Percivall Pott's observation in 1775 that there was an increased occurrence of some forms of cancer among chimney sweeps. Other historical associations include linkage between uranium exposure and lung cancer (1879), links between aniline dyes and bladder cancer (1895), and the sharp foci of lung cancer in the United Kingdom, where high incidence was recorded near shipyards in which workers were exposed to asbestos.

A slightly different type of evidence for the importance of environment in carcinogenesis can be found in Japanese Americans living in Hawaii, who for several generations have had a spectrum of cancer incidence that is very similar to that of other Hawaiian residents but unlike Japanese living in Japan. In fact, more than half of cancer cases are estimated to be related to diet and use of tobacco; therefore, these diseases could be greatly reduced by changes in behavior and habits, such as lowering the intake of fatty foods.

One type of cancer strongly linked to an environmental cause is lung cancer. Causal agents of bronchogenic (cancers that develop from the lung) carcinomas include occupational hazards such as asbestos in shipyard workers; however, most cases result from exposure to tobacco products.

Several different forms of bronchogenic carcinoma occur and can be identified histologically. *Small cell carcinoma* is more common among smokers. Originating in areas of bronchial epithelium, small cell carcinoma tumors often produce hormones and biogenic amines, block the bronchial passages, and often invade other tissues. *Squamous cell carcinoma* often arises in the larger bronchi of the lung and is often preceded by damage and degeneration of the squamous epithelium (often following chronic exposure to pollution or smoking). Ciliated cells are lost and mucosal cells increase in number. It often invades adjacent tissues and nearby lymph nodes, and metastasis is likely. *Adenocarcinoma* is less common; however, among nonsmokers, it accounts for the majority of bronchogenic carcinoma cases. Adenocarcinomas usually originate peripherally in the bronchial tree, exhibiting a glandular form, but are less likely to involve the lymph nodes compared with squamous cell carcinomas. *Large cell carcinoma* also arises peripherally; it, too, commonly affects the lymph nodes. Small cell carcinoma occasionally occurs in a mixture with squamous cell carcinoma or with adenocarcinoma.

Many other cancers have also been linked to environmental risk factors. In Japan, which has a very low incidence of all cancers, stomach cancer is the most frequent form. This may be related to the relatively high intake of salt and nitrites in preserved foods in Japan. Asbestos exposure is linked to the development of a relatively rare form of cancer called mesothelioma, and benzene exposure is linked to increased risks of leukemia.

The International Agency for Research on Cancer has identified a little over one hundred compounds as human carcinogens. For a compound to have been categorized as such, convincing data from both epidemiological and laboratory studies are required. Additional compounds have been identified by that group as probable or possible carcinogens.

GENETIC FACTORS IN CANCER

There are, however, many cases of cancer in which there are clearly genetic factors operating. These cases are generally characterized by the presence of cancer in multiple members of a family, a younger age of onset of cancer, multiple cancer sites (or even the presence of cancer of more than one type) in an individual, and sometimes the association of cancer with other characteristic diseases or conditions.

One example of an often genetically linked cancer is *retinoblastoma*, a childhood cancer of the retina, which occurs primarily in families with a history of the disease. Other examples include *multiple endocrine neoplasia* (which carries increased risks of endocrine cancers), *hereditary nonpolyposis colon cancer*, and *familial adenomatous polyposis syndrome* (which also carries increased risks of colon cancer). 5%–10% of breast cancers, as well as some cases of ovarian cancer, are also thought to be hereditary. Finally, a few families worldwide suffer from *Li–Fraumeni syndrome*, putting them at increased risks for a variety of different cancers. We will discuss the mechanisms behind these apparent increased susceptibilities later in this chapter.

CARCINOGENESIS

THE MUTATIONAL THEORY OF CARCINOGENESIS

Carcinogenesis is a process by which cancer develops in the body. The prevailing theory of carcinogenesis for many years has been the *mutational theory*. This theory describes carcinogenesis as a multistep process, involving three phases beginning with mutational change in one cell and building through an accumulation of factors to the malignant tumor.

According to the mutational theory, carcinogenesis begins with an *initiation* phase, in which damage to DNA (which may be either random or chemically induced) occurs. After undergoing initiation, the cell is considered to have been transformed into a *neoplastic* cell, but remains latent. Possible outcomes for this cell would include remaining in the latent state, succumbing to apoptosis or other forms of cell death, or proliferation.

As the first two options are less likely to result in physiological consequences, it is the third option that we will be concerned with. The second phase of carcinogenesis, *promotion*, involves further cellular changes, which lead to the development of a tumor from the initial neoplastic cell. (Most experimental evidence indicates that tumors are monoclonal in origin.) Generally, this involves the stimulation of proliferation and/or the selection for survival of the altered cells. This stage may be affected by the action of chemical *promoters* that stimulate cell division and produce clonal proliferation. This is a critical step, and in fact, it has been argued that carcinogens requiring very high doses might act through generalized cell damage and the consequential *mitogenesis* (cell proliferation during the repair process) rather than through any specific mechanism.

The third stage is considered to be *progression* and is the stage during which a tumor becomes malignant. The progression to malignancy is characterized by the aggressive nature of the cells, which can attach and penetrate membranes to invade new tissues in a process called *metastasis*. It is also characterized by genetic instability, as evidenced by further mutations and alterations in gene expression, which lead to the heterogeneity of cells within the tumor.

COMPETING THEORIES

Recently, some researchers have suggested alternatives to the mutational theory. Cancer cells typically contain an abnormal number of chromosomes (a condition known as *aneuploidy*), a phenomenon that has been attributed to the general genetic instability seen in malignant cells. It has been suggested, however, that aneuploidy might actually be a cause rather than an effect of neoplastic transformation. For example, studies on individuals with the genetic disease *mosaic variegated aneuploidy* (MVA) indicate that these patients (who show aneuploidy in 25% of their cells) are at greater risks for childhood cancers. Although this study does indicate that aneuploidy can, under some circumstances, be a causative factor in cancer, it does not, of course, demonstrate that it is the only causative factor in tumorigenesis.

There are other theories of carcinogenesis as well. One proposes that cancer is a result of failure of genetic regulation, particularly regulation of genes involved in development. Instead of postulating direct damage to genes, this theory focusses on disruption of epigenetic (non-DNA-related) mechanisms, which then affect gene expression (there will be more about epigenetic mechanisms later in the chapter). Recently, embryonic master regulatory genes *twist* and *slug* have been implicated with the suggestion that networks used in embryonic invaginations could be triggered again in carcinogenesis (Gupta et al., 2005). All theories, however, agree that alterations in gene expression seem to be at the heart of the carcinogenic process.

ONCOGENES AND TUMOR SUPPRESSOR GENES

THE DISCOVERY OF ONCOGENES

The question of how changes to DNA (whether large or small) can result in the neoplastic transformation of a cell is the key to understanding cancer. Initial clues to the answer came with the study of certain RNA tumor viruses, called *acutely transforming retroviruses*, which can cause cancer in animals. Using the enzyme *reverse transcriptase* (encoded in the RNA), the virus can produce DNA from its RNA. The DNA can then be inserted into the genome of the host. When the cancer-causing retroviral genes that were inserted into tumor cells were examined, some of them were found to resemble normal cellular genes of the host. These cancer-inducing retroviral genes were called *oncogenes*, and the analogous (yet apparently inactive) genes in normal cells became known as *proto-oncogenes*. This discovery, which has led to many new molecular approaches to understanding carcinogenesis, earned the Nobel Prize for researchers Michael Bishop and Harold Varmus. There are also oncogene-containing DNA tumor viruses, which typically code for proteins involved in viral replication. These would include *the human papilloma virus*, as well as herpes viruses such as *Epstein–Barr* (which has been shown to be involved in the development of a human lymphoma).

Insertion of a viral oncogene, however, is not the only way in which cancer arises. It appears that cellular proto-oncogenes can be activated to become oncogenes themselves. Proto-oncogene activation to an oncogene can occur in several ways. For example, some cancer viruses lack oncogenes; however, they can exert a similar effect by disrupting a proto-oncogene through *insertional mutagenesis* (that is, by inserting their viral genome within the sequence of a proto-oncogene).

Yet in malignant tumor biopsy samples, activated oncogenes are found even without the involvement of RNA virus. This indicates that other agents are also capable of activating proto-oncogenes. One possibility is that chemically induced mutations in the proto-oncogene or in areas of the genome that control its transcription may initiate activation. *Gene amplification*, where multiple copies of a region of DNA are made, may also trigger activation. Activation can also be associated with chromosome aberration such as the *Philadelphia chromosome* found in *chronic myelogenous leukemia* (CML).

An Example of an Oncogene: The Philadelphia Chromosome

Proto-oncogenes are given three-letter names, e.g., *src*, and those with both viral and eukaryotic homologs are denoted by a prefix as either *v-src* for viral-*src* or *c-src* for cellular-*src*. The Philadelphia chromosome phenomenon involves the reciprocal translocation of large segments of chromosomes 9 and 22. This chromosomal abnormality is seen in almost all cases of CML and results in the disruption of two genes: the *c-abl* proto-oncogene and the *bcr* gene. The piece broken off of chromosome 9 contains *c-abl*, and breakpoints for the translocation on chromosome 22 generally occur within the *bcr* gene. Following translocation, a new hybrid gene is formed that codes for a hybrid protein—the Bcr-Abl fusion protein. The mechanism by which this protein induces leukemia is still not completely clear, but the normal c-Abl protein is a tyrosine kinase and is thought to be involved in the regulation of gene transcription and ultimately control of the cell cycle.

The Role of Proto-Oncogenes in Cell Function

Proto-oncogenes have been mapped throughout the human genome, and they are also found in many other organisms, such as *Drosophila melanogaster*, for example. Their evolutionary conservation suggests important roles in the normal life of the cell. The process of cell growth and division, known as *mitogenesis*, is regulated by biochemical signals that are received, relayed through the cell, and ultimately regulate DNA transcription or replication. Proto-oncogenes typically encode various protein products associated with one of the following processes:

- Signal transduction across the membrane (which involves the interaction between ligands and receptors)
- Signal transduction through the cytoplasm (which involves the interaction between receptors and second messenger systems)
- Regulation of gene expression (which involves binding to DNA and enhancing or blocking the transcription of various genes)

Thus, protein products of proto-oncogenes generally affect interactions between cells, often in pathways with ties to mitogenesis. Evidence suggests that proto-oncogenes may even play a role in the regulation of energy metabolism in tumor cells, perhaps playing a role in the switch to aerobic glycolysis which was discussed earlier in this chapter.

EXAMPLES OF PROTO-ONCOGENES

Among the chemical messengers that promote cell division are small- to medium-sized polypeptides called *growth factors*. One such example is *epidermal growth factor*. Growth factors promote cell division by binding to selective receptors on the outer surface of the cell. Several proto-oncogenes code for proteins that are related to growth factors. For example, the *platelet-derived growth factor* (PDGF) b-chain is coded for by the proto-oncogene *c-sis*, and similar fibroblast growth factor-like proteins are encoded by *int* and *hst* proto-oncogenes.

Some proto-oncogenes code for receptors rather than for ligands. *PDGF receptor* (PDGF-R) is a growth factor receptor with tyrosine kinase activity, as are the products of the oncogenes *c-fms* and *c-kit*. These genes resemble tyrosine kinase genes from retroviral oncogenes, but they all produce proteins composed of three parts: a growth hormone receptor domain on the outer membrane surface, a single hydrophobic domain crossing the membrane, and an obligatory tyrosine kinase domain on the interior.

Binding of growth factors to these growth factor receptors triggers autophosphorylation (self-phosphorylation) of tyrosines on the portion of the receptor inside the cell. The phosphorylation signal is then passed to cytosolic (nonreceptor) tyrosine kinases and other signal-transducing proteins. The signal-transducing proteins wrap tightly around the phosphorylated tyrosine, so that the phosphate groups on the tyrosine associate with SH_2 residues lying too deeply within the signal-transducing molecule to be reached by other phosphorylated amino acid residues (such as serine or threonine).

Mutations that destroy the tyrosine kinase activity also block receptor signaling activity in these proteins. Mutations in the transmembrane region can enhance cell-transforming activity of the proto-oncogenes, suggesting a short circuiting of the signal through an unknown mechanism.

Other proto-oncogenes, including *erb*, *ros*, and *met*, are also associated with growth factor receptor tyrosine kinases. In addition, genes involved in the differentiation of *Drosophila* (such as *sevenless* and *torso*) have the characteristics of this class. Another similar proto-oncogene is *ret*, a gene that has undergone a mutation in individuals with multiple endocrine neoplasia. This mutation leads to increased activity of the RET receptor, which is produced by the gene.

Some cellular functions are mediated through a cascade of protein phosphorylation, beginning with the growth factor receptor autophosphorylation just described. Proto-oncogenes *src*, *abl*, *fgr*, *yes*, *fes*, *fps*, *sea*, *tck*, and *trk* all encode tyrosine kinases that catalyze protein phosphorylation on tyrosine residues. Serine/threonine kinases are produced by *raf*, *mos*, and *pim* proto-oncogenes.

Analysis of the human carcinoma-associated *mas* proto-oncogene has implicated another type of receptor—a neurotransmitter receptor—in mitogenesis.

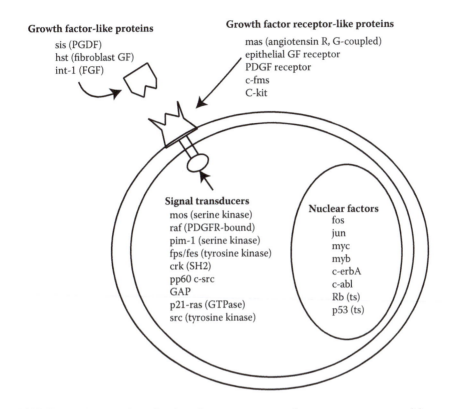

Growth factor-like proteins

sis (PGDF)
hst (fibroblast GF)
int-1 (FGF)

Growth factor receptor-like proteins

mas (angiotensin R, G-coupled)
epithelial GF receptor
PDGF receptor
c-fms
C-kit

Signal transducers

mos (serine kinase)
raf (PDGFR-bound)
pim-1 (serine kinase)
fps/fes (tyrosine kinase)
crk (SH2)
pp60 c-src
GAP
p21-ras (GTPase)
src (tyrosine kinase)

Nuclear factors

fos
jun
myc
myb
c-erbA
c-abl
Rb (ts)
p53 (ts)

FIGURE 6.1   Cellular sites of action of proto-oncogene and tumor suppressor gene (ts) protein products.

This proto-oncogene encodes a receptor protein with seven hydrophobic transmembrane domains and a lack of intrinsic tyrosine kinase activity. It is homologous to the receptor for the small peptide neurotransmitter, angiotensin. Signal transduction from these receptors depends on interaction with G proteins on the inner surface of the membrane. GTPases, such as p21, are products *of N-ras*, *H-ras*, and *K-ras*.

Other proto-oncogene products are found in the cell nucleus. These include transcription factor AP-1, the product of *jun*, which interacts with the protein product of proto-oncogene *fos* to form a protein heterodimer capable of regulating gene transcription. Proto-oncogenes *myc*, *ski*, *ets*, *myb*, and *rel* also generate nuclear proteins. Nuclear oncogenes often contain a zinc finger motif, which is characteristic of proteins capable of interacting with DNA. Figure 6.1 shows a variety of cellular sites where the protein products of proto-oncogenes act.

TUMOR SUPPRESSOR GENES

Tumor suppressor genes are genes that limit cell proliferation and must be overcome in order for a developing tumor to progress. Like oncogenes, their mutated forms were also associated with tumors; therefore, some were originally considered to be oncogenes. Like proto-oncogenes, tumor suppressors are also critical cellular components, as demonstrated by the lack of viable progeny from mice with *Rb-1* selectively deleted. Tumor suppressors can be considered guardians of the body against proliferation of deleterious cells. They can halt cell phase progression, activate postmitotic differentiation, or cause apoptosis (programmed cell death). For example, the *p53* gene generates nuclear protein p53, which monitors for deleterious mutations and can shut down replication of cultured tumor cells when its concentration rises (cells with damaged DNA are arrested in the G1 phase of the cell cycle under p53's influence). The p53 protein also acts to promote apoptosis.

The *p53* gene is, in fact, an excellent example of a tumor suppressor gene. This gene is found to be mutated in many spontaneous cancers and is one of the genes found to be mutated in individuals with Li–Fraumeni syndrome. Mutations in the tumor suppressor genes *BRCA-1* and *BRCA-2* are found in individuals with familial breast and ovarian cancers. Other tumor suppressor genes include *APC* (adenomatous polyposis coli, found to have undergone point or frame-shift mutations in individuals with familial adenomatous polyposis syndrome), *DCC* (deleted in colon carcinoma), *Rb-1* (found in retinoblastoma), and *NF-1* (found in neurofibrosarcoma and schwannoma—cancers of the nervous system).

Development of a tumor most likely requires that functional products of suppressor genes be negated by deletion or destructive mutation of both copies of the gene. It was recently observed, for example, that the tumor suppressor gene *APC* in colorectal tumors was usually mutated on both alleles. An indirect, epigenetic way to negate the action of these genes is to sequester the suppressor protein. For example, the proto-oncogene *MDM2* produces the protein MDM2, which binds to the tumor suppressor gene product p53. Gene amplification of *MDM2* in human sarcomas probably results in high concentrations of the MDM2 protein, which then removes p53 through binding.

The negation of tumor suppressors, as contrasted with the positive activation of proto-oncogenes, is also supported by the observation that there are many different destructive mutation sites in *p53* and *APC*, but there are only a few mutations conferring activation of *ras*. Figure 6.1 shows the sites of action of some tumor suppressor gene protein products.

Single Nucleotide Polymorphisms and Cancer

Single nucleotide polymorphisms (SNPs) are DNA sequence variations that consist of alterations to a single nucleotide. There are, of course, likely to be many (millions of) SNPs in the human genome. A few appear to play a role in genetic susceptibility to cancer. These variations have been found in a variety of different genes coding for proteins involved in metabolism (glutathione-*S*-transferases and cytochrome CYP1A1, for example), as well as DNA repair. The significance and usefulness of SNPs in assessing and predicting individual cancer risk have yet to be fully determined. New techniques such as whole genome sequencing of individuals (which is now becoming both more technically feasible and more affordable) may help in answering some of these questions.

CHEMICAL CARCINOGENS: INITIATION

Although there may be continued debate over the specific molecular mechanisms involved in carcinogenesis, there is no debate over the fact that chemicals can act as carcinogens. Evidence for this comes from both laboratory and epidemiological studies (more on this later in the chapter). Carcinogens that have the ability to bind to and alter the structure of DNA are generally classified as *genetic carcinogens*, whereas carcinogens that bind to and affect other cellular targets, or which do not alter the structure of the DNA, are called *epigenetic carcinogens*. Some substances appear to be intrinsically carcinogenic, whereas others (*procarcinogens*) must undergo bioactivation (such as metabolism by the P450 system) to produce reactive metabolites.

Certain *complete* carcinogens seem to be able to produce malignant tumors without administration of a second chemical. Other carcinogens appear to be only initiators and seem to require subsequent action of a promoter in order to produce cancer. Conversely, many promoters seem to be inactive unless there was prior exposure to an initiator; however, some promoters have induced cancer when used at high doses without an initiator.

Bioactivation

See also:
  Biotransformation
                                Chapter 3

There are both naturally occurring and synthetic carcinogens. Exposure to carcinogens can occur at work, in the diet, or from other environmental sources. Carcinogenesis can result from chronic exposure to low levels of certain carcinogens, and a long latent period may occur before clear manifestation of the disease. Cancer can also arise through exposure to ionizing radiations and certain cell-transforming viruses. Estimating potential carcinogenicity is a major concern in the registration of drugs, food additives, and pesticides, as well as in the control of toxic substances in research and manufacturing.

GENETIC CARCINOGENS

Genetic carcinogens contribute to the process of carcinogenesis through their interactions with the purine and pyrimidine bases found in the DNA molecule. These purine and pyrimidine bases possess the critical nucleophilic sites where chemically induced changes such as *alkylation* (addition of an alkyl group) can occur. This reaction occurs when the nucleophilic atom of a nucleotide, such as guanine *N*-7 or *O*-6, attacks the electrophilic carbon of the alkylating agent (the carcinogen), forming a covalent bond. If a nucleotide base (such as the purine base, guanine, for example) is alkylated to form such an altered DNA molecule (referred to as an *adduct*), the ability of the base to correctly pair with its complementary base (in this case, cytosine) may be impaired. Thus, as the enzyme DNA polymerase catalyzes the synthesis of the complementary strand during DNA replication, an incorrect pyrimidine base (thymine rather than cytosine) might be inserted into the new strand, resulting in mutation (Figure 6.2).

Alkylating agents are one of the best-defined categories of chemical carcinogens. One typical example of an alkylating agent is *mustard gas*, which was employed as a chemical weapon in World War I. This agent was first found to react with nucleotide bases of DNA *in vitro*. Then, as predicted, alkylated DNA was isolated from mice

Alkylation of guanine leads to the misreading of the sequence on the right when its complementary strand is synthesized. The alkyl group (R) allows the substitution of thymine for the normal complement, cytosine.

GTG ⟶ CAC

G*TG ⟶ TAC

Alkylation of guanine

FIGURE 6.2 Alkylation of guanine.

that were exposed *in vivo*. It was also found that the ability of various alkylating agents to induce cancer in mice was dose-dependent and was directly related to the degree of alkylation of mouse thymus DNA *in vivo*. Other studies have demonstrated that alkylation of DNA is better correlated with cancer development than alkylation of RNA or protein. This provides strong evidence that DNA is indeed the target for the reactions that can initiate carcinogenesis.

Another class of alkylating agents is the *polycyclic aromatic hydrocarbons*. Produced during the combustion of organic materials (for example, they are an important component of cigarette smoke), these compounds are metabolized by P450 through epoxidation to yield reactive metabolites. One example is *benzo[a] pyrene*, which, following metabolism to an epoxide, forms adducts with guanine. *N-nitroso compounds* are another example. These compounds are also found in cigarette smoke and are potentially formed in the stomach when ingested nitrates combine with secondary amines.

CONSEQUENCES OF MUTAGENESIS

There are multiple ways in which mutations can alter the structure of the DNA and thus potentially initiate carcinogenesis. Mutations that affect only a single pair of bases in the DNA chain are known as *gene* or *point mutations*. Point mutations generally affect one single codon (the group of three bases that code for a single amino acid in protein synthesis), potentially causing substitution of one amino acid for another in the resulting protein. This change may or may not affect protein function, depending on the location of the substituted amino acid in the molecule and on the similarity of the substituted amino acid for the original amino acid (Figure 6.3). Point mutations may also affect the regulatory regions of DNA that control gene expression.

Mutations that lead to the insertion or deletion of bases, however, disrupt the triplet code and can affect all downstream codons. Such an event is known as a *frameshift mutation* and is likely to block the production of functional proteins by that gene (Figure 6.3).

Are these single-gene changes sufficient to induce carcinogenesis? Again, there is significant debate over this issue. However, mutations can also lead to events that are much larger on the molecular scale. These include changes in the chromosome structure, e.g., breakage and rearrangement, or loss of genetic material (potentially leading to aneuploidy). *Radiation* is an example of a potent genetic carcinogen that can produce changes in the chromosome structure.

Finally, although the focus on initiation in the past has primarily been on genes and on the DNA sequences that control their expression, it is clear that there is more to the story. For example, there are DNA sequences that code for small, noncoding RNA molecules (microRNAs), molecules that play an important role in the regulation of translation. Effects either on the DNA coding for these molecules or on the regulatory sequences that control their transcription can clearly have significant functional consequences. For example, miR-29b suppresses the translation of p21 and Bim, proteins that promote apoptosis. Thus, overexpression of this miRNA may contribute to carcinogenesis, and it can be considered an oncogenic miRNA. Conversely, miR-34 suppresses the transcription of proteins, which stimulate cell division, and may be considered a tumor suppressor miRNA. Underexpression of this miRNA would thus also contribute to carcinogenesis.

Polycyclic Aromatic Hydrocarbons

See also:
The Role of Cytochrome P450
Chapter 3

Endocrine Disrupters and Interference with Hormonal Controls
Chapter 7

Lung Cancer
Chapter 8

PAHs
Appendix

---

miRNA

See also:
Effects of Toxicants on Nucleic Acids
Chapter 4

Other Roles for RNA
Chapter 5

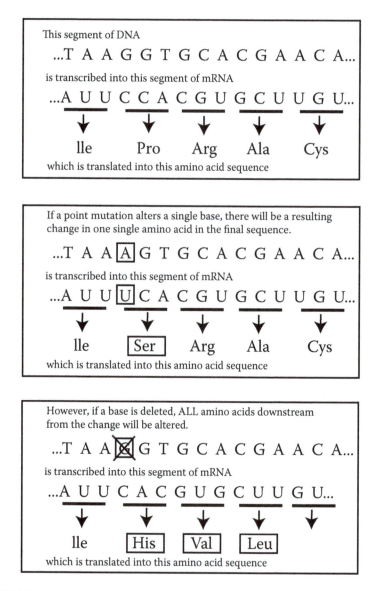

FIGURE 6.3 A comparison between the effects of point and frame-shift mutations in genes on the resulting protein.

CHEMICAL CARCINOGENS: PROMOTION

Although genotoxic carcinogens such as alkylating agents are reactive with DNA, many other carcinogens are believed to produce cancer through epigenetic mechanisms, i.e., by interacting with other parts of the cell. These compounds are generally considered to be acting in the promotion phase.

Stimulation of Cell Growth and Division

One example of a compound that acts as a promoter is dioxin (TCDD). TCDD displays highly specific affinity to a cytosolic receptor protein. Upon binding of TCDD to this receptor, the complex moves to the nucleus, binds to a DNA receptor binding site, and induces the transcription of the gene for the enzyme *aryl hydrocarbon hydroxylase* (AH), along with several other genes. This may lead to the promotion of cell division through actions on various regulating enzymes. TCDD may also act in synergy with carcinogens that require bioactivation, producing an increase in P450 levels and thus leading to an increase in the production of reactive metabolites. Finally, TCDD may also promote development of cancer through its immunosuppressive effects.

Peroxisome proliferators are compounds that are known to be able to produce an increase in the number of peroxisomes, structures within the cell which are involved in the breakdown of fatty acids. Evidence has indicated that some of these chemicals may also act as promoters. The mechanism of action likely involves their interaction with a nuclear receptor, the *peroxisome proliferator-activated receptor alpha* (*PPARα*). Activation of this receptor can lead to both increased cell proliferation and reduction in apoptosis, both of which are events that can play a role in promotion. Other promoters of cell division include the *phorbol esters.*

TCDD

See also:
The Role of Cytochrome P450
Chapter 3

Endocrine Disrupters
and Interference with
Hormonal Controls
Chapter 7

Toxicant-Induced
Immunosuppression
Chapter 13

Pesticides
Chapter 17

TCDD
Appendix

Suppression of Cell Death

Another potential mechanism of promotion would be suppression of apoptosis or other mechanisms of cell death. Apoptosis and autophagy are known to play significant roles in the elimination of cells with genetic and other damage and can thus act as a screening mechanism to eliminate neoplasms. Chemicals that suppress those processes (such as phenobarbital, for example) can allow cells to divide and establish clonal populations in spite of genetic damage.

Stimulation of Cellular Production of Reactive Compounds

Of course, the cell itself produces potentially genotoxic metabolites (free radicals, reactive oxygen species, etc.) through its own normal processes. Thus, compounds that affect normal cellular metabolism may act epigenetically through indirectly increasing the level of potentially DNA-damaging molecules within the cell. Free radicals and reactive oxygen species may also stimulate cell division through their effects on cell growth regulation and signaling pathways.

Hormones as Promoters

Hormones most certainly play a role in many cancers (particularly cancers of the reproductive tracts) and are thought to act through epigenetic mechanisms such as influencing gene expression. Both endogenous and exogenous estrogens have been implicated in the promotion of breast cancer, and therapeutic interventions that antagonize the actions of estrogen have been shown to reduce risks of occurrence or reoccurrence of cancers. One drug that is commonly used to treat estrogen-responsive cancers is *tamoxifen*, which is a competitive inhibitor of the estrogen receptor. However, many tumors become resistant to tamoxifen, and other drugs that interfere with estrogen signaling have been developed. For example, the drug *fulvestrant* produces receptor downregulation, and the drug *anastrozole* blocks estrogen synthesis through inhibition of the enzyme aromatase.

Intriguingly, obesity and type 2 diabetes are well known risk factors for cancer, and several types of tumor cells in culture are insulin-dependent in a way that normal cells are not. Although many normal cells will respond to insulin with growth, the response in cancer cells seems to be amplified. In addition, many cancer cells display elevated numbers of insulin receptors, as well as receptors for insulin-like growth factor (IGF), another growth-related hormone. Thus, some researchers have hypothesized that the hormonal environment resulting from these conditions may act to promote carcinogenesis.

The Role of Cellular Energetics

One early hypothesis of carcinogenesis, the Warburg hypothesis, postulated that defects in energy metabolism were responsible for development of the cancer phenotype in cells. This was based on observations that respiration in cancer cells does, in fact, differ from that in normal cells, in that respiration in cancer cells tends to rely more on aerobic glycolysis (as opposed to more efficient aerobic pathways) than respiration in normal cells does. Nowadays, the general belief is that these metabolic differences are a result (rather than a cause) of the genetic changes that produce neoplastic cells, although the shift in metabolism probably enhances survivability of tumor cells in some fashion (even though it requires large amounts of glucose in order to run). Is this metabolic difference related to the previously discussed differences in insulin response? Does it, in fact, follow rather than precede genetic mutations? Some researchers still think otherwise. Hopefully, future research will tell.

Epigenetic Impacts on Genes and Gene Expression

Epigenetic mechanisms may also be responsible for the gain or loss of entire chromosomes, thus producing aneuploidy. Unlike genetic carcinogens, epigenetic toxicants causing aneuploidy do not act directly on the DNA, but instead act on other cellular components involved in cell division (such as spindle fibers, for example). Other possible epigenetic mechanisms include interactions of carcinogens either with promoters or with products of thousands of repeated elements—short and long interspersed nuclear elements (SINEs and LINEs, respectively), etc.—that inhabit the human genome and are transcribed to RNA polymers. These include active transposable elements that can produce cDNA products via reverse transcriptase and that can be mutagenic, depending on the site of re-integration.

Finally, as a greater appreciation for the role of methylation in control of DNA expression has emerged, more attention has been paid to the potential for epigenetic carcinogens to alter gene expression through effects on methylation patterns. In fact, both hypermethylation and hypomethylation have been seen in a variety of cancers, with hypermethylation generally suppressing gene activity and hypomethylation amplifying expression. Histones and the enzymes that modify them are also potential epigenetic targets.

OTHER MECHANISMS OF PROMOTION

Other potential epigenetic carcinogens include *heavy metals*. The mechanism by which metals act as carcinogens is unclear. Some evidence indicates that they may in fact interact with DNA (a genetic mechanism), but other evidence points toward inhibition of DNA repair mechanisms (an epigenetic mechanism). Cadmium, for example, is nonmutagenic in the Ames test, but has been found to produce overexpression of a number of genes potentially involved in cancer (see the section on oncogenes) and to interfere with DNA repair. Beryllium, in another example, may act through effects on enzymes involved in DNA synthesis and replication, producing an increase in errors associated with those processes. There is also evidence that compounds that interfere with communication between cells (through interference with gap junctions, for example) may contribute to tumor formation.

DNA Repair

See also:
   Protection Against the
   Development of Cancer
                            Chapter 6

PROTECTION AGAINST THE DEVELOPMENT OF CANCER

Fortunately, there are multiple intrinsic mechanisms that act to block the initiation, promotion, and progression of neoplasms. First, there are several enzymes that work to repair damage to DNA. *Methyltransferase enzymes*, for example, cleave methyl groups from guanine. Base or nucleotide *excision repair systems* consist of *endonucleases*, which open the DNA strand to allow removal of the damaged or mispaired base; *polymerases*, which insert the correct base; and *ligases*, which reseal the strand. *Mismatch repair* recognizes and corrects incorrectly paired bases and involves numerous proteins that recognize, bind to, remove, and correct mismatched bases. The quicker that repair occurs, of course, the less likely it is that the original damage will have functional consequences.

Individuals with defects in DNA repair systems are much more susceptible to developing cancer than the general population. For example, individuals with hereditary nonpolyposis colon cancer typically show mutations in one or more genes coding for proteins involved in nucleotide mismatch repair, and individuals with *xeroderma pigmentosum* are extremely susceptible to sunlight-induced skin cancer due to defects in nucleotide excision repair.

The immune system also provides surveillance against neoplastic cells. Abnormal surface proteins (called *tumor-specific antigens*) may appear on the surface of neoplastic cells, marking them for destruction by a type of lymphocyte called a *natural killer cell*. *Cytotoxic T cells* and a class of cells called *tumor-infiltrating lymphocytes* (TILs) also participate in tumor destruction. Evidence for the participation of the immune system in tumor destruction is strong. Individuals with disease-induced or deliberate (as in the case of organ transplantation) immunosuppression have a much higher risk of cancer than the general population. Also, physicians have recently had some success in treating cancer through immune system stimulation.

TESTING COMPOUNDS FOR CARCINOGENICITY

On the basis of the observation that many carcinogens seem to be genetic in their mechanism of action (although this is certainly not true in all cases), initial screening tests for carcinogenicity generally evaluate the mutagenicity of a compound. A good screening test should be simple and replicable, with few false negatives. One test that meets these criteria is an *in vitro* assay called the *Ames test*.

In the Ames test, a mutant form of the bacteria *Salmonella typhimurium* is used. This mutant is unable to grow unless the nutrient histidine is supplied in the growth medium. Reversion to wild-type (which does not require an external supply of histidine) can occur with base-pair substitution or frame-shift mutation at the appropriate location on the bacterial genome. Mutant bacteria are exposed to the potential mutagen/carcinogen and are then cultured on a medium without histidine. Thus, the only colonies that will survive are those that arise from bacteria that have undergone mutation and reverted to wild-type. Bacterial growth is compared between the test group and one or more control groups (which are unexposed to the potential mutagen/carcinogen). A positive control (a group that has been treated with a known mutagen) is often also used.

There are numerous variations on this test, some of which use mammalian cells. For example, Chinese hamster ovary cells are often used in an assay that measures mutations in the gene for the enzyme hypoxanthine–guanine phosphoribosyltransferase (HGPRT). Another type of *in vitro* testing, cytogenetic testing, focuses on visual identification of chromosomal aberrations in populations of exposed cells. (Chinese hamster ovary cells are also used here, as are human lymphocytes.)

One problem with the Ames test and similar *in vitro* tests is that compounds that require metabolic activation to produce mutagenic/carcinogenic metabolites may test negative. To remedy this, preparations of isolated mammalian liver smooth endoplasmic reticulum may be added to the test. This cytochrome P450-containing preparation (called the *S9* component) carries out metabolism of the test compound and thus allows testing of metabolites as well.

There are also a variety of *in vivo* test methods to determine carcinogenicity. One *in vivo* gene mutation assay involves first dosing transgenic mice (whose cells contain an inserted bacterial gene within a bacteriophage DNA sequence) with potential carcinogens. Then, the bacteriophage DNA is extracted from the tissue of interest, used to infect a bacterial host, and assayed for mutations. *In vivo* assays are also available to look for larger changes such as chromosomal damage.

Exposure

FIGURE 6.4   The threshold and nonthreshold models of cancer risk estimation.

Longer term, chronic exposure testing is generally done using mice and rats, with routes of administration of the potential carcinogen chosen to most closely resemble the route by which human exposures would be expected to occur. In a typical chronic study, exposure to the potential carcinogen begins shortly after birth and continues for 1–2 years. At the end of the study, animals are examined, and control and treated groups are compared with respect to survival, number of tumors, types of tumors, and onset time to development of tumors. Because incidence of background tumors may be high, however, high dosages and large group sizes may be necessary to demonstrate statistically significant differences between the treated and control groups.

The relationship between exposure to carcinogens and actual risk of developing cancer is controversial. Although there is evidence that the relationship may be linear at high exposure levels, it cannot be assumed that this is also true at low exposure levels, and the limits in sensitivity of testing methods make experimental verification difficult, if not impossible.

Some scientists believe that the exposure–risk curve is, in fact, linear; others, however, believe that there is a threshold exposure below which risk is negligible. Proponents of the *threshold model* argue that our food, water, and environment are replete with carcinogens and that without a threshold, cancer rates would be much higher than they are. They also point to defense mechanisms such as detoxification reactions, DNA repair, and immune surveillance that should be able to cope with low-level exposures. Obviously, regulatory decision-making involving carcinogens is heavily influenced by whether the participants accept a threshold or nonthreshold model of risk. A comparison of both models is shown in Figure 6.4.

CRITIQUES OF STRATEGIES IN CANCER RESEARCH

Some researchers dispute the mutational theory and argue that direct evidence for the cause-and-effect relationship sought for many years continues to be lacking, despite several decades of experiments. Interested readers should consult the references (Li et al., 1997; Prehn, 2005). A detailed investigational criticism of established cancer research, including the relative lack of experimental investigation of the process of metastasis, has been published (Leaf, 2004).

CARCINOGENESIS: A COMPLEX PROCESS

At this point, although much has been discovered about the process of carcinogenesis, much more remains unknown. Most scientists would agree that cancer is triggered by alterations in the DNA structure and function, by either direct (genetic) or indirect (epigenetic) means. It also appears clear that activation of oncogenes, inactivation of tumor suppressor genes, and changes in expression of many other genes involved in the cell cycle or signal transduction in general play a role in the development of the neoplastic phenotype. However, it is clear that there are many paths to cancer, and recent estimates are that there are as many as 140 genes, mutations of which may play significant roles in the process of carcinogenesis. It is estimated that the process involves up to eight mutations developing over the course of two or more decades, with the end result being a heterogeneous tumor of clonal origin.

Gene Arrays

See also:
Genomics and New Genetics in Toxicology
Chapter 5

Hopefully, future research will not only clarify the remaining questions, but also lead to better methods of prevention and treatment. In fact, the use of gene arrays to provide more precise and specific information as to the proteins being produced in cancer cells is at the forefront of new strategies being developed to fight cancer. Currently, much cancer chemotherapy focusses on destroying cells that are actively undergoing cell division. This, of course, targets cancer cells, but it also targets normal cell populations such as the bone marrow. Also, cancer cells tend to develop resistance to many traditional chemotherapeutic agents. A better understanding of the mechanisms of gene expression in cancer cells will hopefully allow the development and use of more sophisticated methods (involving, for example, specific manipulation of levels and activity of some of the DNA, RNA, and protein molecules described in this chapter) to both selectively destroy cancer cells and successfully negotiate the problem of resistance.

CASE STUDY   Predicting Carcinogenesis Based upon Chemistry (QSAR)

It would be very useful to be able to predict the carcinogenic potential of a new chemical to which humans might be exposed (intentionally or otherwise), simply by examining its chemical structure. Manufacturers and regulators of drugs and pesticides are especially interested in developing these chemistry-based methods to predict biological responses, as accurate predictions have the potential to identify both potentially useful compounds and compounds that might be problematic in terms of carcinogenesis or other toxicity issues. This modeling of biological response based on mathematical descriptions of a series of chemicals is known as the *quantitative structure–activity relationship* (QSAR).

The approach of QSAR begins with determining a biological response (BR) to a *congeneric series* of chemicals (chemicals with a similar basic structure). This is usually a series in which a core molecule is derivatized (altered) by attaching various substituent atoms or groups. The BR that is measured could be a toxic, mutagenic, carcinogenic, or allergenic response, or other response of interest. This is followed by describing each molecule in arithmetic descriptors that represent the *hydrophobic* (or lipophilic), *electronic* (charge), and *steric* (shape or size) properties of the molecule. Descriptors can be theoretically computed (*a*) or empirically measured (*x*). In the simplest form, a regression equation is then calculated from the plot of the logarithm of the BR versus a quantity representing the chemical structure, as estimated from the sum of the descriptors:

$$\log \mathrm{BR} = a_h x_h + a_e x_e + a_s x_s + \text{constant}.$$

In this equation, the first term on the right refers to the hydrophobic properties, the second to the electronic properties, and the third to the steric properties. This general approach of making a regression model was developed independently in 1964 by Hansch and Fujita and by Free and Wilson.

The hydrophobic properties of substituents are based on the estimated contribution that substituents make to the total hydrophobicity of the molecule.

Electronic properties of substituents are often based on the Hammett equation, which gives an estimate of sigma (σ), a quantity related to the effects of a substituent as measured by electron-withdrawing or electron-donating influence on a functional group in the molecule. An example is the electron-withdrawing character of a halogen substituted for hydrogen *para* (across the ring) to a carboxyl functional group.

Steric properties are estimated from the van der Waals radii of substituents or by considering the three-dimensional size and shape of the molecule using STERIMOL size parameters as introduced by Verloop.

BRs to small molecules often involve some interaction with a lipid bilayer membrane not present in a chemical reaction in a test tube; therefore, the QSAR approach can be strengthened by adding terms to estimate transport into or through a membrane to the site of action. This is accomplished by adding terms to model a bilinear or parabolic relationship based on a descriptor of the hydrophobicity of the compound, *P* (the octanol/water partition coefficient). The value of *P* can either be measured as the coefficient for partitioning of the chemical of interest between layers of *n*-octanol and water in a separatory funnel, or it can be calculated from the chemical formula.

This addition to the model results in the equation

$$\log \mathrm{BR} = a_h x_h + a_e x_e + a_s x_s - a_1 (\log P)^2 + a_2 \log P + \text{constant}.$$

QSAR analysis is performed with a set of assumptions. First, the logarithm of the BR is assumed to correlate with the free energy changes associated with binding of the chemical to the molecular site of action. From this assumption, QSAR describes the biological process in the same way that a physical organic chemist would describe chemical reactions by finding rate constants or chemical equilibria.

Another assumption is that substituents of the core structure contribute independently and additively to the response. It is also assumed that additive descriptors representing hydrophobic, electronic, and steric properties can be used to estimate a quantity related to the response.

In application to cancer biology, QSAR is most reliable for modeling mutagenicity. Predictions based on present QSAR models are more accurate for genotoxic carcinogens than for nongenotoxic carcinogens. Finally, it was found that models deteriorate with addition of marginally significant descriptors, so that a few good descriptors constitute the optimal model in most cases.

BIBLIOGRAPHY

Aaronson, S.A., Growth factors and cancer, *Science*, 254, 1146, 1991.
Balmain, A. and Brown, K., Oncogene activation in chemical carcinogenesis, *Adv. Cancer Res.*, 51, 147, 1988.
Bock, G. and Marsh, J., *Proto-oncogenes in Cell Development*, John Wiley & Sons, Chichester, U.K., 1990, p. 295.
Boguski, M.S., Bairoch, A., Attwood, T.K., and Michaels, G.S., Proto-vav and gene expression, *Nature*, 358, 113, 1992.
Bourne, H.R. and Stryer, L., G proteins: The target sets the tempo, *Nature*, 358, 541, 1992.
Choueiri, T.K., Alemany, C.A., Abou-Jawde, R.M., and Budd, G.T., Role of aromatase inhibitors in the treatment of breast cancer, *Clin. Ther.*, 26(8), 1199, 2004.
Cohen, S.M. and Ellwein, L.B., Cell proliferation in carcinogenesis, *Science*, 249, 1007, 1990.
Cuperlovic-Culf, M., Belacel, N., and Ouellette, R.J., Determination of tumor marker genes from gene expression data, *Drug Discov. Today*, 10(6), 429, 2005.
Dawson, M.A. and Kouzarides, T., Cancer epigenetics: From mechanism to therapy, *Cell*, 150, 12, 2012.
Downward, J., Signal transduction: Rac and Rho in tune, *Nature*, 359, 273, 1992.
Duesberg, P. and Rasnick, D., Aneuploidy, the somatic mutation that makes cancer a species of its own, *Cell Motil. Cytoskel.*, 47, 81, 2000.
Farmer, G., Bargonetti, J., Zhu, H., Friedman, P., Prywes, R., and Prives, C., Wild-type p53 activates transcription *in vitro*, *Nature*, 358, 83, 1992.
Franke, R. and Gruska, A., General introduction to QSAR, in *Quantitative Structure–Activity Relationship (QSAR) Models of Mutagens and Carcinogens*, Benigni, R., Ed., CRC Press, Boca Raton, FL, 2003.
Gupta, P.B., Kuperwasser, C., Brunet, J.P., Ramaswamy, S., Kuo, W.L., Gray, J.W., Naber, S.P., and Weinberg, R.A., The melanocyte differentiation program predisposes to metastasis after neoplastic transformation, *Nat. Genet.*, 37, 1047, 2005.
Grimm, D., Disease backs cancer origin theory, *Science*, 306, 389, 2004.
Harlow, E., Retinoblastoma: For our eyes only, *Nature*, 359, 270, 1992.
Harris, C.C. and Liotta, L.A., *Genetic Mechanisms in Carcinogenesis and Tumor Progression*, John Wiley & Sons, New York, 1990, p. 235.
Hemminki, K., Rawal, R., Chen, B., and Bermejo, J.L., Genetic epidemiology of cancer: From families to heritable genes, *Int. J. Cancer*, 111, 944, 2004.
International Agency for Research on Cancer, Monographs on the Evaluation of Carcinogenic Risks to Humans, Vol. 100, IARC, World Health Organization, Lyon, France, 2012.

Israelsen, W.J. and Vender Heiden, M.G., ATP consumption promotes cancer metabolism, *Cell*, 143, 669, 2010.

Kaplan, D.R., Hempstead, B.L., Martin-Zanca, D., Chao, M.V., and Parada, L.F., The trk proto-oncogene product: A signal transducing receptor for nerve growth factor, *Science*, 252, 554, 1991.

Kilpivaara, O. and Aaltonen, L.A., Diagnostic cancer genome sequencing and the contribution of germline variants, *Science*, 339, 1559, 2013.

Klaunig, J.E., Chemical carcinogenesis, in *Casarett and Doull's Toxicology*, Klaassen, C.D., Ed., McGraw-Hill, New York, 2013, chap. 8.

Koren, I. and Kimchi, A., Promoting tumorigenesis by suppressing autophagy, *Science*, 338, 889, 2012.

Kurzrock, R., Kantarjian, H.M., Druker, B.J., and Talpaz, M., Philadelphia chromosome-positive leukemias: From basic mechanisms to molecular therapeutics, *Ann. Intern. Med.*, 138, 819, 2003.

Lane, D.P., Cancer: p53, guardian of the genome, *Nature*, 358, 15, 1992.

Leaf, C., The war on cancer, why we're losing the war on cancer—And how to win it, *Fortune*, March 22, 2004.

Levine, A.J. and Puzio-Kuter, A.M., The control of the metabolic switch in cancers by oncogenes and tumor suppressor genes, *Science*, 330, 1340, 2010.

Li, R., Yerganian, G., Duesberg, P., Kraemer, A., Willer, A., Rausch, C., and Hehlmann, R., Aneuploidy correlated 100% with chemical transformation of Chinese hamster cells, *Proc. Natl. Acad. Sci. USA*, 94, 14506, 1997.

Lichtner, R.B., Estrogen/EGF receptor interactions in breast cancer: Rationale for new therapeutic combination strategies, *Biomed. Pharmacother.*, 57, 447, 2003.

Loktionov, A., Common gene polymorphisms, cancer progression and prognosis, *Cancer Lett.*, 208, 1, 2004.

Marsh, D.J. and Zori, R.T., Genetic insights into familial cancers: Update and recent discoveries, *Cancer Lett.*, 181, 125, 2002.

Oliner, J.D., Kinzler, K.W., Meltzer, P.S., George, D.L., and Vogelstein, B., Amplification of a gene encoding a p53-associated protein in human sarcomas, *Nature*, 358, 80, 1992.

Pavletich, N.P. and Pabo, C.O., Zinc finger-DNA recognition: Crystal structure of a Zif268-DNA complex at 2.1 Å, *Science*, 252, 809, 1991.

Petsko, G.A., Signal transduction: Fishing in Src-infested waters, *Nature*, 358, 625, 1992.

Pitot, H.C., III and Dragan, Y.P., Chemical carcinogenesis, in *Casarett and Doull's Toxicology*, Klaassen, C.D., Ed., McGraw-Hill, New York, 2001, chap. 8.

Plass, C., Cancer epigenomics, *Hum. Mol. Genet.*, 11, 2479, 2002.

Powell, S.M., Zilz, N., Beazer-Barclay, Y., Bryan, T.M., Hamilton, S.R., Thibodeau, S.N., Vogelstein, B., and Kinzler, K.W., APC mutations occur early during colorectal tumorigenesis, *Nature*, 359, 235, 1992.

Prehn, R.T., The role of mutation in the new cancer paradigm, *Cancer Cell Int.*, 5, 9, 2005.

Preston, R.J. and Hoffmann, G.R., Genetic toxicology, in *Casarett and Doull's Toxicology*, Klaassen, C.D., Ed., McGraw-Hill, New York, 2013, chap. 9.

Ries, L.A.G., Eisner, M.P., Kosary, C.L., Hankey, B.F., Miller, B.A., Clegg, L., Mariotto, A., Feuer, E.J., and Edwards, B.K., Eds, *SEER Cancer Statistics Review, 1975–2002*, National Cancer Institute, Bethesda, MD, 2004, available at http://weer.cancer.gov/csr/1975_2002.

Roberts, L., More pieces in the dioxin puzzle, *Science*, 254, 377, 1991.

Saglio, G. and Cilloni, D., Abl: The prototype of oncogenic fusion proteins, *Cell. Mol. Life Sci.*, 61, 2897, 2004.

Schneider, G. and So, S.-S., *Adaptive Systems in Drug Design*, Landes Bioscience, Georgetown, TX, 2002.

Shen, H. and Laird, P.W., Interplay between the cancer genome and epigenome, *Cell*, 153, 38, 2013.

Simon, M.I., Strathmann, M.P., and Gautam, N., Diversity of G proteins in signal transduction, *Science*, 252, 802, 1991.

Solomon, E., Colorectal cancer genes, *Nature*, 343, 412, 1990.

Solomon, E., Borrow, J., and Goddard, A.D., Chromosome aberrations and cancer, *Science*, 254, 1153, 1991.

Suva, M.L., Riggi, N., and Bernstein, B.E., Epigenetic reprogramming in cancer, *Science*, 339, 1567, 2013.

Taub, G., Unraveling the obesity–cancer connection, *Science*, 335, 28, 2012.

U.S. Cancer Statistics Working Group, United States Cancer Statistics: 1999–2001, incidence and mortality Web-based report version, Department of Health and Human Services, Centers for Disease Control and Prevention, and National Cancer Institute, 2004, available at www.cec.gov/cancer/npcr/uscs.

Vogelstein, B., Papadopoulos, N., Velculescu, V.E., Zhou, S., Diaz, L.A., Jr., and Kinzler, K.W., Cancer genome landscapes, *Science*, 339, 1546, 2013.

Waisberg, M., Joseph, P., Hale, B., and Beyersmann, D., Molecular and cellular mechanisms of cadmium carcinogenesis, *Toxicology*, 192, 95, 2003.

Waksman, G., Kominos, D., Robertson, S.C., Pant, N., and Baltimore, D., Crystal structure of the phosphotyrosine recognition domain SH2 of V-src complexed with tyrosine-phosphorylated peptides, *Nature*, 358, 646, 1992.

zur Hausen, H., Viruses in human cancers, *Science*, 254, 1167, 1991.

7 Reproductive Toxicology and Teratology

INTRODUCTION

The functions of reproduction and development are complex and involve many relatively unique cellular-level processes. As such, the effects of toxicants on the process of reproduction and on developing organisms may be quite different from those of the same toxicant on other systems in the adult organism. This chapter first reviews a few basic concepts in reproduction and development and then considers the effects of toxicants on reproductive function (the production of eggs in the female and sperm in the male). It concludes with an examination of the effects of toxicants on developing organisms.

BASIC PROCESSES IN REPRODUCTION AND DEVELOPMENT: CELL DIVISION

THE CELL CYCLE AND MITOSIS

Cell division plays a major role in both reproduction and development. The two basic types of cell division are (1) *mitosis*, in which a single cell divides to form two identical daughter cells, and (2) *meiosis*, in which daughter cells are produced that have only one instead of two copies of each chromosome.

During their lifetime, somatic cells (cells not involved in the formation of eggs or sperm) move through various stages in what is known as the *cell cycle* (Figure 7.1). The cell cycle consists of two distinct phases: *interphase* and the *mitotic phase*. Each of these phases can be subdivided on the basis of the activities being carried out in the cell. During the G_1 *phase* of interphase, duplication of organelles and other cytoplasmic constituents occurs in preparation for cell division. Cells may spend from only a few hours to several months in this phase.

During the next phase, the *S phase* (which lasts several hours), the cell makes a copy of its DNA. DNA in eukaryotic cells is, of course, found in the form of *chromatin*, long strands of DNA wrapped around proteins called *histones*. During DNA replication, the hydrogen bonds that bind together the two complementary strands of the DNA in each chromosome are disrupted, and the two strands separate. Enzymes called *DNA polymerases* and *ligases* then link together free DNA nucleotides to form a new complementary strand for each of the original strands (Figure 7.2). The result is that the cell now contains two copies of each stretch of chromatin.

After a short period of additional protein synthesis (the G_2 *phase*), the cell enters the first stage of mitosis, which is *prophase* (Figure 7.3). At the beginning of

123

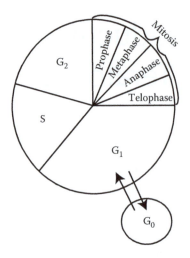

FIGURE 7.1 The stages of the cell cycle, including the G_0 (nondividing) phase, from which cells may move in and out over their lifetime.

prophase, the DNA coils up tightly to form the visible structures known as *chromosomes*. A chromosome consists of the two identical copies of the DNA made during the S phase (each of which is called a *chromatid*), held together at a point called the *centromere*. Also during prophase, the two pairs of microtubular structures called *centrioles* move to opposite ends of the cell, and the mitotic spindle (another microtubular structure that serves as a scaffolding for mitosis) forms between them.

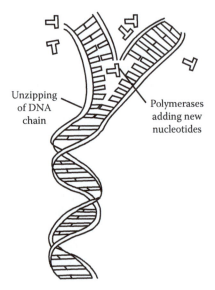

Unzipping of DNA chain

Polymerases adding new nucleotides

FIGURE 7.2 Semiconservative replication of DNA. (For the chemical formula of DNA, see Figure 6.1.)

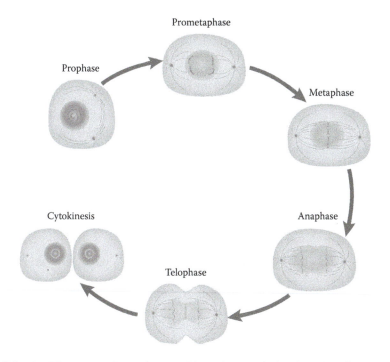

FIGURE 7.3 The stages of mitosis. (Modified with permission from http://www.shutterstock.com/pic-112083344/stock-vector-cell-division-mitosis-vector-scheme.html?src=&ws=1. Accessed on June 22, 2014.)

With the disappearance of the nuclear envelope, the cell moves into the second stage of mitosis, *metaphase*. During metaphase, the chromosomes attach to the mitotic spindle and are moved to the center of the cell, where they line up across a plane called the metaphase plate. Then, in the third stage, *anaphase*, chromatids of each pair separate and are pulled by the mitotic spindle toward opposite ends of the cell. In the final stage, *telophase*, the nucleus reappears, the chromosomes uncoil, and the process of *cytokinesis*, the division of the cytoplasm, is completed. The end result is two daughter cells with the identical genetic makeup of the parent cell.

Movement of cells through the cell cycle is controlled by an elegant system of molecular checkpoints. There are several of these checkpoints spread throughout the cell cycle, and whether a cell proceeds past a checkpoint is determined by the proteins called *cyclins*, whose levels within the cell fluctuate with time.

For example, one important checkpoint is located at the end of the G_2 phase. The levels of cyclins rise during G_2, and the newly synthesized cyclins form associations with proteins known as *cyclin-dependent kinases* (Cdks). One cyclin–Cdk complex is known as *maturation promoting factor (MPF)*. When MPF levels become high enough, passage of the cell from interphase into mitosis occurs. MPF then turns itself off later in mitosis by destroying its cyclin component and resuming its inactive Cdk form.

Another checkpoint occurs during G_1. If cells cannot pass the G_1 checkpoint (which is also regulated by cyclins and Cdk proteins), they enter a nondividing phase known as G_0. Many cells spend most of their lives in G_0, only returning to the cell cycle if prompted to do so by the correct molecular signals.

The final cell cycle checkpoint is located during mitosis, at the beginning of anaphase. At this checkpoint, chromosomes that remain unattached to the mitotic spindle will block further progression into mitosis. The mechanism of the signal is not yet well understood, but even one unattached chromosome has been shown to block passage.

Further elucidation of cell cycle control mechanisms and identification of compounds that can influence the cell cycle are important research directions, with applications to a variety of biomedical problems. Examples are cancer research, where reduction in cell division rates is the goal, and neurodegenerative disease research, where stimulation of cell division would be beneficial.

MEIOSIS

Cells that are to form eggs or sperms must go through a different process. Human somatic cells contain 23 pairs of chromosomes: 22 *homologous* pairs of *autosomal* (non-sex-related) chromosomes and 1 pair of chromosomes that determine the sex of the individual. Each of the two chromosomes that make up a homologous pair of autosomal chromosomes contains the same basic genes, and yet the two are not identical. This is because any given gene may exist in two or more variations called *alleles*. Thus, the copy of a gene that is on one chromosome of a homologous pair may or may not be the same allele that is found on the other chromosome of the homologous pair. In the case of the sex chromosomes, there are two distinctive forms: the X chromosome and the Y chromosome. Individuals with two X chromosomes have a female genotype, whereas individuals with one X and one Y chromosome have a male genotype. The total assortment of alleles possessed by an individual is known as that individual's *genotype*.

Cells that contain these two complete sets of chromosomes, such as most of the cells that make up the human body, are described as *diploid*. It is easy to see that in order for the fusion of one egg and one sperm to produce a normal diploid zygote, each gamete must contribute half the normal complement of chromosomes. Egg and sperm cells are, in fact, *haploid* (i.e., containing only one of each chromosome). Thus, when they combine together to form a diploid zygote, one chromosome of each diploid pair is contributed by the egg and the other is contributed by the sperm.

But where do these haploid gametes come from? They are created from diploid cells through the process of meiosis (Figure 7.4). Meiosis consists of two separate divisions: meiosis I and meiosis II. Following duplication of cellular constituents (including DNA), such as precedes mitosis, the cell enters meiosis I. The stages of meiosis I are similar to those of mitosis, except that when the chromosomes line up along the midline of the cell in metaphase I, they line up as homologous pairs, which link together to form a structure called a *tetrad* (as there are two copies of each of the two chromosomes, for a total of four chromatids). During this time, homologous chromosomes may exchange genetic material in a process called *crossing over*, which can result in new combinations of alleles on a chromosome. Then, during anaphase I, the homologous chromosomes separate and move to opposite ends of the cell. In telophase I and cytokinesis, the cytoplasm of the cell divides, giving rise to

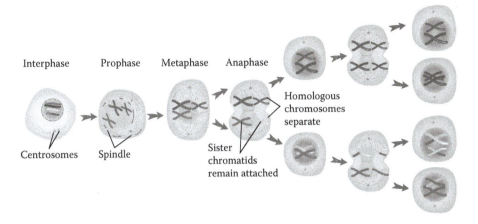

FIGURE 7.4 The stages of meiosis. (Modified with permission from http://www.shutterstock.com/pic-116607409/stock-vector-vector-diagram-of-the-meiosis-phases.html?src=&ws=1. Accessed on June 22, 2014.)

two haploid daughter cells. Each chromosome in these daughter cells does, however, still have two sister chromatids, and in meiosis II, each of the newly formed daughter cells divides again through a process virtually identical to mitosis. The end result is a total of four haploid cells.

From the molecular perspective of the RNA world, sexual reproduction is a means of rescuing the DNA blueprint of the species from an invasion of RNA information in transposable elements and other forms. The offspring of sexual reproduction can have either less or more RNA-based information; however, the offspring of asexual reproduction will have at least as much RNA information as its parent, and more if invasion or amplification by copy-and-paste transposition has occurred.

Cloning

Cloning of Dolly the sheep, mules, horses, cats, and other mammals is based on a patented process of reversing differentiation. First, somatic cells (often mammary cells) of the individual animal to be cloned are cultured. During the G or G_0 phase of the donor cell cycle, a nucleus obtained from the donor cell is used to replace the nucleus of a surrogate egg (arrested at metaphase II). During this process, the donor cell nucleus undergoes reprogramming, a process that reverses the state of differentiation of the nucleus. This process, as of yet, is not completely understood, and failure of reprogramming may be responsible for the low percentage of viable clones produced from somatic nuclear transfer. Note that the process, as currently practiced, uses the mitochondria, including the circular DNA of the mitochondria, of the surrogate, not of the donor, so the resultant clone is a clone of only the nuclear DNA and not all the genomic DNA of the individual.

New applications of veterinary cloning continue to appear. Increasingly, mules, the sterile hybrid offspring of a cross of a mare (female horse) and a jack (male donkey), are valued for many varied competitive activities. Several clones of a champion mule have been produced, cloning being the only means of reproduction for mules. Similarly, a champion gelding (neutered male horse) was cloned in Italy, with the intent of standing the clone at stud.

Opponents of cloning point to the incidence of defective offspring produced by the process. Cloning of humans is banned in the United States, but has been practiced in Republic of Korea to produce an embryo from which stem cells were harvested. The ethics of using embryonic stem cells, producing embryos for the purpose of harvesting stem cells, employing cloning techniques for human reproduction, and related activities are highly controversial. With increasing success of such techniques, geneticists in this era of biotechnology must confront philosophical issues of good and evil in the application of new technology, just as physicists have had to face these issues in the era of nuclear physics.

THE MALE REPRODUCTIVE SYSTEM

Development of cells, which will form the gonads, begins early in human embryos (around the fourth week), although the structures characteristic of the male reproductive system do not begin to become apparent until around the seventh week. A protein coded for by the SRY gene on the Y chromosome triggers development of the male phenotype, and the absence of this protein during development ultimately confers a female phenotype upon the developing fetus.

The formation of the male gamete, the sperm, occurs within organs called the *testes*, in tubules called *seminiferous tubules* (Figure 7.5). The process of *spermatogenesis* (sperm formation) begins at puberty and is regulated by the steroid hormone *testosterone*, which is produced by *interstitial cells* that lie between the tubules.

FIGURE 7.5 The testes, with seminiferous tubules. (Modified with permission from http://www.shutterstock.com/pic-176748164/stock-vector-illustration-of-the-anatomy-of-the-male-reproductive-system-on-a-white-background.html?src=&ws=1. Accessed on June 22, 2014.)

Secretion of testosterone is itself regulated by the pituitary hormone *luteinizing hormone* (LH). Another pituitary hormone, *follicle-stimulating hormone* (FSH), is also important in spermatogenesis. Release of these pituitary hormones is, in turn, regulated by *gonadotropin-releasing factor* (GnRF), which is released by a section of the brain called the hypothalamus. The presence of testosterone produces a *negative feedback effect*, inhibiting the release of GnRF by the hypothalamus and thus blocking testosterone synthesis. When testosterone levels decline, though, GnRF release occurs and testosterone production is stimulated.

The process of spermatogenesis begins when diploid cells called *spermatogonia* divide by mitosis, forming daughter cells. Some of these cells will remain as spermatogonia, and some will mature into *primary spermatocytes*. Each primary spermatocyte undergoes meiosis I to form two haploid *secondary spermatocytes*, each of which completes meiosis II, leading to the formation of a total of four *spermatids*. Cytokinesis during these divisions is not completed; however, the spermatids remain connected through their cytoplasm. Under the influence of cells called *Sertoli cells*, the spermatids separate and mature into *spermatozoons*, which then migrate to the *vas deferens*, where they continue to mature and eventually are stored for release. The whole process takes about 80 days in humans.

Sperm stored in the vas deferens can remain active for several weeks. During the process of ejaculation, these sperm are ejected through the urethra, along with fluid from glands such as the seminal vesicles, bulbourethrals, and prostate. Typically, a few milliliters of this *semen* is released, containing upwards of 100 million spermatozoa. Normal reproductive function is often assessed by examining the concentration of sperm in semen, as well as the motility and histological appearance of individual spermatozoons.

THE FEMALE REPRODUCTIVE SYSTEM

The same basic hormonal system that controls reproduction in the male also controls reproduction in the female. GnRF is released by the hypothalamus, initiating secretion of LH and FSH by the pituitary. In the female, though, the effects of LH and FSH are to promote the syntheses of the *estrogens* and *progestins*. These hormones then participate in a complex feedback system that in turn regulates the release of GnRF, LH, and FSH in the same type of negative feedback regulatory cycle that operates in males.

The pair of organs in which egg production takes place is the *ovaries* (Figure 7.6). Early in development (around the fifth month), several million oogonia develop in each ovary. These oogonia barely begin the process of meiosis and then stop in prophase I. At this stage, they are called *primary oocytes*. They are still diploid, and they remain in an arrested state until many years later when puberty begins. These primary oocytes, along with the cells that surround them, make up the *primary follicles*. Many primary oocytes degenerate, so that by the time of puberty, each ovary probably contains only a few hundred thousand primary follicles. The belief has always been that these initial primary oocytes are the only oocytes that will be formed in a woman's lifetime. However, experimental work has indicated that, at least in mice, proliferative germ cells may continue to produce additional oocytes postnatally. This is still an open question, as evidence based on lineage analysis

FIGURE 7.6　The ovaries, fallopian tubes, uterus, and cervix. (Modified with permission from http://www.shutterstock.com/pic-182553398/stock-vector-uterus-and-ovaries-organs-of-female-reproductive-system.html?src=&ws=1. Accessed on June 22, 2014.)

(tracking the origin of individual oocytes) as well as evidence for the existence of female germline stem cells can be interpreted in a variety of ways. Also, even if postnatal oogenesis is eventually convincingly demonstrated in other species, whether or not the process occurs to humans may still remain an open question.

At the onset of puberty, the release of the hormone FSH each month (as part of the menstrual cycle) stimulates some follicles to grow into larger *secondary follicles*. One secondary follicle will continue to grow each month, while the primary oocyte within it completes the first part of meiosis to become a *haploid secondary oocyte*. (During this first division, the secondary oocyte receives most of the cytoplasm; the other, much smaller daughter cell is called a *polar body* and ultimately disintegrates.)

The secondary oocyte continues to grow along with the follicle, which fills with fluid secreted by the follicular cells. The structure is now known as a *tertiary follicle*. A surge in the hormone LH at about the 14th day of the menstrual cycle stimulates release of the secondary oocyte and some of its surrounding follicular cells. The oocyte then enters the *fallopian tubes* (where fertilization, if it is to occur, takes place) and begins the trip to the *uterus*. The empty follicle becomes the *corpus luteum*, secreting estrogen and progesterone until it eventually decays. If pregnancy occurs, a hormone secreted by the developing embryo (*human chorionic gonadotropin*) maintains the corpus luteum until the placenta develops and takes over the secretion of these hormones. If the oocyte is not fertilized, the corpus luteum begins to degenerate within about 10 days, resulting in a drop in progestins and initiating menstruation.

The uterus is a muscular, pear-shaped organ where development of the fertilized egg occurs. Under the influence of estrogens and progestins, the lining of the uterus (the endometrium) undergoes a monthly cycle of changes, first proliferating to produce a thick zone where the developing zygote can implant, then, if fertilization does not occur, shedding this newly developed tissue during menstruation. Menstruation is a process that is unique to primates; however, rats, mice, and other mammals have *estrous cycles*, in which the endometrium is reabsorbed rather than shed.

THE EFFECTS OF TOXICANTS ON THE MALE AND FEMALE REPRODUCTIVE SYSTEMS

Throughout the process of gametaogenesis, there are several points at which toxicants may act. We will look briefly at an endogenous protective mechanism and then at three major functional categories of toxicants: those that interfere with cell division, those that act directly on reproductive cells, and those that interfere with hormonal control of reproduction.

PROTECTIVE MECHANISM: THE BLOOD–TESTIS BARRIER

In the male reproductive system, the testes are afforded some degree of protection against toxicants by what is termed the blood–testis barrier. In humans, this barrier begins developing in childhood and is completed at puberty. Tight junctions between the Sertoli cells in the seminiferous tubules form a barrier that prevents many substances from entering the areas where spermatozoons are developing.

The testes also have metabolic capability in the form of cytochrome P450 activity, and the cells that will give rise to spermatozoa have at least some DNA repair capabilities. The ovaries also have some capacity for biotransformation.

INTERFERENCE WITH CELL DIVISION

Toxicants that interfere with cell division, such as alkylating agents (which damage DNA) and antimetabolites (which inhibit nucleotide biosynthesis), can directly inhibit sperm production. Likewise, physical agents such as x-rays and other forms of ionizing radiation can also cause decreases in spermatogenesis through effects on dividing cells.

These agents can also potentially interfere with oogenesis, because even though the process of cell division is thought to be initiated prenatally, it is not completed until just prior to ovulation.

Lead

See also:
Anemias, Hemolysis, and
Related Disorders
Chapter 9

Myelinopathies
Other Neurotoxicants
Chapter 10

Toxicant-Induced
Immunosuppression
Chapter 13

Metals
Chapter 17

Lead
Appendix

Cadmium

See also:
Vascular Spasms and Blood
Pressure
Chapter 9

Damage to the Proximal
Tubule
Chapter 12

Metals
Chapter 17

Cadmium
Appendix

CYTOTOXICITY AND INFERTILITY

Carbon Disulfide

See also:
  Atherosclerosis
                          Chapter 9

  Distal Axonopathies
                          Chapter 10

  Carbon Disulfide
                          Appendix

Smoking

See also:
  Genetic Carcinogens
                          Chapter 6

  Chronic Obstructive
    Pulmonary Disease:
    Bronchitis and Emphysema
  Lung Cancer
                          Chapter 8

  Atherosclerosis
  Effects of Toxicants on
    Hemoglobin
                          Chapter 9

  Tobacco
                          Appendix

Polycyclic Aromatic Hydrocarbons

See also:
  The Role of Cytochrome P450
                          Chapter 3

  Genetic Carcinogens
                          Chapter 6

  Lung Cancer
                          Chapter 8

  PAHs
                          Appendix

Some toxicants may exert their effects through direct action on cells of the reproductive system. For example, the pesticide *dibromochloropropane* (DBCP) caused destruction of the seminiferous tubule epithelium in exposed male workers. The inhibition of spermatogenesis produced by DBCP may have been caused either through effects on primary spermatogonia or through effects on Sertoli cells. The mechanism of action of DBCP in either case may be through inhibition of oxidative phosphorylation. Interestingly, DBCP does not seem to produce reproductive effects in females.

Other toxicants that may affect energy metabolism in the testes include *m-dinitrobenzene (m-DNB)*, *dinitrotoluene* (DNT), and various *phthalates* (plasticizing compounds). *m*-DNB and the phthalates appear to exert their effects primarily on the Sertoli cells, whereas the chemical ethylene glycol monomethyl ether (EGME) appears to affect not only Sertoli cells but also spermatocytes themselves. Heavy metals such as *lead* and *cadmium* are also well-known reproductive toxicants in males. Exposure to lead has been associated with infertility as well as with chromosomal damage in sperm. Cadmium can cause testicular necrosis, probably by decreasing blood flow to the testes. Ethanol causes delays in testicular development and may affect supporting cells. Other male reproductive toxins include the pesticides *kepone* and *DDT*, the solvent *carbon disulfide*, and even tobacco (smokers have higher percentages of abnormal sperm than nonsmokers).

In females, cytotoxic substances such as *antineoplastic agents*, *heavy metals*, *polycyclic aromatic hydrocarbons* (PAHs), or *radiation* may damage oocytes. The effects of such agents depend on which stage of oocyte development is affected. Effects on mature secondary oocytes will lead to temporary infertility, with fertility being restored as new secondary oocytes develop. Partial destruction of primary oocytes, however, may lead not to immediate infertility, but to early onset of menopause (which occurs when the total pool of primary oocytes in the ovary falls below a minimum number). The total destruction of primary oocytes, though, will lead to infertility as well as to premature menopause. For example, studies indicate that cigarette smoke, known to produce impairment in female fertility, depletes follicles and triggers oocyte apoptosis, most likely through oxidative stress. Nonlethal effects of these agents on oocytes include DNA damage, which may lead to genetic defects in offspring.

Some of these cytotoxic toxicants (PAHs, for example) must be metabolized in order to produce toxicity. This activation can occur in the ovary, as cytochrome P450 and other enzymes involved in xenobiotic metabolism are found in ovarian tissues. Toxic PAH metabolites destroy primary oocytes, which is one possible explanation for the observation that exposure to cigarette smoke (which contains PAHs) may lead to premature menopause.

Many other toxicants, including some *pesticides*, *chlorinated hydrocarbon solvents*, and *aromatic solvents*, have also been reported to interfere with female reproductive capacity.

ENDOCRINE DISRUPTERS AND INTERFERENCE WITH HORMONAL CONTROLS

A topic of much discussion in recent years has been that of *endocrine disrupters*. The attention in this category was initially focussed on *environmental estrogens*, compounds that are present in the environment in low concentrations and that possess estrogenic activity, but has now been broadened to include chemicals with the capability of disrupting endocrine function through other mechanisms. Typically, mechanisms of endocrine disruption would include interference with hormone levels (through impact on production, secretion, binding, metabolism, or clearance); upregulation or downregulation of receptors; or effects on hormone/receptor interactions. Examples of endocrine disrupters would include DDT and other organochlorine pesticides, polychlorinated biphenyls (PCBs), and dioxins (such as TCDD).

There are a number of examples in which interference with the secretion of hormones such as GnRF, LH, FSH, testosterone, estrogens, or progestins have been shown to have an impact on the reproductive process. For example, in males, estrogens and progestins block spermatogenesis by

Organochlorine Pesticides

See also:
Effects of Toxicants on Electrical Conduction
Chapter 10

Material Cycling in Ecosystems
Chapter 14

Pesticides
Chapter 17

Organochlorine Pesticides
Appendix

Polychlorinated Biphenyls (PCBs)

See also:
Toxicant-Induced Immunosuppression
Chapter 13

Other Organic Compounds
Chapter 17

Polychlorinated Biphenyls
Appendix

TCDD

See also:
The Role of Cytochrome P450
Chapter 3

Stimulation of Cell Growth and Division
Chapter 6

Toxicant-Induced Immunosuppression
Chapter 13

Pesticides
Chapter 17

TCDD
Appendix

suppressing LH and FSH and thus suppressing testosterone secretion. Another group of compounds that can interfere with hormonal control of reproduction in the male is the *anabolic steroids*. These are synthetic drugs that were developed in an attempt to separate the anabolic (muscle-building) effects of steroids from the androgenic (reproductive) effects. This separation is, in fact, unachievable, as both reproductive and muscle tissues seem to contain the identical type of androgen receptor. These synthetic steroids may, however, lower testosterone levels (probably through feedback inhibition of GnRF, LH, and FSH) and suppress spermatogenesis. Other undesirable side effects of anabolic steroids include hepatotoxicity, behavioral changes, and potential shortening of stature in prepubertal males through premature termination of long bone growth.

These hormonal compounds, along with others that either block binding of testosterone to the androgen receptor or inhibit enzymes involved in testosterone synthesis, have been suggested as potential male birth control agents. Unfortunately, it is difficult to completely block spermatogenesis reliably. In addition, many of these drugs also produce unacceptable side effects such as irreversibility of effects, depression of libido, or toxicity to other organ systems.

There is particular concern with endocrine disruptors in the environment, as many of these compounds are quite lipophilic and may accumulate over time in fatty tissues in humans and other organisms. In fact, studies have established relatively firm links between exposure to environmental estrogens such as DDT and reproductive dysfunction in wildlife. The related question, of course, is whether there are also measurable effects on the human population. There have been a variety of researchers who have begun to address this issue. For example, the possibility of a link between exposure to endocrine disrupters and decreased male fertility has been raised. One study (Carlson et al., 1992) has indicated a significant worldwide decline in sperm count; other studies have not detected any decline. In fact, sperm counts may be too highly variable both in location and in time to serve as a good overall indicator of male reproductive health.

Toxicants that interfere with the hormonal control of reproduction in the male can, of course, also impair fertility in the female. Anesthetics, analgesics, and other drugs that interfere with either neuronal or hormonal control of hypothalamic or pituitary function can prevent ovulation. In fact, *birth control drugs* such as oral contraceptives, as well as injectable (Depo-Provera®) and implantable (Norplant®) contraceptives, also act on this hormonal system. These drugs contain a mixture of estrogen and progestins that inhibit the release of FSH and LH and thus inhibit ovulation. Side effects associated with these drugs include some increase in the risk of thromboembolism (obstruction of a blood vessel by a blood clot), myocardial infarction, and stroke. These risks increase with age and with the presence of contributing factors such as smoking and underlying cardiovascular disease.

Another potential issue of concern with endocrine disrupters (in both males and females) is their effect on the timing of the onset of puberty. The average age of the onset of puberty in humans has decreased in many parts of the world in recent years, and several scientists have put forward the hypothesis that endocrine disrupters may be involved. The issue is confounded, however, by an accompanying trend of increasing body weight in children. Studies have noted that obese girls tend to enter puberty earlier, and there is a complex relationship between body weight, hormone levels, and development. Other studies have, however, shown possible associations between exposure to chemicals and both accelerated and delayed puberty. Exposure to DDT and phthalates, to give two examples, have been associated with accelerated puberty; on the other side of the coin, exposure to PCBs has been linked to pubertal delay in boys, as has exposure to dioxins in both boys and girls.

Concerns have also been raised over reports of increasing rates of reproductive abnormalities and cancers. At this point in time, more research is required to make any definitive determination as to whether there is a link between exposure to these chemicals and adverse effects on human reproductive health.

THE PROCESS OF DEVELOPMENT

The next stage in the reproductive process begins with the process of *fertilization*, where egg and sperm combine to form a single-celled diploid *zygote*. This step generally occurs in the fallopian tubes when a spermatozoon penetrates the outer covering of the oocyte and activates it. At that time, the second stage of meiosis is completed, leading to the formation of an oocyte and also a second polar body. Following penetration, the nuclei of the spermatozoon and the oocyte fuse, forming the zygote.

It should be noted here that although it was long thought that the genetic contributions of the egg and the sperm to the genetic makeup of the zygote were independent of the source (i.e., it did not matter if a particular allele came from the maternal or paternal genome), this turns out not to be the case. In point of fact, the expression of certain genes in the zygote has been found to depend explicitly on whether they are of maternal or paternal origin. This phenomenon is termed *imprinting* and appears to result from maternal and paternal differences in DNA methylation and/or chromatin conformation, both of which, of course, can influence gene expression.

Following fertilization, then, the single-celled zygote enters into a period of rapid cell division, or *cleavage*, and eventually forms a hollow ball of cells called a *blastocyst* (Figure 7.7). By this time (a few days after fertilization), the blastocyst has passed from the fallopian tubes into the uterus, where it implants in the uterine lining. The outer cells of the blastocyst (called the *trophoblast*) divide, grow, penetrate, and break down the endometrial tissues, releasing nutrients that can be used by the *inner cell mass* from which the embryo will develop.

Meanwhile, the inner cell mass separates from the trophoblast, and a fluid-filled cavity (the amniotic cavity) forms between them. The cells of the inner cell mass form an oval sheet called the *blastodisc*. By the end of the second week, the *differentiation* (adoption of different developmental pathways) of cells has begun. The mechanism behind this routing of genetically identical cells to different developmental fates is one of the central questions studied by researchers in the field of development. Differentiation appears to have its basis in differences in cytoplasmic constituents (resulting from unequal distribution of components in the egg) as well as differences in cellular environments involving interaction between the cells themselves. The combination of these factors leads to the formation of three distinct layers in a developing animal embryo: the *ectoderm*, which will form the epidermis as well as the epithelial linings of oral, nasal, anal, and vaginal cavities, nervous tissue, and some endocrine organs; the *mesoderm*, which will form muscle and connective tissues, vascular

FIGURE 7.7 The development of the blastocyst. (Modified with permission from http://www.shutterstock.com/pic-132798476/stock-vector-fertilized-egg-development-isolated-on-a-white-background.html?src=&ws=1. Accessed on June 22, 2014.)

endothelium and lymph vessels, the lining of some body cavities, the reproductive and urinary systems, and some other endocrine organs; and the *endoderm*, which becomes the epithelial lining of the gastrointestinal, respiratory, and urinary tracts.

Some of these tissues will also become the *extraembryonic membranes*, serving to protect, nourish, and support the developing embryo (Figure 7.8). One of these

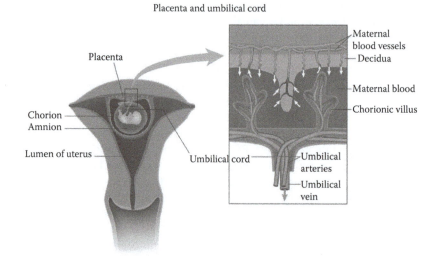

FIGURE 7.8 The placenta and the arrangement of the extraembryonic membranes. (Modified with permission from http://www.shutterstock.com/pic-163605584/stock-photo-placenta-labeled.html?src=&ws=1. Accessed on June 22, 2014.)

membranes is the *yolk sac*, which provides nourishment in some species and in humans produces blood cells and future gametes. The *amnion* is an ectodermal and mesodermal membrane that encloses the developing embryo and the amniotic fluid that cushions it, and the *allantois* stores metabolic waste and is involved in the formation of fetal blood vessels, blood cells, and the bladder. The *chorion* will form the fetal portion of the *placenta*. The placenta is not completely developed until about 12 weeks postfertilization. Blood flows into the placenta from the fetus through the umbilical arteries and returns through the umbilical vein. In the placenta, close juxtaposition of branches of the umbilical arteries with the maternal blood that is circulating through cavities in the placenta allows the interchange of oxygen, nutrients, and waste materials through the process of diffusion. The placenta also produces estrogen, progesterone, and other hormones that help maintain the uterine lining and thus the pregnancy.

By the fourth week, the longitudinal axis of the embryo develops in the form of a *primitive streak*, a thickened band along the midline of the blastodisc. During the next few weeks, the rudiments of the nervous system develop, arm and leg buds develop, and the basic structures of most organ systems are formed. At this time, the production of testosterone by the embryonic testes initiates the steps that will result in the development of a male phenotype. In the absence of testosterone, a female phenotype will develop regardless of genotypic sex. By the end of 8 weeks, the embryo is quite well developed and is referred to as a fetus. Development and growth continue throughout the fetal period, and for some systems (such as the nervous system, for example), development even continues postnatally.

EMBRYOGENESIS AND DEVELOPMENTAL GENETICS

Systems Biology

See also:
  Monitoring Transcription:
  Gene Expression and
  Microarrays
Systems Toxicology
                                        Chapter 5

The study of developmental genetics has been enhanced by the application of techniques for observing the sequential activation of genes using microarrays of oligonucleotide probes of the genes cataloged in the Human Genome Project. Beginning with maternal effect genes expressed in the egg, embryogenesis as described earlier proceeds through cascades of *transcription factors*. Master genes activate expression of target genes, which include subordinate transcription factors in a hierarchical network. Some genes are activated whereas others are inactivated. A web of positive or negative interactions can be described as in an engineering diagram using the new principles of systems biology.

During embryogenesis, the fate of cells is determined in the process of differentiation; e.g., a cell in the embryonic *neural crest* can become a melanocyte or a neuron, depending on which transcription factor is active. A mutation in just one transcription factor can result in the failure of an organ to form; e.g., deletion of the microphthalmic-associated transcription factor can result in diminished eyes in mice.

Master transcription factors are similar over widely diverse phyla, suggesting high conservation of the primary pathways of development. Transgenic introduction of mouse *PAX6* and its expression in fruit flies induces the formation of *ectopic eyes*; i.e., eyes are formed on legs and other structures of the insect. Conversely, the *eyeless* gene of fruit fly can induce ectopic eyes in the frog, *Xenopus*.

Protein transcription factors encoded by master genes regulate target genes by binding to specific *response elements* in the promoter/operator region upstream of the start codon. Usually, this binding occurs as part of an assembled complex of protein cofactors and an *RNA polymerase*. For example, the site of action of the chemicals *TCDD* and *TCDF* is likely the protein encoded by the gene *AhR*, which makes a dimer with the product of *Arnt*. The dimer translocates to the nucleus and becomes a component of a nuclear transcription factor complex. A similar dimer is formed by *fos/jun* oncogene products.

Embryonic stem cells are nondifferentiated cells considered to be *pluripotent*, i.e., capable of differentiating into any type of tissue, depending on the triggering signal received. Insects, following embryonic development and hatching to larvae, continue to carry small clusters of semidifferentiated cells called *imaginal discs*, which begin to differentiate into eye, antenna, leg, wing, and other structures during metamorphosis to the imago (adult). Each and every eye imaginal disc cell can become any of the diverse types of cells in the eye: eight different photoreceptor cells, cone cells, corneal cells, pigment cells, or others. The process is driven by the timing of concentration gradients and interactions of diffusing growth factors and transcription factors. Photoreceptor 8 cells are first to differentiate as triggered by a gradient of the protein product of *atonal* and its interaction with the product of *hedgehog*. Once formed, photoreceptor 8 cells begin to influence adjacent cells as additional protein factors sequentially replace atonal and hedgehog in the program. Humans also possess *adult stem cells*, which are semidifferentiated to the fate of a certain organ or type of tissue.

Understanding the impact of toxicants on the genetic basis of embryonic development is an aim of many researchers. One current trend is the application of microarray technology in the study of gene expression and its perturbation by exposure to environmental toxicants. *Knockout* technology (targeted disruption of a gene) and *antisense* technology (using probes that are complimentary to messenger RNA to block gene expression) are also useful techniques. These can be useful in determining both the normal role of genes in development and the potential consequences of interference with that role.

EFFECTS OF TOXICANTS ON DEVELOPMENT: TERATOGENS AND TERATOGENESIS

Teratology is the study of birth defects—structural or functional abnormalities that are present at birth. In humans, it is estimated that the presence of abnormalities leads to the spontaneous abortion of somewhere between 25% and 50% of all pregnancies. In addition, birth defects occur in up to 10%–15% of all live births. Of course, many of these problems may be caused by random genetic errors, but some may also be caused by environmental factors.

A *teratogen* is an environmental agent that produces birth defects. Examples of effects of teratogens on developing organisms include structural malformations, growth retardation, and death. Whether or not exposure to a teratogen produces a birth defect depends on several different factors. Two of the most important factors are dose or exposure level and timing of exposure. There are, of course, other factors involved as well, many of which we do not yet understand. For example, prenatal exposure of a litter of rat pups to a teratogen may produce severe defects in some pups, milder defects in others, and no effects at all in a few. Reasons for this may include microdifferences in intrauterine conditions (causing, for example, some pups to be exposed to higher levels of the teratogen than others), variations in the state of development of different pups in the litter, or even genetic variations in susceptibility between pups.

EFFECTS OF DOSE OR EXPOSURE LEVEL ON TERATOGENICITY

First, as with any toxicant, the actual dose or exposure level is an important factor. During pregnancy, such physiological changes as increased absorption of substances through the gastrointestinal tract, increases in the lung tidal volume, and increases in the blood flow to the skin all may enhance absorption of environmental toxicants. Also, as the maternal blood volume increases, concentrations of the plasma protein albumin decrease, leading to fewer binding sites for toxicants and a greater tendency for those toxicants to enter tissues. Counteracting this, however, is an increase in the rate of renal excretion.

It is important to note that the placenta itself fails to provide much of a barrier to transfer of toxicants between the maternal and fetal compartments. Substances that are lipid soluble, small, and neutral in charge diffuse easily through the placenta. Other substances may cross by means of facilitated diffusion or active transport mechanisms. Metals (cadmium, for example) may also accumulate in the placenta.

Xenobiotic metabolism of toxicants is thought to occur mostly in maternal tissues or the placenta, as the levels of many enzymes that participate in both phase I and phase II biotransformation are quite low in developing organisms (in humans, P450 levels at

> **Xenobiotic Metabolism**
>
> *See also:*
>   *Biotransformation*
>   *Chapter 3*

midgestation are less than half the adult levels). Prenatal levels of P450 do, however, increase on exposure to inducers such as phenobarbital. Placental biotransformation activities, although low, also are inducible (exposure to PAHs, as contained in cigarette smoke, significantly increases P450-A1 activity, for example).

With teratogens in general, as exposure levels increase, so does the severity of the teratogenic effect. Some teratogens produce only structural defects at low levels of exposure, but may be lethal at higher levels. Others may produce a range of effects, from structural defects to lethality, at the same exposure level. Maternal toxicity may or may not occur at exposure levels sufficient to produce birth defects. Thus, in many cases, exposures that would not threaten the health

> **Threshold Model**
>
> *See also:*
>   *Testing Compounds for*
>   *Carcinogenicity*
>   *Chapter 6*

of the mother may be quite hazardous to the developing child. However, the relationship between dose and response as it relates to teratogenicity is complex, and the same debate concerning the presence or absence of a safe "threshold" of exposure (below which the risk of adverse events is virtually zero) exists in teratogenicity as it does in carcinogenicity. In particular, the ability of embryonic tissues to recover from damage at the cellular level is not well understood.

EFFECTS OF TIMING OF EXPOSURE ON TERATOGENICITY

Because a variety of events occur at so many different times during the prenatal period, it stands to reason that the timing of exposure to a teratogen is critical in determining the potential effects. Exposure during the early stages (prior to implantation), for example, is most likely to lead to embryonic death. Exposure during the late stages (in humans, the third trimester) is most likely to lead to growth retardation. It is during the middle stage, organogenesis, that exposure is most likely to lead to structural defects. Exposure to a teratogen during the *critical period* when a particular organ system is forming may lead to malformations in that system. For example, exposure to the rubella virus during the first 8 weeks of pregnancy frequently produces defects of the visual and cardiovascular systems, whereas exposure during weeks 8–12 leads to hearing impairment. Critical periods in humans may vary in length from as long as several weeks to as short as a day.

EXAMPLES OF TERATOGENS

Thalidomide

See also:
  Thalidomide
                    Appendix

Diethylstilbestrol

See also:
  DES
                    Appendix

One of the earliest identified teratogens was, in fact, a biological agent. This was the rubella, or German measles virus, and it was identified when an Australian ophthalmologist named Norman Gregg observed that an epidemic of cases of congenital cataracts closely followed an epidemic of rubella. This case also illustrates the importance of critical periods, as the type of malformations produced was closely related to the time of exposure. Eye malformations were observed to correspond most closely to exposure during the first 8 weeks of development, and deafness and heart problems were seen to correspond to exposure between the 8th and 12th weeks of development. Another case that also illustrates the importance of critical periods is the case of the drug thalidomide (see the case study at the end of this chapter).

An historic example of a chemical teratogen was the use of the drug *diethylstilbestrol* (DES) to prevent potential miscarriage (ironically, a use for which it has since been shown to be ineffective). It was administered to women considered to be at high risk for miscarriage (typically because of a prior history of early miscarriage) and was given to several million women in the United States alone between the late 1940s and early 1970s. The problems associated with DES use were first noted when an unusually large number of cases of a rare vaginal cancer (a cancer most often seen in older women) were seen in young women at Massachusetts General Hospital. Interference of this drug with normal reproductive tract development (particularly during weeks 6–16) led to structural and functional abnormalities of the reproductive tract in both DES-exposed daughters and sons (and now possibly even in children of DES-exposed daughters). This case heightened awareness of the possibility of what has come to be called *transplacental carcinogenesis*. Since that time, other agents (other chemicals and radiation) that may also increase the risk of cancer in prenatally exposed offspring have also been identified.

For some systems, such as the nervous system, critical periods extend throughout development. One teratogen with significant effects on this system is *ethanol. Fetal alcohol syndrome* was identified in the early 1970s and is a group of related effects, including craniofacial abnormalities, growth retardation, and mental retardation, all of which result from intrauterine exposure to ethanol. Severity of the abnormalities seems to increase with increases in exposure levels. It is not certain whether there is a "safe" level of alcohol consumption during pregnancy, but risk of craniofacial and neurological abnormalities rises with the consumption of 2 oz. of ethanol per day and risk of growth retardation rises with consumption of 1 oz. per day. Cocaine use has also been associated with various abnormalities, including neurological problems and developmental deficits that may persist throughout life.

Ethanol

See also:
> *Other Enzymes Carrying Out Oxidation*
> > *Chapter 3*

> *Cardiomyopathies and Other Effects on Cardiac Muscle*
> > *Chapter 9*

> *Other Neurotoxicants*
> > *Chapter 10*

> *Fatty Liver*
> *Liver Cell Death: Necrosis and Apoptosis*
> *Fibrosis and Cirrhosis*
> > *Chapter 11*

> *Forensic Toxicology and Alcohol Use*
> > *Chapter 16*

> *Ethanol*
> > *Appendix*

A controversial prescription drug currently on the market is the drug *isotretinoin* (trade name Accutane®). Although effective in the treatment of severe cystic acne in adults, it is also a potent teratogen, causing craniofacial, thymus, cardiac, and neurological defects. In spite of label warnings, educational programs, and pregnancy testing, some affected children are born each year. Although many people would prefer to see this drug taken off the market, its effectiveness in dermatological treatment makes such a decision difficult. This is an excellent example of a regulatory risk–benefit decision.

There are many other compounds on the list of known or suspected teratogens. Other pharmaceuticals include the antiepileptic drugs *valproic acid* and *phenytoin*, as well as the tetracycline antibiotics. Heavy metals that are known to be teratogenic include methylmercury and lead, both of which produce neurological dysfunction. Various herbicides (such as Agent Orange, a mix of the herbicides 2,4-D and 2,4,5-T) have been suspected by some of being teratogenic; however, the evidence is not completely clear. Finally, smoking during pregnancy (or even being exposed to secondhand smoke) has been associated with increased risks of spontaneous abortion, sudden infant death syndrome, low birth weight, and a variety of neurological disorders.

MECHANISMS OF TERATOGENICITY

Not much is known about the biochemical mechanism of action of many teratogens, perhaps because there are so many gaps in our knowledge of the biochemical and cellular aspects of development. Of course, any agent that interferes with cell division is likely to damage developing organisms, where rates of cell division are very high. High levels of such compounds during organogenesis may produce organ malformations; lower levels may result in the development of structurally normal, but smaller organs.

Probably more is known about the mechanisms involved in the production of *cleft palate* than for any other structural abnormality. The palate is formed by growth and fusion of the maxillary and palatine processes, an event involving cell division, migration, programmed cell death, and other complex processes. Cleft palate occurs when this event is disrupted, leaving a gap where fusion was to occur. One toxicant thought to be capable of producing cleft palate is TCDD, and it appears to do so by binding to proteins in the cytosol and blocking the programmed cell death necessary for normal palatal development. Exposure to high levels of *glucocorticoids* also causes cleft palate. Interaction of glucocorticoids with the glucocorticoid receptors in the maxillary cells inhibits cell growth and thus blocks normal palate formation.

Other teratogens also appear to interact directly with cells in the developing organism. Thalidomide has been hypothesized to directly damage developing limb tissue or to interfere with communication between that tissue and surrounding tissues. DES seems to lead to failure of tissues from a temporary structure called the Mullerian duct to either transform into normal tissues (in women) or degenerate (in men).

Neurological effects of teratogens may be produced by many different mechanisms. Because the increased permeability of the fetal blood–brain barrier allows greater access to toxicants, the fetal brain may be susceptible to a wider range of insults than the adult brain. In addition, there are many specialized steps in neurological development that can potentially be disrupted by toxicants, including the development of neurotransmitters and receptors. Evidence indicates that these systems must be functioning properly in order for innervation to proceed correctly.

Cocaine, for example, potentiates the effects of norepinephrine (a neurotransmitter found in the sympathetic branch of the autonomic nervous system) by blocking reuptake. This may have a direct effect on the cells of the developing brain, or because norepinephrine stimulates blood vessels to contract, it may lead to a decrease in the blood flow to the fetus, causing hypoxia. Norepinephrine also stimulates contraction of the uterine smooth muscle, perhaps causing the tendency for premature labor observed in cocaine-use cases.

Finally, advances in the molecular biology of development have led to a greater understanding of the potential of teratogens to interact with the genetic mechanisms that drive development. Retinoids, for example, may exert their effects through interaction with a set of developmental genes known as *hox* genes. Recently, the potential for *epigenetic* effects of teratogens has also been recognized. These effects would include impact on gene expression without direct alteration of the DNA structure. Ethanol, to name one example, has been hypothesized to exert at least some of its teratogenicity through epigenetic mechanisms. In terms of epigenetic mechanisms, attention has been recently focused on the regulatory role of histone acetylation and deacetylation in gene expression and on the possibility that some teratogens (including valproic acid) may act through inhibition of enzymes involved in this process.

TESTING FOR REPRODUCTIVE AND DEVELOPMENTAL TOXICITY

HUMAN ASSESSMENT

Reproductive and developmental toxicity assessment often involves the identification of problems in humans. In humans, reproductive history, sperm count (the concentration of sperm in the semen) and normality of sperm, and hormone levels in the blood are typically used to assess male reproductive functioning. In females, reproductive history is also an important tool for assessing human fertility. In addition, x-rays of the uterus and fallopian tubes can identify structural abnormalities, and measurement of serum hormone levels, changes in body temperature, and other indicators can be used to evaluate for the presence or absence of ovulation.

Epidemiological studies have, in the past, been useful in identifying links between exposure and reproductive or teratogenic outcomes, but several factors complicate their usefulness. First of all, there are many confounding factors that complicate the identification of such an association. Secondly, the relatively high background rate of spontaneous abortion and birth defects makes it difficult to identify many common birth defects as being related to a toxic exposure. Thus, epidemiological studies must not only be carefully designed, but typically must also incorporate a large number of subjects in order to produce clear evidence as to whether or not there is an association between exposure to a toxicant and reproductive or developmental effects.

TESTING OF LABORATORY ANIMALS: GENERAL PRINCIPLES

A number of factors must be considered in developing tests for reproductive toxicity and teratogenicity involving laboratory animals. Because of interspecies differences in developmental pathways, xenobiotic metabolism, and so on, the choice of species to be used in the test is critical. Hamsters, mice, and rats are common choices, due in part to their short gestation times (a purely practical advantage), but rabbits are also used, as are primates.

The route of administration for a toxicant should be similar to any route for human exposure and may include injection, intubation, inhalation, or delivery in food or water. Due to variations in food and water consumption, however, it is difficult to deliver a precise dose through this last route. Normally, multiple dose levels and a control (as well as solvent control, if necessary) are used, with dosages ranging from near the no-effect level to near the lethal level. In the case of teratology studies, exposure should continue throughout the length of the gestational period, especially during organogenesis.

There are a number of different ways in which fertility is evaluated in the research laboratory in both male and female animals. In males, one of the most direct methods is to assess sperm count, sperm motility, and sperm morphology (abnormalities in head shape, tail length, etc.). The ability of sperm to penetrate and fertilize an egg *in vitro* can also be studied. Another measure that has been proven to be effective is histological evaluation of the testes, although more general measures such as weights of testes, epididymis, prostate, and other glands can also be used. Blood levels of various hormones can be measured, and evaluation of hormone/receptor binding *in vivo* or *in vitro* can also be carried out. Ultimately, reproductive success as measured by the number of viable young produced can be assessed. Fertility profiles can be developed through regular mating of a toxicant-exposed male with a number of females, followed by calculation of the percentage of females impregnated. During these matings, reproductive behavior can also be observed. Offspring from these matings can then be studied for evidence of genetic defects.

In female laboratory animals, similar tests are used. Organ weights (ovaries and uterus) can be evaluated, and histological examination of the ovaries can indicate whether ovarian toxicity has occurred. As in males, hormonal levels and studies of hormone/receptor binding can be undertaken. Finally, production of viable offspring can also be assessed. Fertility profiles would involve mating of treated females, observation of mating behavior, assessment of the outcome of mating, and possibly evaluation of offspring.

Testing *In Vitro*

Very young rat or mouse embryos (from conception up to the point where placental formation occurs) can be maintained in culture, exposed to teratogens, and observed for changes in normal development. Organs removed during organogenesis can also be cultured, as can cells or groups of cells (such as the rat embryo limb bud, to give one example). Mouse embryonic stem cell lines have also been developed and can be used in testing both cell viability and development (mEST assays). Some nonmammalian cell culture systems are also used in research and testing. These include cells derived from *Drosophila* (fruit fly) eggs, hydra cells, and *Xenopus* (an amphibian) embryos. One such well-established test system is the Frog Embryo Teratogenesis Assay—*Xenopus*, or FETAX test system, which is a 96 h developmental toxicity test using *Xenopus* embryos. Increasingly, zebrafish models have also been used in developmental assays.

Established Procedures for Testing

With the goal of standardizing requirements for product testing, a group known as the International Conference of Harmonization of Technical Requirements for Registration of Pharmaceuticals for Human Use (the ICH) has developed a typical set of guidelines for teratogenicity testing that has been well accepted worldwide. This relatively recent set of tests includes a fertility protocol (involving treatment of both male and female animals for several weeks prior to mating), tests for prenatal/postnatal development and maternal function (involving treatment of the pregnant female through the end of lactation), and tests for effects on embryo/fetal development (involving treatment of the pregnant female from implantation through the end of organogenesis in two species).

For endocrine disruptors, the EPA's Endocrine Disruptor Screening and Testing Advisory Committee has proposed a series of tests to determine whether or not compounds have the potential to be endocrine disruptors. Tests in this protocol assess effects of chemicals on rats *in vivo*, looking at effects on male and female reproductive tissues in both intact animals and animals that have had their ovaries or testes removed. Tests for effects on onset of puberty in female rats are also performed.

To truly assess both reproductive and developmental effects, however, multigenerational studies assessing impact at a number of life stages would be an expensive and time-consuming, but optimal, solution.

CASE STUDY   Thalidomide

Perhaps, the most notorious teratogen in history is the drug *thalidomide*. Synthesized in 1954, with a chemical formula similar to barbiturates, it was developed as a sedative/hypnotic. It also proved to be effective as an *antiemetic*, easing the nausea (morning sickness) that often plagues early pregnancy. With a very low acute toxicity, it was rapidly approved for use and was marketed in Germany in 1956 and in England and Canada in 1958. In fact, it was even available in England over the counter (in other words, without a doctor's prescription). There were a few reports that adults taking thalidomide over several months had developed peripheral neuropathies, but knowledge of these problems was not generally widespread in the medical community.

In late 1961, however, reports began to surface of a link between thalidomide exposure and the birth defects *phocomelia* (drastic shortening of the limbs; fingers attached at the shoulder, for example) and *amelia* (lack of limbs). These effects were accompanied by malformations of the cardiovascular, renal, and other systems, but it was the sudden increase in the frequency as well as the severity of the previously quite rare limb defects that attracted the attention of medical personnel and allowed the link between exposure to the drug and these teratogenic effects to be uncovered. By 1961, thalidomide had been withdrawn from European markets, and by 1962, it had also been withdrawn from Canadian markets. Estimates of the number of children affected range from 6000 to 10,000, with perhaps 20% of exposed children developing defects.

Altogether, thalidomide was sold in about 30 countries worldwide. It was not, however, sold in the United States due to inadequacies in the safety data submitted to the FDA. The fact that the drug was not released in the United States is primarily due to an FDA employee, Dr. Frances Kelsey, who repeatedly rejected the application even under pressure from the manufacturer to approve the drug. Nonetheless, 1200 US doctors were sent samples of thalidomide, which the FDA had to go to great lengths to recover. Still, there were only a handful of thalidomide cases in the United States—a few resulting from distribution of the samples and a few resulting from acquisition of the drug in a country where it had been approved.

A number of hypotheses as to the mechanism of action for thalidomide's teratogenic effects have been put forward. One leading hypothesis is that because thalidomide has been shown to be an inhibitor of *angiogenesis* (the formation of new blood vessels), interference with the development of blood vessels in the limbs as they formed might be the cause of phocomelia and amelia. However, thalidomide has numerous other cellular effects as well, any one of which might have contributed to limb malformation. For example, thalidomide has also been shown to block the production of the lymphokine tumor necrosis factor alpha (TNF-α), which is involved in the inflammatory response, as well as to inhibit production of other lymphokines such as interleukins and interferons. Finally, there is some evidence that some of thalidomide's effects may be due to increased production of free radicals leading to either oxidative damage to DNA or alterations in gene expression mediated through altered transcription factor binding.

The obvious question about thalidomide is why was it ever allowed on the market? Certainly, evidence indicates that safety testing for the compound was, in fact, inadequate, but there were also a number of contributing factors that would have made accurate testing and risk assessment difficult even under the best of conditions. First of all, the critical period for the development of phocomelia and amelia in humans was quite short, corresponding to a 2-week period between days 21 and 35 of gestation. Also, although the drug is a potent teratogen in humans, with doses as little as 100 mg/day (around 1 mg/kg maternal weight) producing severe defects, it is much less potent in other species. In fact, mice and rats are relatively resistant to the teratogenic effects of thalidomide. Only rabbits and primates have been proven susceptible to developing the limb defects that are so characteristic in humans. All of these factors made the developmental problems associated with thalidomide usage quite difficult to accurately identify. In fact, even with the more stringent testing measures in place today, it is not at all certain that a drug such as thalidomide could not still clear the regulatory hurdles.

The final chapter on thalidomide has actually yet to be written. In 1998, the FDA did, in fact, approve thalidomide for use in the United States for the treatment of the skin condition erythema nodosum leprosum (ENL), which is a complication of leprosy. Thalidomide also shows some promise for treatment of some forms of arthritis, for Crohn's disease (an inflammatory bowel disease), and for multiple myeloma and other cancers, although side effects such as thrombosis and peripheral neuropathies have been noted. The drug is tightly regulated, and only physicians participating in an FDA educational program are permitted to prescribe it. In addition, women of childbearing age who receive the drug must undergo weekly or monthly pregnancy testing. These restrictions are similar to those placed upon the prescription of Accutane, an anti-acne retinoid drug that was discussed earlier in this chapter. In both cases, there has been considerable debate over whether the benefits gained by patients using these drugs are worth the teratogenic risks taken by making the drugs available.

BIBLIOGRAPHY

Beckman, D.A. and Feuston, M., Landmarks in the development of the female reproductive system, *Birth Defects Res. B*, 68, 137, 2003.

Carlson, E., Giwercman, A., Keiding, N., and Skakkebaek, N.E., Evidence for decreasing quality of semen during the past 50 years, *Br. Med. J.*, 305, 609, 1992.

Christian, M.S., Test methods for assessing female reproductive and developmental toxicology, in *Principles and Methods in Toxicology*, Hayes, A.W., Ed., Taylor & Francis, Philadelphia, 2001, chap. 29.

Claudio, L., Bearer, C.F., and Wallinga, D., Assessment of the U.S. Environmental Protection Agency Methods for Identification of Hazards to Developing Organisms. Part I. The reproduction and fertility testing guidelines, *Am. J. Indus. Med.*, 35, 543, 1999.

Claudio, L., Bearer, C.F., and Wallinga, D., Assessment of the U.S. Environmental Protection Agency Methods for Identification of Hazards to Developing Organisms. Part II. The developmental toxicity testing guideline, *Am. J. Indus. Med.*, 35, 554, 1999.

Clegg, E.D., Perreault, S.D., and Klinefelter, G.R., Assessment of male reproductive toxicity, in *Principles and Methods in Toxicology*, Hayes, A.W., Ed., Taylor & Francis, Philadelphia, 2001, chap. 28.

Collins, T.F.X., Current protocols in teratology and reproduction, in *Safety Evaluation of Drugs and Chemicals*, Lloyd, W.E., Ed., Hemisphere Publishing Corporation, Washington, D.C., 1986, chap. 13.

Daston, G.P., Cook, J.C., and Kavlock, R.J., Uncertainties for endocrine disrupters: Our view on progress, *Toxicol. Sci.*, 74, 245, 2003.

Dimopoulos, M.A. and Eleutherakis-Papaiakovou, V., Adverse effects of thalidomide administration in patients with neoplastic diseases, *Am. J. Med.*, 117, 508, 2004.

Foster, P.M.D. and Gray, L.E., Jr., Toxic responses of the reproductive system, in *Casarett and Doull's Toxicology*, Klaassen, C.D., Ed., McGraw-Hill, New York, 2013, chap. 20.

Franks, M.E., Macpherson, G.R., and Figg, W.D., Thalidomide, *Lancet*, 363, 1802, 2004.

Johnson, J., Canning, J., Kaneko, T., Pru, J.K., and Tilly, J.L., Germline stem cells and follicular renewal in the postnatal mammalian ovary, *Nature*, 428, 145, 2004.

Kalter, H., Teratology in the 20th century. Environmental causes of congenital malformations in humans and how they were established, *Neurotoxicol. Teratol.*, 25, 131, 2003.

Loose-Mitchell, D.S. and Stancel, G.M., Estrogens and progestins, in *Goodman and Gilman's: The Pharmacological Basis of Therapeutics*, Hardman, J.G. and Limbird, L.E., Eds, McGraw-Hill, New York, 2001, chap. 58.

Mangelsdorf, I., Buschmann, J., and Orthen, B., Some aspects relating to the evaluation of the effects of chemicals on male fertility, *Regul. Toxicol. Pharmacol.*, 37, 356, 2003.

Marty, M.S., Chapin, R.E., Parks, L.G., and Thorsrud, B.A., Development and maturation of the male reproductive system, *Birth Defects Res. B*, 68, 125, 2003.

McKinnell, R.G. and Di Berardino, M.A., The biology of cloning: History and rationale, *Bioscience*, 49, 11, 1999.

Menegola, E., Cappalletti, G., and DiRenzo, F., Epigenetic approaches and methods in developmental toxicology: Role of HDAC inhibition in teratogenic events, *Meth. Mol. Biol.*, 889, 373, 2012.

Miller, R.K., Perinatal toxicology: Its recognition and fundamentals, *Am. J. Ind. Med.*, 4, 205, 1983.

Miller, R.K., Kellogg, C.K., and Saltzman, R.A., Reproductive and perinatal toxicology, in *Handbook of Toxicology*, Haley, T.J. and Berndt, W.O., Eds, Hemisphere Publishing Corporation, Washington, D.C., 1987, chap. 7.

Notarianni, E., Reinterpretation of evidence advanced for neo-oogenesis in mammals, in terms of a finite oocyte reserve, *J. Ovarian Res.*, 4, 2011, available at http://www.ovarianresearch.com/content/4/1/1.

Radike, M., Reproductive toxicology, in *Industrial Toxicology*, Williams, P.L. and Burson, J.L., Eds, Van Nostrand Reinhold Company, New York, 1985, chap. 16.

Reizel, Y., Itzkovitz, S., Adar, R., Elbaz, J., Jinich, A., Chapal-Ilani, N., Maruvka, Y.E. et al., Cell lineage analysis of the mammalian female germline, *PLoS Genet.*, 8, e1002477, doi:10.1371/journal.pgen.1002477, 2012.

Rideout, W.M., III, Eggan, K., and Jaenisch, R., Nuclear cloning and epigenetic reprogramming of the genome, *Science*, 293, 5532, 2001.

Rogers, J.M., Developmental toxicology, in *Casarett and Doull's Toxicology*, Klaassen, C.D., Ed., McGraw-Hill, New York, 2013, chap. 10.

Rogers, J.M. and Kavlock, R.J., Developmental toxicology, in *Casarett and Doull's Toxicology*, Klaassen, C.D., Ed., McGraw-Hill, New York, 2001, chap. 10.

Safe, S., Endocrine disruptors and human health: Is there a problem?, *Toxicology*, 205, 3, 2004.

Snyder, P.J., Androgens, in *Goodman and Gilman's: The Pharmacological Basis of Therapeutics*, Hardman, J.G. and Limbird, L.E., Eds, McGraw-Hill, New York, 2001, chap. 59.

Sobinoff, A.P., Beckett, E.L., Jarnicki, A.G., Sutherland, J.M., McCluskey, A., Hansbro, P.M., and McLaughlin, E.A., Scrambled and fried: Cigarette smoke exposure causes antral follicle destruction and oocyte dysfunction through oxidative stress, *Toxicol. Appl. Pharmacol.*, 271, 156, 2013.

Thomas, M.J. and Thomas, J.A., Toxic responses of the reproductive system, in *Casarett and Doull's Toxicology*, Klaassen, C.D., Ed., McGraw-Hill, New York, 2001, chap. 20.

Toppari, J. and Juul, A., Trends in puberty timing in humans and environmental modifiers, *Mol. Cell. Endocrinol.*, 324, 39, 2010.

Whorton, M.D., Bedinghaus, J., Obrinsky, D., and Spear, P.W., Reproductive disorders, in *Occupational Health*, Levy, B.S. and Wegman, D.H., Eds, Little, Brown, & Company, Boston, 1983, chap. 20.

Wilmut, I., Beaujean, N., de Sousa, P.A., Dinnyes, A., King, T.J., Paterson, L.A., Wells, D.N., and Young, L.E., Somatic cell nuclear transfer, *Nature*, 419, 583, 2002.

Woods, D.C., Telfer, E.E., and Tilly, J.L., Oocyte family trees: Old branches or new stems? *PLoS Genet.*, 8, e1002848, doi:10.1371/journal.pgen.1002848, 2012.

8 Respiratory Toxicology

FUNCTION OF THE RESPIRATORY SYSTEM

The primary functions of the respiratory system are to deliver oxygen to the bloodstream where it can be routed throughout the body to every cell, and to remove the waste product of metabolism—carbon dioxide. Mitochondria within cells require oxygen to carry out oxidative phosphorylation, the series of reactions whereby energy contained in chemical bonds in food is repackaged into the bonds in the molecule ATP (a form of energy the cell can directly use). Although some cells in the body can function without oxygen for a short time, many cells (such as heart or brain cells) are absolutely dependent on an adequate supply of oxygen in order to survive. The respiratory system also plays a role in the process of speech, the defense of the body, and the regulation of body pH. It is also a rapid route by which volatile xenobiotics can reach the brain.

ANATOMY AND PHYSIOLOGY OF THE RESPIRATORY SYSTEM

RESPIRATORY ANATOMY

The respiratory system can be divided into two basic parts (Figure 8.1). The first part, the *conducting portion*, is responsible for carrying air to and from the second part, the *respiratory portion*. The respiratory portion is where the process of *gas exchange*, the movement of oxygen into and carbon dioxide out of the bloodstream, occurs.

The conducting portion of the respiratory system begins with the *nose*. The external portion of the nose consists of cartilage covered with skin. The nostrils open into the internal portion of the nose, the *nasal cavity*, which is bounded below by the hard palate and above and to the sides by other cranial bones. The nasal cavity is divided into halves by the nasal septum. Scroll-like bones called *turbinates* project into the interior of the nasal cavity.

The nose is lined with epithelial tissue consisting of both column-shaped epithelial cells covered with cilia and cells called *goblet cells* that secrete mucus. Underneath the epithelial layer is a layer of connective tissue that contains many blood vessels. This combination of epithelium and connective tissue is called a *mucous membrane*.

The nose functions both to filter and to condition inhaled air. As air passes through the nose, any entering particles can become entrapped in the cilia and mucus. Also, the temperature of the incoming air is raised and moisture is added as the air passes over the warm, moist surfaces of the nasal mucous membranes. The nose also contains receptors for the sense of smell and serves as a resonating chamber for the voice. Some animals are obligate nose breathers (in other words, they cannot inhale

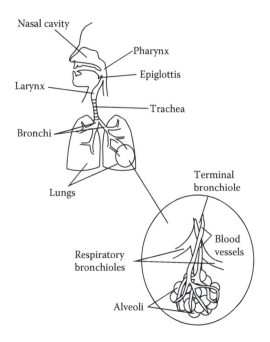

FIGURE 8.1 The anatomy of the respiratory system, with inset showing terminal bronchiole with respiratory bronchioles and alveolus, along with associated capillaries.

through the oral cavity); humans, of course, are not. In fact, this is one of the many factors that may need to be considered when it comes to the choice of animal models for studying respiratory exposure to toxicants. There is also significant capacity for xenobiotic metabolism in the nasal mucosa of most species.

Air moves next from the nose into the *pharynx*. This chamber functions as a passageway between the nose and the *larynx* (which opens to the trachea) and also between the mouth and the esophagus (which leads to the stomach). The larynx (or voice box) is composed of cartilage lined with a mucous membrane. Folds of this membrane extend into the open center of the larynx and vibrate as air passes over them. These vibrating folds are the *vocal cords*. Muscles in the larynx control the tension of the cords, as well as the size of the opening into the larynx, allowing the production of both high- and low-pitched sounds. A flexible flap called the *epiglottis* closes down over the top of the larynx during swallowing, preventing food or drink from entering the larynx and passing on into the lungs. Irritation of the larynx produces a reflex action called a *cough*. In coughing, the opening to the larynx is temporarily closed, air is forced upward, pressure builds, and the larynx then opens again to allow the blast of air out (hopefully along with the irritant that provoked the cough). Severe irritation, however, may cause the larynx to clamp shut in a life-threatening spasm.

The larynx opens into the *trachea* (windpipe). This flexible tube is also lined with a mucous membrane and is supported around the outside by C-shaped rings of cartilage. These rings keep the trachea from collapsing from the changes in air pressure that accompany breathing. At its base, the trachea branches into the right and left *primary bronchi*, which then lead to the right and left lungs, respectively. Within the lungs, the primary bronchi branch out into *secondary bronchi*, each of which leads to a different segment of the lung. Secondary bronchi continue to branch out, forming smaller tubules called *bronchioles*. The amount of cartilage in the airways decreases as the bronchi branch out and become smaller, until it finally disappears in the bronchioles. Smooth muscle content, however, increases, and bronchioles are completely ringed by a layer of smooth muscle. Spasms of that smooth muscle, in fact, are what produce the condition called *asthma*.

Bronchioles continue to branch out, forming *terminal bronchioles*, each of which branches into several *respiratory bronchioles*, which then terminate in sacs called *alveoli*. These respiratory bronchioles and alveoli make up the respiratory portion of the respiratory system—in other words, the area where gas exchange takes place. In fact, the extensive branching of bronchioles and the expanded sacs of the alveoli serve to dramatically increase the surface area across which gas exchange can occur.

Respiratory bronchioles and alveoli are composed of one thin layer of epithelial tissue, with a thin layer of elastic fibers underneath. The epithelial tissue contains small cells called *Clara cells* (where xenobiotic metabolism occurs), thin flat *type I* cells, and cuboidal *type II* cells. Both the overall levels and the specific isozymes of cytochrome P450 in Clara cells vary between species, with other Phase I and II enzymes occurring as well. There are also polymorphisms (genetically different forms) of many of these enzymes, some of which may play a role in individual susceptibility to toxicants. Type II cells also have some cytochrome P450 activity, can divide to produce new type I cells, and can manufacture a substance called *surfactant*. Surfactant is a lipid-rich material that decreases surface tension in the alveoli, allowing the sacs to inflate properly and to remain inflated during the process of breathing. *Alveolar macrophages*, cells that digest and destroy debris, are also found in the alveoli, as are a variety of other types of cells, including fibroblasts.

Xenobiotic Metabolism

See also:
*   *Biotransformation*

Chapter 3

In order for gases to be exchanged between the lungs and the blood, there must be an adequate concentration of blood vessels in the area. Thus, the lungs are highly vascularized. Blood enters the lungs through the *pulmonary arteries*, which branch out into arterioles and finally capillaries. Networks of capillaries (which are composed of one layer of thin, flat epithelial cells, often called the *endothelium*) surround each terminal bronchiole and its respiratory bronchioles and alveoli, allowing gas exchange to take place. The capillaries then merge together to form venules, which merge to form the *pulmonary veins*, which carry oxygenated blood back to the heart.

PULMONARY VENTILATION

The lungs are located in the thoracic (chest) cavity. A membrane called the *parietal pleura* covers the surface of each of the lungs, and a membrane called the *visceral pleura* lines the walls of the thoracic cavity. The small space between these two membranes is called the *pleural cavity*, and it is filled with fluid. The fluid acts as a lubricant and also holds the two membrane surfaces together.

Breathing in, or *inspiration*, is initiated by contraction of two muscles. The primary muscle involved is a dome-shaped sheet of skeletal muscle called the *diaphragm*, which forms the floor of the thoracic cavity and separates it from the abdominal cavity. As the diaphragm contracts, it flattens out and increases the size of the thoracic cavity. At the same time, a set of muscles called the *external intercostals* (which extend from each rib to the rib below) contracts, thus moving the ribs up and out and also contributing to the increase in the size of the thoracic cavity.

As the thoracic cavity expands, the visceral pleura (which, as you remember, lines that cavity) is pulled outward, pulling the parietal pleura (to which it is attached) outward and thus expanding the volume of the lungs and inflating the alveoli. As the lung volume increases, pressure in the lung decreases and air is pulled in through the conducting airways into the lung. A hole in the visceral or parietal pleura, however, can allow air into the pleural cavity and break the seal between the membranes. This allows the two membranes to separate, so that the lung no longer inflates with expansion of the thoracic cavity. This situation is referred to as a *pneumothorax*, or *collapsed lung*, and may result from disease or traumatic injury to the thoracic cavity.

In contrast to inspiration, breathing out, or *expiration*, is usually a passive process. This is because of the elastic properties of the lungs. During expiration, the diaphragm and the external intercostal muscles relax, and the volume of the thoracic cavity, and thus the lungs, decreases. When the lung volume decreases, pressure in the lungs increases and air is forced out of the lungs through the conducting airways. Abdominal muscles as well as a set of rib cage muscles called the *internal intercostals* can be used, though, to effect a more forceful expiration.

The amounts of air moved by the lungs are called *respiratory volumes* and can be measured using an instrument called a *spirometer*. The amount of air breathed in or out during normal quiet breathing is called the *tidal volume* (TV). The additional volume of air that can be inhaled with effort is the *inspiratory reserve volume* (IRV). The additional volume of air that can be exhaled with effort is the *expiratory reserve volume* (ERV). TV + IRV + ERV together are called the *vital capacity*. No matter how hard you try, though, you can never expel all the air from your lungs, and the volume of air that always remains in your lungs is called the *residual volume* (RV). RV + vital capacity together make up what is called the *total lung capacity*. The respiratory volumes are illustrated in Figure 8.2.

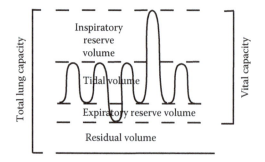

FIGURE 8.2   The respiratory volumes.

The rate of respiration can also be measured, and the number of inspirations per minute is called the *respiratory rate*. Respiratory rate multiplied by tidal volume gives a measure called the *minute volume*, which is the volume of air moved in and out per minute. Another common measurement is the *FEV1*, the volume of air that can be forcibly exhaled in a period of 1 second, following a maximum inhalation.

Many pathological conditions of the lungs affect respiratory volumes and rates. *Restrictive conditions* result from decrease in elasticity of the lungs and cause decreases in lung volumes. *Obstructive conditions* result from blockage or narrowing of the airways and cause decreases in airflow rates (which can be measured by an FEV1).

GAS EXCHANGE

In the respiratory bronchioles and alveoli there are few barriers to the diffusion of gases. To move between alveoli and the bloodstream, gases need only cross a thin, flat type I alveolar cell and a thin, flat capillary endothelial cell. Also, the large amount of available surface area of both alveolar and endothelial surfaces facilitates rapid diffusion of gases between the alveolar space and the bloodstream.

The force driving the diffusion of gases is quite simply the difference in concentration of the gas between alveoli, blood, and tissues. The concentration of a gas is reflected by its *partial pressure*, the pressure exerted by that particular gas in a given situation. For example, atmospheric pressure is the sum of all the partial pressures of the gases that make up the atmosphere. Gases dissolved in liquids also have partial pressures and tend to diffuse from areas of high partial pressure to areas of lower partial pressure.

Air that has just arrived in the alveoli (in other words, air that has just been inhaled) has a relatively high partial pressure of oxygen. The partial pressure of oxygen in the bloodstream, however, is low, as oxygen has been used up by the cells in the process of oxidative phosphorylation. In contrast, the partial pressure of carbon dioxide is much higher in the bloodstream (which has picked it up as a waste product from cells) than it is in inhaled air. Thus, in the lungs, partial pressures dictate that oxygen diffuses from the alveoli into the bloodstream and carbon dioxide diffuses from the bloodstream into the alveoli. This process of gas exchange in the lung is diagrammed in Figure 8.3.

The story in the rest of the body, however, is a little different. In the various tissues of the body, where oxygen is being consumed and carbon dioxide produced, oxygen partial pressure is low and carbon dioxide partial pressure is high. Blood traveling to these tissues, though, has just exited to the lungs and so has a high partial pressure of oxygen and a low partial pressure of carbon dioxide. Thus, in the tissues, oxygen leaves the bloodstream and diffuses into the tissues, whereas carbon dioxide leaves the tissues and diffuses into the bloodstream.

Most of the oxygen carried in the bloodstream, however, is not found dissolved in the blood fluid. Instead, it is carried by an iron-contaning protein called *hemoglobin*, which is found in red blood cells (*erythrocytes*) and which will be discussed further in Chapter 9. Most carbon dioxide is also carried in

Hemoglobin

See also:
  The Blood
                                              Chapter 9

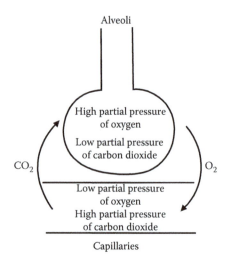

FIGURE 8.3 The process of gas exchange as it takes place in the alveoli and capillaries of the lung.

red blood cells. When CO_2 is picked up from the tissues, some is bound to hemoglobin (at different sites than where oxygen is carried), but most combines with water to form *carbonic acid*, in a reaction that is catalyzed by an enzyme called *carbonic anhydrase*. Carbonic acid breaks down in the red blood cells into a hydrogen ion and a *bicarbonate* ion. The bicarbonate ion leaves the red blood cells and moves into the plasma in exchange for chloride (a mechanism called the *chloride shift*), whereas the H^+ binds to hemoglobin and stimulates the molecule to release oxygen, which can then diffuse into the oxygen-poor tissues. When blood reaches the lungs and carbon dioxide levels decline, the bicarbonate diffuses back into the red blood cells and the series of reactions runs in reverse to liberate carbon dioxide (Figure 8.4). This system is a major part of the buffer system that regulates blood pH.

CONTROL OF RESPIRATION

Rate and depth of respiration are controlled by the *respiratory centers*, located in an area of the brain called the brain stem. These centers control normal respiration and also make the necessary responses to any changes in physiological status. Receptors located in the arteries monitor the partial pressures of both oxygen and carbon dioxide in the blood, and receptors in the brain monitor the partial pressure of carbon dioxide in the cerebrospinal fluid. Actually, changes in carbon dioxide levels are not measured directly by these receptors; instead, they respond to the accompanying changes in pH. As carbon dioxide levels go up, carbonic acid levels go up, and hydrogen ion levels go up, so pH goes down. Likewise, as carbon dioxide levels go down, carbonic acid levels go down, and hydrogen ion levels go down, so pH goes up.

It is these changes in pH (reflecting changes in carbon dioxide levels) that stimulate the greatest responses from the respiratory centers. If carbon dioxide levels go up, the rate of respiration is stimulated, leading to a more rapid release of carbon

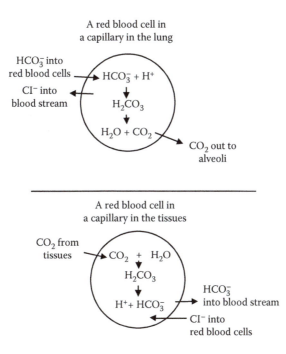

A red blood cell in
a capillary in the lung

HCO_3^- into
red blood cells

CI^- into
blood stream

$HCO_3^- + H^+$

H_2CO_3

$H_2O + CO_2$

CO_2 out to
alveoli

A red blood cell in
a capillary in the tissues

CO_2 from
tissues

$CO_2 + H_2O$

H_2CO_3

$H^+ + HCO_3^-$

HCO_3^-
into blood stream

CI^- into
red blood cells

FIGURE 8.4 The carbon dioxide/carbonic acid/bicarbonate buffer system. The top figure shows the removal of carbon dioxide from the red blood cells to the alveoli in the lung, and the bottom figure shows the entry of carbon dioxide into the red blood cells from the tissues.

dioxide from the lungs. If carbon dioxide levels go down, the rate of respiration is slowed.

Receptors also monitor expansion of the lungs, preventing overextension or collapse during forced inspiration and expiration. In addition, other receptors respond to changes in blood pressure, increasing the rate of respiration when blood pressure goes down and decreasing the rate of respiration when blood pressure goes up.

EFFECTS OF TOXICANTS ON THE RESPIRATORY SYSTEM: GENERAL PRINCIPLES

Just as the respiratory system is an important route of exposure for various toxicants, it is also a target for many of them. For some toxicants, the lungs are a target only when the route of exposure to that toxicant is inhalation. Other toxicants, however, produce effects on the lungs even when exposures occur through ingestion or absorption through the skin.

Toxicants that affect the respiratory system following inhalation can be divided into two general categories: gases and particulates. The chemical and physical properties of these toxicants determine how they will be distributed in the respiratory system. Exposure to these toxicants can be either acute or chronic, and effects due to exposure may be either immediate or delayed. Of course, the respiratory system also has several defenses against injury by these toxicants. We will discuss defense mechanisms first, then the types of toxicants and their properties, then immediate effects, and then delayed effects. Finally, we will discuss laboratory testing of respiratory toxicants.

DEFENSE MECHANISMS OF THE RESPIRATORY SYSTEM

The respiratory system may be particularly vulnerable to exposure to toxicants, but it also has several defense mechanisms that help protect it. First of all, the cilia and the mucus found in the mucous membranes of the upper airways help trap particles and prevent them from penetrating further into the lungs. Particles trapped in the mucus are moved along by motion of the cilia, in what has been termed the *mucociliary escalator*, upward toward the mouth to be swallowed. Particles from the lower reaches of the lungs may be consumed by macrophages, which then move onto the mucociliary escalator for elimination. Particles may also leave the lungs through dissolution or absorption into the bloodstream or lymphatic system.

These clearance mechanisms may, however, be altered by exposure to toxicants. In heavy smokers, for example, the rate of clearance of particles from the respiratory system is significantly slowed. This slowing might be due to inhibition of ciliary motion, changes in mucous viscosity, damage to macrophages, or a combination of factors. Damage to macrophages may not only slow clearance, but also lead to an increased risk of infection because macrophages are important in the destruction of pathogens. One inherited disease that affects viscosity of mucus is *cystic fibrosis*. In cystic fibrosis, defects in chloride ion channels lead to the production of overly thick mucus, which blocks the airways and shuts down the mucociliary escalator, leading to difficulties with ventilation complicated by frequent infections. Other organs, such as those of the digestive system, are frequently affected in cystic fibrosis as well.

Immune System

See also:
Effects of Toxicants on the Immune System
Chapter 13

There are also many cells of the immune system located within the lungs, ready to respond to invaders. Some of these cells produce antibodies against foreign antigens, and others release endogenous chemicals that mediate allergic responses (such as symptoms of *asthma* or bronchitis). Thus, as well as being protective, the presence of these cells means that exposure to some toxicants can trigger an allergic attack or chronic inflammation. Also, remember that Clara cells contain cytochrome P450 and are capable of carrying out xenobiotic metabolism, which is frequently protective.

If, in spite of other defense mechanisms, damage to alveolar cells occurs, some repair is possible. When type I cells are damaged, type II cells can undergo mitosis, proliferate, and replace the damaged cells.

EXPOSURE TO RESPIRATORY TOXICANTS

MEASURING EXPOSURE LEVELS

Both gases and particles suspended in gases can be inhaled easily. Because the amount of a gas or particle that is inhaled and retained is difficult to measure exactly, exposures can be estimated on the basis of the concentration of the gas or particle in the environment and the length of exposure. Of course, other factors, such as breathing rate and depth, can also influence exposure, but are often much less easily quantified.

Typically, concentrations of gases are expressed as *parts per million* (ppm). This unit expresses concentration as the volume of the gas per million volumes of air. Concentrations of gases and suspended particles may also be expressed in a weight per volume manner, usually as *milligrams per cubic meter of air* (mg/m^3).

Based on laboratory and epidemiological studies, the *American Conference of Governmental and Industrial Hygienists* (ACGIH) has developed a list of allowable exposures in the occupational setting for various respiratory toxicants. These *threshold limit values* (TLVs) specify the average maximum allowable concentrations to which workers can be exposed without undue risk. The *TLV-TWA* (time-weighted average) gives the maximum allowable concentration for exposure averaged over an 8 h day. The *TLV-STEL* (short-term exposure limit) gives the maximum allowable concentration for a 15 min period, and the *TLV-C* (ceiling) gives the concentration limit that should never be exceeded. Some representative TLVs are shown in Table 8.1.

Deposition of Gases

Deposition of a gas in the respiratory system depends primarily on the water solubility of the gas. Water-soluble gases are likely to dissolve quickly into the watery mucus secreted by the cells lining the upper parts of the respiratory tract, whereas gases that are less water soluble are more likely to continue deeper into the respiratory tract.

Deposition of Particulates

For particulates, size is the main factor that influences deposition in the respiratory system. Fibers, by benefit of their length, may be *intercepted* by physical contact with the airway surface (the longer the fiber, the greater the chance of interception). Very large particles (greater than 5 μm in diameter) are likely to *impact* on the walls of the nasal cavity or pharynx during inspiration. Medium-sized particles (1–5 μm in diameter) tend to *settle as sediment* in the trachea, bronchi, or bronchioles as air velocity decreases in these smaller passageways. Particles less than 1 μm in diameter typically move by *diffusion* into alveoli. Of special concern are *nanoparticles*,

TABLE 8.1
Examples of Some Threshold Limit Values (TLVs)

Substance	TLV-TWA (ppm)	TLV-STEL (ppm)
Ammonia	25	35
Benzene	0.5	2.5
Diethyl ether	400	500
Nitrogen dioxide	3	5
Trichloroethylene	50	100

particles with diameter of less than 100 nm. For further discussion of these, see the case study at the end of this chapter.

Nanoparticles

See also:
*   *Case Study: Nanoparticles*
        Chapter 8

*   *Other Additives, Including*
    Indirect Additives
        Chapter 19

Particulate matter dispersed in air (also known as an *aerosol*) can be described by its average size and the source from which it is generated. Starting from *dusts*, which are generated by mechanical grinding and have an average diameter of less than 1 μm, progressively smaller particulates are referred to as *fumes* (generated by vapor condensation), *smoke* (generated by combustion), *mists* (generated by spraying of liquids), and *fogs* (involving condensation of water vapor).

Of course the shape of a particle can also affect its aerodynamic performance and thus its site of deposition. Physiological factors influence particle deposition as well. The narrower the airways, the more deposition will occur. Rapid inhalation of deep breaths (those that occur during exercise) also increases exposure and deposition.

IMMEDIATE RESPONSES TO RESPIRATORY TOXICANTS: MECHANISMS

Many compounds tend to produce their effects on the respiratory system within a few minutes to a few hours following exposure, often through direct damage to the cells of the respiratory tract. There are several different mechanisms by which toxicants produce their damage: three major mechanisms are irritation, involvement of the immune system, and free radical-induced damage. In all three cases, the degree of reversibility of immediate responses depends directly on the degree of damage produced in the respiratory tissue.

THE IRRITANT RESPONSE

Inflammation

See also:
*   *The Inflammatory Response*
        Chapter 13

Some gases and particulates act as physical or chemical irritants. Injury to the epithelial cells lining the respiratory tract by these compounds may produce an inflammatory response, characterized by an increase in the permeability of blood vessels and accumulation of immune system cells in the area of the damage. The increase in blood vessel permeability leads to an accumulation of fluids or *edema* in the airways. Cell death, or necrosis, may also result.

Exposure to irritants can also promote contraction of the ring of smooth muscle surrounding bronchioles, an effect known as bronchoconstriction. Bronchoconstriction can result from impact of the irritant on the muscle directly or may be mediated through effects on receptors and nerves. Some irritants not only trigger bronchoconstriction themselves, but also produce an increase in sensitivity of bronchiolar smooth muscle to other agents (other irritants or perhaps even endogenous substances). It is likely that this is mediated through increased sensitivity of receptors to cholinergic stimulation. Finally, irritants can also affect other nerves in the respiratory system, producing sensory irritation in the upper respiratory tract or alterations in breathing patterns.

INVOLVEMENT OF THE IMMUNE SYSTEM

Some toxicants may produce their effects on the respiratory system through their interactions with the immune system. In an *immune response*, the immune system reacts to the presence of specific molecules (*antigens*) with the production of proteins (*antibodies*) designed to neutralize and destroy the perceived threat. In an *allergic response*, the immune system overresponds, reacting against molecules that are generally harmless and that do not provoke an immune response in most individuals.

> **Allergic Reaction**
>
> *See also:*
> > *Toxicant-Induced Allergies*
> > > *Chapter 13*

There are a number of proinflammatory molecules that appear to be important in the production of immune-related damage to the lung. These would include *high-mobility group protein B1 (HMGB1)*, whose presence is associated with the development of edema. Molecules such as *histamine* and *prostaglandins* that are released during an allergic response can also produce edema, increase in the mucus secretion, and bronchoconstriction in the respiratory system.

FREE RADICAL-INDUCED DAMAGE

Finally, some respiratory toxicants appear to produce damage through production of free radicals, which can interact with membranes and other cellular constituents to produce significant cellular damage. Free radicals would include reactive oxygen-containing compounds (often called *reactive oxygen species* or ROS) such as the *superoxide anion* (O_2^-) that is produced normally during the process of oxidative phosphorylation (as well as during other normal cellular processes). Another reactive molecule, hydrogen peroxide (H_2O_2), is formed when the superoxide anion is further oxidized in a reaction catalyzed by the enzyme *superoxide dismutase*. H_2O_2 can also be broken down to form another free radical, the hydroxyl radical ($\bullet OH$). In a high-oxygen environment such as the alveoli, these reactions would be expected to occur frequently, and compounds that stimulated these normal metabolic processes might be expected to also stimulate free radical production.

> **Free Radicals**
>
> *See also:*
> > *Primary Biotransformation (Phase I Reactions): Reduction*
> > > *Chapter 3*
> >
> > *Effects of Toxicants on Lipids*
> > > *Chapter 4*

> **Nitric Oxide**
>
> *See also:*
> > *Effects of Toxicants on Synaptic Function*
> > *Effects of Toxicants Directly on Neurons and Glial Cells*
> > > *Chapter 10*

An additional source of free radicals would be those generated through cytochrome P450 metabolism of certain xenobiotics.

One other free radical that may play a significant role in respiratory damage is *nitric oxide*. This compound is produced in the lung by the action of the enzyme *nitric oxide synthase*, which has been found in a variety of lung cell types, including type II cells, macrophages, and endothelial cells. Nitric oxide probably plays a role in regulating the blood flow, but levels of nitric oxide synthase in the lung have been shown to increase following exposure to toxicants such as asbestos and ozone. However, as the effects of nitric oxide can be protective as well as damaging (it does have some anti-inflammatory actions), its role in mediating pulmonary damage is not yet clear.

IMMEDIATE RESPONSES TO RESPIRATORY TOXICANTS: EFFECTS ON UPPER AND LOWER AIRWAYS

IMMEDIATE RESPONSES: UPPER AIRWAY EFFECTS

Sulfur Dioxide

See also:
  *Sulfur Oxides and Nitrogen
    Oxides*
                    Chapter 17

  Sulfur Dioxide
                    Appendix

Formaldehyde

See also:
  Toxicant-Induced Allergies
                    Chapter 13

  Formaldehyde
                    Appendix

Exposure to water-soluble irritants such as *sulfur dioxide* (SO_2) produces swelling and edema in the upper airways, causing narrowing of the passageways and making breathing more difficult. This may be accompanied by an increase in the secretion of mucus. Studies have shown that exposure to as little as 5 ppm sulfur dioxide can affect airways. Sulfur dioxide is also a potent bronchoconstrictor. *Formaldehyde* is another upper airway irritant. Individuals with pre-existing respiratory diseases such as asthma may be particularly susceptible to these two irritants. Another irritant of both upper and lower airways is *acrolein*, a volatile compound that is released during heating of cooking oils, as well as being found in mainstream and sidestream smoke and automobile exhaust. There is evidence that acrolein may also be involved in the production of chronic obstructive pulmonary disease (COPD) and lung cancer (conditions that are described later in this chapter).

IMMEDIATE RESPONSES: LOWER AIRWAY EFFECTS

Nitrogen Dioxide

See also:
  *Sulfur Oxides and Nitrogen
    Oxides*
                    Chapter 17

  Nitrogen Dioxide
                    Appendix

Irritant gases and particles that are less water soluble, such as *nitrogen dioxide* (NO_2) and *ozone* (O_3), produce accumulation of fluid in the alveoli. This fluid interferes with gas exchange, acting as an additional barrier to diffusion of gases. Exposure of rats to as little as 1 ppm ozone has caused cell death and edema. Both these gases are oxidants and so are likely to damage membranes through the process of lipid peroxidation. Although relatively insoluble in water, nitrogen dioxide is actually slightly absorbed all along the respiratory tract, thus producing both upper and lower airway irritation.

One very unique respiratory toxicant is the herbicide *paraquat*. Paraquat was an important herbicide for cleaning up fields, roadsides, and rights of way because it possesses a very broad spectrum of activity. With an LD_{50} of 30 mg/kg, however, it is quite toxic. What is unique about paraquat is the way in which it accumulates in and damages the lungs no matter what the route of absorption: respiratory, oral, or dermal. Paraquat seems to accumulate in type II cells by an active transport process. Paraquat produces damage through lipid peroxidation, most likely generating free radicals and depleting NADPH as it is alternately oxidized and reduced within the cell.

Ozone

See also:
  Hydrocarbons and the Formation of Secondary Pollutants (Including Ozone)
                                  Chapter 17

Ozone
                                  Appendix

DELAYED AND CUMULATIVE RESPONSES TO RESPIRATORY TOXICANTS

Repeated (chronic) exposure to respiratory toxicants often leads to long-term changes in respiratory function, some of which may not occur until some time after exposure to the toxicant begins and others that may accumulate gradually before noticeable changes occur. These changes may lead to obstructive or restrictive lung diseases or lung cancer and are typically irreversible.

Paraquat

See also:
  Pesticides
                                  Chapter 17

Paraquat
                                  Appendix

ASTHMA AND IMMUNE-RELATED CHRONIC CONDITIONS

Asthma is an acute effect that is characterized by increased sensitivity of bronchial smooth muscle, leading to repeated episodes of bronchoconstriction that may range from mild to severe. It can be induced through exposure to either allergens or irritants. The underlying mechanisms by which asthma is produced are not clear, but may involve injury to airway epithelial cells by free radicals. This may be a result of increased production of ROS by immune system cells that are part of the chronic inflammation response. A chronic predisposition to asthma may develop following exposure to toxicants such as the chemical *toluene diisocyanate* (TDI). Not only do individuals with TDI-induced asthma react to even very low levels of TDI, but many also suffer the generalized increase in sensitivity of airway smooth muscle mentioned earlier.

Toluene Diisocyanate

See also:
  Toxicant-Induced Allergies
                                  Chapter 13

TDI
                                  Appendix

Exposure to cotton dust can produce a condition called *byssinosis* (also sometimes called brown lung), which is also characterized by bronchoconstriction. Symptoms of byssinosis seem to be most severe when a worker in a cotton mill returns to work after a day or two off. For this reason, it is also termed Monday morning sickness. The cause of byssinosis is not clear—it may be an allergic reaction to microorganisms on the dust particles or a simple reaction to an irritant in the cotton dust itself.

Another type of allergic reaction is *hypersensitivity pneumonitis*. Symptoms of this problem are shortness of breath, fever, and chills. Hypersensitivity pneumonitis results from exposure to organic materials that trigger an immune response localized primarily in the lower airways. Exposure to moldy hay, for example, can lead to a condition called farmer's lung, whereas exposure to fungus found on cheese particles may produce cheese washer's lung. Continued exposure can result in permanent lung damage in the form of fibrosis (as discussed later in this chapter).

CHRONIC OBSTRUCTIVE PULMONARY DISEASE: BRONCHITIS AND EMPHYSEMA

Smoking

See also:
  Genetic Carcinogens
                Chapter 6

  Fibrosis and Pneumoconioses
  Lung Cancer
                Chapter 8

  Atherosclerosis
  Effects of Toxicants on
    Hemoglobin
                Chapter 9

  Tobacco
                Appendix

An individual showing a combination of symptoms of chronic cough and *dyspnea* (difficulty in breathing) will most likely be classified as having *chronic obstructive pulmonary disease* (COPD). These patients typically suffer from some combination of *chronic bronchitis* and *emphysema* and sometimes asthma and show airflow reduction as measured by FEV1. It has been estimated that around 6% of the adult population of the United States may have COPD. Risk for developing COPD is strongly correlated with exposure to respiratory toxicants, with smoking as the primary risk factor. Particularly in developing countries, exposure to smoke generated by cooking and heating with open fires indoors is also a significant risk factor.

In *chronic bronchitis*, excessive secretion of mucus results from increases in the mucus gland size as well as increases in numbers of goblet cells in the respiratory tract. Along with this, inflammation leads to narrowing of the airways. There is also a decrease in the rate of mucociliary clearance. These factors result in a chronic cough and increased susceptibility to infection. *Emphysema* is an obstructive condition characterized by breakdown of walls of alveoli and loss of elasticity. This leads to both reduction in elasticity and decrease in the rate of gas exchange (due to the decrease in alveolar surface area).

The causes of emphysema are complex, involving (among other factors) *proteases*, which are enzymes that break down proteins. Several studies have indicated that an increase in protease activity or a decrease in antiprotease activity may lead to the destruction of alveolar tissue that is typical of emphysema. Evidence for this hypothesis includes the observation that individuals with genetic antiprotease deficiencies are known to be at increased risk for emphysema. There are also studies demonstrating that delivery of the protease *elastase* to the lungs can produce emphysema in animal models. *Metalloproteinases* may also play a role in the development of this condition. Smoking, the major environmental risk factor for emphysema, may act through increasing protease activity.

FIBROSIS AND PNEUMOCONIOSES

A number of different toxicants that produce irritation and inflammation in the lower respiratory system may, after some years of exposure, lead to a restrictive condition

called *fibrosis*. Fibrosis occurs when repeated activation of macrophages leads to chronic inflammation of an area. This results in the recruitment of
fibroblasts: cells that proliferate and produce the
rigid protein *collagen*. The accumulation of collagen interferes with ventilation (by reducing elasticity) and with blood flow within the lung. Abnormal
cross-linking between collagen fibers may also contribute to the stiffness associated with fibrosis.

Fibrosis

See also:
 Stress, Repair, and Recovery
Chapter 4

Fibrosis and Cirrhosis
Chapter 11

Fibrosis is a major characteristic of the diseases called *pneumoconioses*, which are
diseases associated with dust exposure. Among the dusts that produce fibrosis are the
crystalline *silicates*. In *silicosis*, one of the most widespread and serious occupational lung
diseases, alveolar macrophages ingest the inhaled silica crystals and may be damaged or
destroyed in the attempt. This results in the release of cytokines that attract and stimulate
fibroblasts and lead to the laying down of collagen in the area. Silica crystals thus accumulate in the lungs, surrounded by areas of inflammation characterized by collagen nodules.
Silicosis is a potential hazard for anyone whose occupations involve mining, quarrying,
blasting, grinding, or other types of stoneworking.

Asbestosis, a similar condition, is caused by exposure to *asbestos*, itself a fibrous
silicate. There are several different forms of asbestos, including serpentine forms (a
group to which the most commonly used type, chrysotile asbestos, belongs) and
amphibole forms. From the 1940s to the 1960s, a significant number of workers were
exposed to asbestos and later developed asbestosis.

Experimental evidence indicates that potential
damage to DNA as well as other cellular constituents is produced by asbestos fibers in two ways:
first, reactive oxygen species (ROSs) are generated
by direct chemical interactions involving the surface of fibers, and secondly, additional ROSs may
be generated by cells such as macrophages as they

Cytokines

See also:
 Function of the Immune
 System
Chapter 13

phagocytize the fiber. These ROSs then produce upregulation of cytokines such as
tumor necrosis factor alpha (TNF-α) in macrophages, and the TNF-α then induces
the production of other cytokines that then recruit fibroblasts and other cells involved
in inflammation. Other cytokines that may also be involved in the development of
asbestosis include *transforming growth factor* (TGF) and *interleukins* 1 and 6.

Currently, concern over asbestos focuses on whether or not there are significant
risks associated with exposure of the general public to fibers that may be shed from
asbestos-containing products such as insulation, brake linings, and other materials.
There is considerable debate in the research community as well over whether the
different forms of asbestos are equally dangerous, particularly because the results
of exposure to the different forms of asbestos in laboratory animals appear to differ
from what is seen in humans. The answers to these questions will prove significant
as decisions are made on whether to attempt to remove existing asbestos in buildings
(an expensive and difficult process).

One other well-known occupational lung disease is black lung or *coal worker's
pneumoconiosis* (CWP). Caused by exposure to coal dust, CWP is characterized by the

presence of black nodules in the lungs, along with widespread fibrosis and emphysema. Also, American veterans of the war in Kuwait and Iraq are being examined after complaining of delayed illness following inhalation of smoke from massive petroleum fires.

LUNG CANCER

Carcinogenesis

See also:
The Mutational Theory of Carcinogenesis
Chapter 6

Oncogenes and Tumor Suppressor Genes

See also:
Oncogenes and Tumor Suppressor Genes
Chapter 6

Polycyclic Aromatic Hydrocarbons

See also:
The Role of Cytochrome P450
Chapter 3

Genetic Carcinogens
Chapter 6

Interference with Cell Division
Chapter 7

PAHs
Appendix

Once rare, lung cancer has now become one of the leading causes of cancer deaths. Lung cancers typically originate from airway epithelial cells either in the center (*squamous cell carcinoma*) or periphery (*adenocarcinoma*) of the lung. Along with a third type of cancer, *large cell carcinoma*, these types make up the majority of lung cancers. A fourth type, *small cell carcinoma*, is less common and also more rapid in growth. Lung cancers most likely develop in response to DNA damage by reactive oxygen species and other free radicals, radiation, or other reactive compounds. Chromosomal changes have been seen in a variety of lung cancers, and there is evidence that loss of *tumor suppressor genes* (genes such as p53 or p16 that suppress cancer) occurs in many tumors. Activation of *oncogenes* (genes that may contribute to the development of cancer) may also play a role. Examples of such genes, which have been found to play important roles in lung cancer, include *EGFR* and *ALK*, both of which code for growth factor receptors.

Of course, the greatest risk factor for lung cancer is exposure to tobacco smoke. It has been well established that smokers have a 10–20 times greater risk of developing lung cancer than nonsmokers and that smoking interacts in an additive or, in some cases, synergistic manner with other risk factors for lung cancer (such as asbestos). Lately, research has focused on the risks of secondhand cigarette smoke to nonsmokers. *Sidestream smoke* (from the end of the cigarette) makes up a significant amount of secondhand smoke and may have even higher concentrations of toxicants than inhaled smoke as well as smaller average particle size. Studies have shown that children and nonsmoking spouses of smokers are more likely to suffer from respiratory problems and lung cancer, respectively, than children and spouses of nonsmokers.

Out of the estimated 4000 or so compounds in cigarette smoke, there are approximately 69 known carcinogens. These include polycyclic aromatic hydrocarbons such as benzo(a)pyrene, nitrosamines, heterocyclic amines, formaldehyde, and benzene; pesticides such as DDT and vinyl chloride; and metals such as nickel, chromium, cadmium, and lead.

Other chemicals have also been implicated as causative agents in lung cancer. Exposure to asbestos, for example, is linked to development of not only the more common form of lung cancer, but also a relatively rare form of cancer called *mesothelioma*. There can be an extremely long latent period (as much as 40 years) between exposure to asbestos and development of mesothelioma.

INHALATION STUDIES

In the laboratory, toxicologists use *inhalation chambers* to study effects of airborne toxicants. An inhalation chamber consists of one or more areas in which animals are held for exposure, along with some apparatus for delivery of the toxicant to be tested. In *static test systems*, the toxicant is simply introduced and mixed into the atmosphere in a closed chamber. Although this method is relatively simple, disadvantages include the tendency for oxygen to be depleted and carbon dioxide to accumulate in the chamber and the constantly decreasing concentration of the toxicant in the atmosphere as it settles out or is absorbed. One way around these difficulties is to use a *dynamic test system*. In this system, air is constantly circulated through the exposure chamber, with the toxicant being introduced into the entering airstream. Gases may be directly mixed in with incoming air; particles may be introduced either as a dry dust or suspended in droplets of water. Concentration of gases and concentration and size of particles can be monitored by sampling within the chamber, and the level of exposure can be adjusted by altering either flow rate through the chamber or rate of addition of the toxicant to the airstream.

The chambers in which the animals are exposed may vary also. The whole body of the animal may be exposed to the toxicant or just the head or neck. In the latter systems, restraint of the animal may pose a problem, but the problems of deposition of toxicant on the animal's coat and subsequent ingestion by licking are solved. Also, if the chamber containing the body can be sealed, it can be adapted as a *plethysmograph*, so that pressure changes within the chamber can be used to estimate lung volumes. Toxicants may also be injected directly into the trachea.

Along with measuring respiratory rates and volumes (vital capacity, minute volume, FEV1, etc.), other parameters such as oxygen and carbon dioxide levels and blood pH can also be used to assess respiratory function in test animals. In addition to *in vivo* studies, washing of the lungs with physiological saline (a technique called *bronchoalveolar lavage*) can supply cells for *in vitro* analysis of cellular function. This technique is particularly useful for studying macrophages.

CASE STUDY   Nanoparticles

Very small particles, or nanoparticles, are being increasingly used in a variety of industrial processes. Also termed *ultrafine* particles, the small size and high surface area to volume ratio of these particles can alter the physical and chemical mechanisms through which they interact with other substances in the environment. This, of course, raises obvious concerns about their potential impact on human health. In particular, questions have been raised about their effects on the respiratory system, which is the most likely avenue through which absorption of these particles may take place (although some absorption may also take place through the skin or gastrointestinal tract). Because of their small size, these particles would be expected to travel into the lower reaches of the lungs, where they could diffuse into alveoli. Thus, they may not only have an impact on the respiratory system itself, but, as recent evidence indicates, may be able to travel through the bloodstream to other organs and systems as well. Much work remains to be done on the kinetics of both absorption and excretion for these particles.

There are several different types of nanoparticles that may be of concern in terms of human exposures. To give one example, there are very small carbonaceous nanoparticles that are able to escape the filters and other control mechanisms which are typically a part of internal combustion engines. Exposure to these nanoparticles is potentially a public health concern, as high levels of exposure have been associated with increased incidence of both respiratory and cardiovascular diseases in epidemiological studies.

Nanoparticles are also used in a variety of industrial processes. *Carbon nanotubes* are long tubular structures made of sheets of carbon that are one atom thick. Their strength and flexibility, conductive properties, and ability to slide along within each other as a nested set of tubes show great promise in numerous practical applications; however, there is some evidence that the elongated shape of nanotubes may be problematic, leading to increased toxicity compared with other nanoparticles. This is reminiscent of some of the issues seen with asbestos and other fibers (many of which have diameters in the 100–200 nm range), where toxicity is at least partially related to the shape rather than to chemical composition of the fibers. In fact, mesotheliomas, the cancer most associated with asbestos exposure, have been induced in mice following exposure to carbon nanotubes, and carbon nanotubes have also been shown to produce DNA damage in mesothelial cells in culture. Inflammatory responses have been noted in mice, as well.

Finally, nanoparticles have also found potential uses in biomedical applications. Their small size allows a wide distribution into tissues, and their high surface area allows the attachment of many molecules to their surfaces. This has led to their usage in both drug and gene delivery systems and as diagnostic probes. Gold nanoparticles have been used for these purposes, as have silica nanoparticles, polymer-based nanoparticles, and even *quantum dots*. Quantum dots are nanoparticles that act as semiconductors and can be used as fluorescent probes. They are brighter than traditional probes, and thus the signal they send can be more easily observed. They are also more stable and so can be used over much longer time spans. However, surface coatings are necessary to provide adequate water solubility and to modify their tendency to interact with cellular macromolecules, a characteristic that can, of course, potentially lead to toxicity.

In terms of the cellular level mechanisms by which nanoparticles produce their effects, *in vitro* studies have shown that exposure of endothelial cells to carbon nanoparticles leads to decreases in cell viability and increased incidence of apoptosis. Evidence also indicates that nanoparticles (including quantum dots, in particular) may stimulate the production of reactive oxygen species. This can then lead to damage to membranes (lipid peroxidation) and resulting negative impact on energy metabolism, transport processes, protein synthesis, and a host of other cellular mechanisms. Nanoparticles may also physically damage membrane integrity, if they are unable to be absorbed and digested through the normal lysosomal systems. There is also evidence that nanoparticles may also interfere directly with protein function, perhaps through effects on folding and conformation. Some nanoparticles even appear to be small enough to penetrate the nucleus, potentially allowing direct interaction with DNA, as well as indirect effects mediated through generation of reactive oxygen species and subsequent damage to the genome.

BIBLIOGRAPHY

Armstrong, B., Hutchinson, E., Unwin, J., and Fletcher, T., Lung cancer risk after exposure to polycyclic aromatic hydrocarbons: A review and meta-analysis, *Environ. Health Perspect.*, 112, 9, 2004.

Barnes, P.J., Chronic obstructive pulmonary disease, *N. Engl. J. Med.*, 343, 4, 269, 2000.

Belvisi, M.G. and Bottomley, K.M., The role of matrix metalloproteinases (MMPs) in the pathophysiology of chronic obstructive pulmonary disease (COPD): A therapeutic role for inhibitors of MMPs?, *Inflammation Res.*, 52, 95, 2003.

Elsaesser, A. and Howard, C.V., Toxicology of nanoparticles, *Adv. Drug Deliv. Rev.*, 64, 129, 2012.

Henderson, R.F. and Nikula, K.J., Respiratory tract toxicity, in *Introduction to Biochemical Toxicology*, Hodgson, E. and Smart, R.C., Eds, Elsevier, New York, 2001, chap. 24.

Hoffmann, D. and Hoffmann, I., The changing cigarette: Chemical studies and bioassays, in *Risks Associated with Smoking Cigarettes with Low Tar Machine-Measured Yields of Tar and Nicotine*, Smoking and Tobacco Control Monograph 13, U.S. Department of Health and Human Services, Public Health Service, National Institutes of Health, National Cancer Institute, 2001.

Hogg, J.C., Pathophysiology of airflow limitation in chronic obstructive pulmonary disease, *Lancet*, 364, 709, 2004.

Jaurand, M.-C., Renier, A., and Daubric, J., Mesothelioma: Do asbestos and carbon nanotubes pose the same health risk?, *Particle Fibre Toxicol.*, 6, 16, 2009.

Leifauf, G.D., Toxic responses of the respiratory system, in *Casarett and Doull's Toxicology*, Klaassen, C.D., Ed., McGraw-Hill, New York, 2013, chap. 15.

Li, N., Hao, M., Phalen, R.F., Hinds, W.C., and Nel, A.E., Particulate air pollutants and asthma. A paradigm for the role of oxidative stress in PM-induced adverse health effects, *Clin. Immunol.*, 109, 250, 2003.

Manning, C.B., Vallyathan, V., and Mossman, B.T., Diseases caused by asbestos: Mechanisms of injury and disease development, *Int. Immunopharmacol.*, 2, 191, 2002.

Mannino, D.M., COPD: Epidemiology, prevalence, morbidity and mortality, and disease heterogeneity, *Chest*, 121(5), 121S, 2002.

Marshall, E., Involuntary smokers face health risks, *Science*, 234, 1066, 1986.

National Library of Medicine, Hazardous Substances Data Bank, available at http://toxnet.nlm.nih.gov/cgi-bin/sis/htmlgen?HSDB.

Pedata, P., Bergamasco, N., D'Anna, A., Minutolo, P., Servillo, L., Sannolo, N., and Balestrieri, M.L., Apoptotic and proinflammatory effect of combustion-generated organic nanoparticles in endothelial cells, *Toxicol. Lett.*, 219, 307, 2013.

Reasor, M.J., The composition and dynamics of environmental tobacco smoke, *J. Environ. Health*, 50, 20, 1987.

Simon-Deckers, A., Gouget, B., Mayne-L'Hermite, M., Herline-Boime, N., Reynaud, C., and Carriere, M., *In vitro* investigation of oxide nanoparticle and carbon nanotube toxicity and intracellular accumulation in A549 human pneumocytes, *Toxicology*, 253, 137, 2008.

Valentine, R. and Kennedy, G.L., Inhalation toxicology, in *Principles and Methods of Toxicology*, 4th edn, Hayes, A.W., Ed., Taylor & Francis, Philadelphia, 2001.

Wang, U. and Chen, L., Quantum dots, lighting up the research and development of nanomedicine, *Nanomed. Nanotechnol. Biol. Med.*, 7, 385, 2011.

Witschi, H.R., Pinkerton, K.E., Van Winkle, L.S., and Last, J.A., Toxic responses of the respiratory system, in *Casarett and Doull's Toxicology*, Klaassen, C.D., Ed., McGraw-Hill, New York, 2008, chap. 15.

Yildirimer, L., Thanh, N.T.K., Loizidou, M., and Seifalian, A.M., Toxicological considerations of clinically applicable nanoparticles, *Nano Today*, 6, 585, 2011.

Zeidler, P.C. and Castranova, V., Role of nitric oxide in pathological responses of the lung to exposure to environmental/occupational agents, *Redox Rep.*, 9, 1, 7, 2004.

9 Cardiovascular Toxicology

FUNCTION OF THE CARDIOVASCULAR SYSTEM

The basic function of the cardiovascular system is transport. This is the system that is responsible for carrying gases, nutrients, waste products, cells, hormones, and other substances from one part of the body to another. As such, it plays a major part in homeostasis through its role in regulating the composition of both intracellular and extracellular fluids. The system is also critical in temperature regulation. Finally, it is the circulatory system that carries defensive elements (cells and molecules of the immune system) to areas of the body that require them. Physically, the system consists of a pump (the heart), a network of tubes (the vascular system), and a transport fluid (the blood). All three components can be affected by toxicants, and we will consider each of them in turn.

ANATOMY AND PHYSIOLOGY OF THE HEART

The heart (Figure 9.1) is a hollow muscular organ located in the thoracic cavity. The bulk of the heart, the *myocardium*, is composed of cardiac muscle tissue. The outside of the heart is covered by a connective tissue sac called the *pericardium*, while the inside of the heart is lined by a layer of epithelial and connective tissue called the *endocardium*. The heart contains four hollow spaces, or chambers: the *right atrium*, the *right ventricle*, the *left atrium*, and the *left ventricle*. The right and left sides of the heart are separated by a wall of tissue called a *septum*, while the upper and lower chambers on each side are separated by flaps of tissues called *valves* that constrain blood flow to a single direction (from the atria into the ventricles) and prevent backflow of blood into the atria.

Blood that is low in oxygen and high in carbon dioxide enters the heart from the *inferior vena cava* and *superior vena cava*—the two major veins that collect blood from all body tissues. This deoxygenated blood enters into the right atrium and then passes through the *tricuspid valve* into the right ventricle. From the right ventricle, the blood is pumped through the *pulmonary semilunar valve* into the pulmonary arteries, which carry blood to the lungs to replenish oxygen and release carbon dioxide. The oxygenated blood returns to the heart from the lungs through the *pulmonary veins*, which empty into the left atrium. Blood passes from the left atrium through the *bicuspid (mitral) valve* and into the left ventricle. From here, the oxygenated blood is pumped through the *aortic semilunar valve* into the aorta, through which it is distributed to the rest of the body.

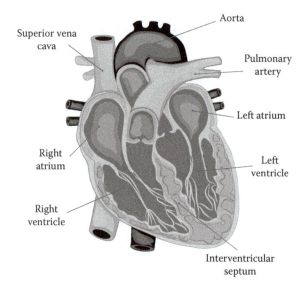

FIGURE 9.1   A coronal section through the heart, showing the four chambers and the vessels that enter and exit those chambers. (Modified with permission from http://www.shutterstock.com/pic-159497261/stock-vector-heart-anatomy-gray-scale-background.html?src=&ws=1. Accessed on June 21, 2014.)

The force required to move blood through these pathways is supplied by the beating action of the heart. During a single heartbeat or *cardiac cycle*, the atria contract together, pushing blood into the ventricles, then the atria relax while the ventricles contract and push blood to the lungs and the rest of the body.

Contractions of the various chambers are produced by the synchronized contraction of cardiac muscle cells. These specialized cells, called *myocytes*, make up roughly 25% of the cells in the heart. Myocytes contain structures called *filaments*, which are made up of proteins. *Thick filaments* are built from the protein *myosin*, and *thin filaments* are built from the protein *actin*. These proteins overlap in a distinct, visible pattern, with projecting heads on the myosin filaments fitting into slots on the actin filaments. Contraction of the muscle involves the release and rebinding of the myosin heads to another open slot further along the actin filament, followed by bending of the myosin heads to rachet the thick fibers further along the thin fibers. Energy in the form of ATP is required for the binding and conformational change in myosin, and calcium is required to pull away an inhibitory protein (*troponin C*) from the binding site on actin.

Myocytes have the ability to contract spontaneously, a property known as *automaticity*. The basis for this excitability lies in the distribution of ions across the cell membranes. At rest, the interior of a cardiac cell is about 80–90 mV more negative than the exterior. This difference is called a *membrane potential* and is produced by unequal distribution of ions (Na^+, K^+, Ca^{++}, and others) across the membrane. A membrane pump (a *Na^+/K^+ ATPase*) maintains a gradient with a high concentration of sodium outside the cell and a high concentration of potassium within the cell.

Small shifts in ionic currents in the cells (such as a small inward sodium leak) can cause a gradual *depolarization*. In other words, the membrane potential becomes less negative and eventually a threshold is reached at which membrane channels specific for sodium open. This allows sodium to rush into the cell (down its concentration gradient), which then alters the membrane potential from negative to positive. As the membrane potential swings to a positive value, other ion channels open, including a calcium channel. This influx of calcium ions, along with release of intercellular calcium, triggers contraction of the muscle fiber. When the potential reaches slightly over 0 mV into the positive range, the sodium (and eventually the calcium) channels close and potassium channels open, allowing an outward movement of potassium. This returns the membrane potential to the original negative potential (a process called *repolarization*). The continuing action of the Na^+/K^+ ATPase rapidly rebuilds the original gradient, and after a brief *refractory period*, the cell will soon be ready to contract again. These changes in membrane potential during depolarization are shown in Figure 9.2. This physiology is similar to depolarization of the nerve axon, which is also rich in sodium ion channels.

For the heart to function efficiently, though, it is not enough for individual cells to contract on their own; cells must be able to communicate and contract in *synchrony*. Cardiac muscle cells are joined together at special communicating junctions called *intercalated discs*. These junctions allow excitatory impulses to pass from one cardiac muscle cell to the next. Rates of contraction are normally controlled by a group of cells in the upper part of the right atrium called the *sinoatrial* (SA) *node*. These cells have the most rapid rate of spontaneous depolarization and thus initiate an impulse before other cells have a chance to initiate. Impulses from these pacemaker cells spread rapidly throughout the atria (causing them to contract simultaneously) and eventually reach a second group of specialized cells in the lower part of the right atrium called the *atrioventricular* (AV) *node*. Here the impulse is delayed briefly and then is sent down a bundle of special muscle fibers that run down the septum between the two ventricles. From this bundle, fibers called *Purkinje fibers* spread out, carrying the impulse to contract to all ventricular muscle cells. The electrical activity of the heart may be viewed on an *electrocardiogram*, which measures spread

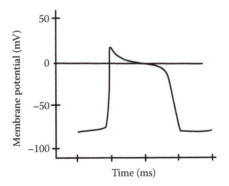

FIGURE 9.2 Changes in membrane potential in a cardiac muscle cell during a contraction (depolarization and repolarization).

of electrical activity across the heart. Many potential cardiac problems can be diagnosed by variations and changes from the normal electrocardiogram pattern.

The SA node controls heart rate through its own spontaneous automaticity, but the actual rate of depolarization is also affected by a branch of the nervous system called the *autonomic nervous system*. This system is under dual positive control of neurohormones eliciting opposite actions through specific receptors. Stimulation of the *sympathetic branch* of the autonomic nervous system causes an increase in heart rate, while stimulation of the *parasympathetic branch* of the autonomic nervous system causes a decrease in heart rate.

Autonomic Nervous System

See also:
   Anatomy and Physiology of
   the Nervous System
             Chapter 10

There is evidence that myocytes have some capacity for cell division, but typically cardiac growth, whether during the normal developmental process or following injury, is due to an increase in size of existing myocytes (hypertrophy) rather than cellular proliferation. Fibroblasts, which are another cell type found in the heart, can and will proliferate following injury, however.

EFFECTS OF TOXICANTS ON THE HEART

ARRHYTHMIAS

One way in which toxicants can interfere with cardiovascular function is through interference with the electrochemical system (as described previously) that regulates contraction of the heart. Abnormalities in this system lead to irregularities in heartbeat, or *arrhythmias*. There are many different types of arrhythmias, with perhaps the most serious being completely asynchronous contraction of muscle cells, or *fibrillation*. Because the ventricles must pump blood with so much more force than the atria (which for the most part empty almost passively into the ventricles), ventricular arrhythmias are typically much more serious than atrial arrhythmias (which may be nearly asymptomatic).

Arrhythmias can be produced indirectly through effects of the autonomic nervous system on the SA node. Excessive sympathetic stimulation can lead to rapid heartbeat (*tachycardia*, defined as over 100 bpm), while excessive parasympathetic stimulation can lead to a slowed heartbeat (*bradycardia*, defined as under 60 bpm). For example, the neurotransmitter *epinephrine* (also known as *adrenaline*), as well as synthetic drugs such as *isoproterenol*, interact with receptors called beta one receptors to increase the heart rate. Other drugs interact more weakly with beta one receptors, but may still produce tachycardia at high enough doses. These drugs would include *ephedrine*, a compound found in herbal preparations of the plant genus *Ephedra*, sales of which were banned by the FDA in 2004 (the ban was later struck down in 2005 and then reinstated in 2006). Another drug with similar mechanism is *pseudoephedrine*, a drug that is often found in over-the-counter decongestant medications designed to treat the common cold.

Halogenated hydrocarbons (chloro- and fluorocarbons) are another class of compounds that can affect activity of the SA node. These compounds suppress activity of the SA node cells and at the same time reduce the refractory period of Purkinje cells. This makes the ventricles in particular more sensitive to the effects of catecholamines (the neurotransmitters released by the sympathetic nervous system), thus increasing the ability of sympathetic stimulation to produce tachycardia and arrhythmias.

Damage to cells of the SA node can also produce arrhythmias, often by interfering with their automaticity. If SA node cells are unable to perform, cells of the AV node (the cells with the next most rapid spontaneous depolarization rate) will attempt to compensate, setting the rhythm of the heart. In the case of extremely rapid atrial depolarization (*atrial flutter*, occurring at over 200 bpm), the ventricles (which have a longer refractory period) cannot keep up, and only about one-third to one-half of the atrial depolarizations result in a ventricular depolarization. Atrial fibrillation (characterized by completely unsynchronized contraction of atrial cells) also leads to ventricular rate being set by the AV node. *Heart block* is a condition that occurs when the signal fails to pass correctly between the atria and ventricles.

When arrhythmias occur elsewhere in the atria or ventricles, it is usually due to underlying damage to the ventricular muscle cells, or *myocarditis*. This damage can be a result of *myocardial infarction* (heart attack), *ventricular hypertrophy*

Halogenated Hydrocarbons

See also:

Primary Biotransformation
(Phase I Reactions):
Reduction
Chapter 3

Other neurotoxicants
Chapter 10

Liver Cell Death: Necrosis
and Apoptosis
Fibrosis and Cirrhosis
Chapter 11

Damage to the Proximal
Tubule
Chapter 12

Halogenated Hydrocarbons
Appendix

(enlargement of the heart), or other disease processes. Myocarditis can also result from direct action of toxicants or can be due to inflammation resulting from hypersensitivity (allergic) reactions to drugs such as penicillin.

Some toxicants can produce alterations in automaticity in non-pacemaker cardiac cells, frequently through effects on ion channels. Cells in the Purkinje network, or even normal atrial or ventricular cells, may be induced to spontaneously depolarize, producing extra beats, or *ectopic beats*. Some toxicants, such as the alkaloid *aconitine* (found in the plant monkshood), keep sodium channels open, preventing repolarization and leading to the repeated generation of impulses. The cardiac glycosides *digoxin* and *digitoxin* (found in the foxglove plant) are Na^+/K^+ ATPase inhibitors that seem to also make resting membrane potential less negative, making ectopic beats more frequent.

In other cases, impulse generation may be inhibited by toxicants. For example, drugs including the *tricyclic antidepressants* suppress activity in Purkinje cells, probably through blockade of sodium channels. This leads to an increase in duration of impulses and increase in the refractory period, thus delaying conduction to the ventricles. At high enough exposures, complete blocks may result (particularly in the AV node, where blocks are most common). Other sodium channel blockers such as the biological toxins *tetrodotoxin* and *saxitoxin* have similar effects; however, nervous system effects usually overshadow the cardiovascular effects.

Ironically, *antiarrhythmic drugs* themselves may produce arrhythmias. Class I antiarrhythmic agents such as *procainamide* or *phenytoin* delay the opening of sodium channels and thus slow conduction. This, however, may lead to the development of *reentrant rhythms*, a situation where if conduction is slow enough, a delayed impulse may be seen as a new impulse and initiate a new wave of depolarization both forward and backward. Beta blockers are classified as Class II agents; the primary arrhythmia which they generate is bradycardia, which is produced through blockade of beta adrenergic receptors. Class III agents such as *amiodarone* block potassium channels and can prolong the duration of impulses. This can lead to *early after depolarizations*, depolarizations that occur during the repolarization process. Class IV agents are calcium channel blockers and can induce bradycardia. Some calcium channel blockers may, however, induce a reflex tachycardia secondary to the peripheral vasodilation which they also produce.

One very specific type of arrhythmia is termed *QT prolongation*, referring to effects seen on a specific interval on an electrocardiogram. This alteration reflects changes in the action potential of ventricular cells and is associated with dysfunction of specific sodium and potassium channels involved in the production of the cardiac action potential. QT prolongation is associated with increased risk for sudden cardiac death and can either be congenital, or a result of tissue damage, ion imbalances, or drug exposure. The drug *quinidine*, for example, used historically for treatment of fevers (as well as treatment of arrhythmias) produces QT prolongation as a result of its effects as a sodium channel blocker. Tricyclic antidepressants have also been associated with QT prolongation, as have some antibiotics (including some of the newer fluoroquinolones).

CARDIOMYOPATHIES AND OTHER EFFECTS ON CARDIAC MUSCLE

Damage to cardiac muscle cells produces a condition called *cardiomyopathy*. Toxicant effects on myocytes can be sublethal, producing changes in myocyte functionality. One example of this would be effects on *contractility*, the ability of cardiac muscle to contract. Decreases in contractility can result from toxicant-induced damage to cardiac muscle cells as well as other factors, such as disruption of oxygen supply. Agents that diminish the availability of calcium (such as heavy metals, including *lead* or *cadmium*) can produce decreases in contractility. The drugs digoxin and digitoxin (which were mentioned before), on the other hand, enhance contractility through increasing calcium levels inside cardiac muscle cells. If you recall, these drugs inhibit the Na^+/K^+ ATPase, which would lead to increased levels of sodium within the cell. This sodium is then available to participate in a Na^+/Ca^{2+} exchange mechanism, raising intracellular calcium levels.

Decreases in contractility can lead to *congestive heart failure*, a condition in which the heart is unable to pump sufficiently to supply blood to all tissues. Individuals with congestive heart failure may suffer from fatigue and edema (accumulation of fluid in tissues) as well as hypertrophy of heart muscle.

As with other cells, minor damage to cardiac myocytes can be repaired. However, if damage is severe enough, cell death may occur. Both apoptosis and necrosis have been observed in toxicant-damaged cardiac tissue. Also, as in other cells, the determination of whether badly damaged cells undergo apoptosis or necrosis may depend on a number of factors, including whether the cells have sufficient energy resources to complete the apoptotic program. Assessment of apoptotic vs. necrotic cell death in cardiac myocytes can be studied in the laboratory using differential immunohistochemical staining techniques. Assays such as the TUNEL or "nick end labeling" assay identify apoptotic myocytes through binding of a labeled probe to the exposed 3′ DNA ends (*overhangs* or *sticky ends*) produced by endonucleases during apoptosis. DNA fragments resulting from necrosis, in contrast, tend to have blunt ends. Myocytes themselves can be distinguished from other cardiac tissues through actin staining.

Recent studies point to an important role of protein kinases in mediating the response to toxicants in myocytes. These kinases would include the *mitogen-activated protein kinases (MAPKs)*, which play an important role in stress responses and are likely to be involved in the initiation of apoptosis. Another important group of kinases is the *protein kinase C* group of isozymes, which are involved in numerous important signaling pathways and which may play a role in cardiomyopathies.

Exposure to *cobalt*, a heavy metal that may block calcium channels, can lead to cardiomyopathy. Although problems with exposure to cobalt occur primarily in the workplace, the association between cobalt and heart disease was first noted in individuals who consumed beer containing 1 ppm cobalt as a foam-stabilizing agent. Of course, long-term *ethanol* exposure can also produce a cardiomyopathy characterized by both degeneration of myocytes and decrease in protein content of surviving myocytes. Alcoholic cardiomyopathy is, in fact, a significant problem in many populations. There is evidence that the molecular mechanism behind this effect is an inhibition of the initiation or elongation steps of protein synthesis, by either ethanol itself or its metabolite acetaldehyde. There is also some evidence that genetic polymorphisms in a mitochondrial aldehyde dehydrogenase (ALDH2) may influence susceptibility to alcohol-induced cardiac complications. In fact, suggestions have been made that stimulation of the activity of this enzyme may even prove to be of therapeutic benefit in protecting patients not only from ethanol cardiomyopathy, but from other forms of damage to cardiac tissues, as well.

Some drugs, including the antitumor drug *doxorubicin*, also produce cardiomyopathy. The drug is metabolized by cytochrome P450 to a free radical, which may then either bind to and damage nucleic acids or bind to and damage mitochondria. Doxorubicin may also disrupt calcium handling by the cell. More recently, long-term use of *reverse transcriptase inhibitors*, drugs used in the treatment of AIDS, has been associated with the development of cardiomyopathy. This cardiomyopathy appears to be secondary to mitochondrial dysfunction involving inhibition of a DNA polymerase involved in mitochondrial replication. Myocardial cells are likely to be affected both because of their many mitochondria and because of their possession of an enzyme that can phosphorylate (and thus activate) the reverse transcriptase inhibitors. Effects have been seen in skeletal muscle, probably for the same reasons, as well as in some other tissues. It is not yet clear, however, why this impacts some patients more significantly than others, but genetic factors may play a role.

Following the development of cardiomyopathy, hypertrophy of the heart muscle can occur. This generally involves an increase in the size of remaining myocytes, but also can involve fibrosis. The fibrosis (as is seen in other tissues following injury) is characterized by accumulation of collagen and other extracellular components. Of course, hypertrophy in and of itself is not necessarily problematic, and can result, for example, from the increased demand that follows from exercise. Unlike exercise-induced hypertrophy, however, the hypertrophy that follows toxic insult does not appear to be physiologically adaptive.

Long thought to be irreversible, some cardiomyopathies now appear to have some degree of reversibility. Recently, cardiac stem cells have been identified, and there is some evidence that they may play an important role in recovery from cardiac damage. There is also some evidence that fibrosis may be reversible. One other essential key to repair of cardiomyopathy is *angiogenesis*, the capacity to form new blood vessels, so as to supply any new cells with oxygen and nutrients. Obviously, the ability of the heart to reverse damage depends upon the functionality of each of these cardiac repair mechanisms. Unfortunately, some toxicants (adriamycin, for example) may not only trigger apoptosis, but may also inhibit regeneration of myocytes, thus blocking a potential recovery mechanism.

MYOCARDIAL INFARCTIONS

Myocardial infarction, or heart attack, is clearly capable of producing a great deal of damage to the heart. Although generally not directly induced by toxicants, the underlying condition that is the major cause of heart attacks (*atherosclerosis* leading to *ischemic heart disease*; for more details, see the following sections in this chapter) can be influenced by chemical exposure. In addition, the cellular mechanisms by which heart attack-induced damage occurs are of interest to toxicologists.

Generally, myocardial infarctions occur when blood supply to an area of the heart is dramatically reduced or cut off, producing the condition of *ischemia*, or lack of oxygen. (Again, the underlying causes of this ischemia will be addressed in the following sections.) Cardiac myocytes are dependent on ATP for contraction and contain many mitochondria in order to supply their intensive energy demands. Myocytes can also store energy through the *phosphocreatine reserve system*, in which the enzyme *creatine kinase* can both phosphorylate creatine in the presence of an excess of ATP and also rapidly transfer a phosphate from phosphocreatine to ADP to restore depleted ATP stores. In the absence of oxygen, however, aerobic respiration cannot proceed and both ATP and phosphocreatine can become quickly depleted. The metabolic changes the cell undergoes as it switches from aerobic to anaerobic metabolism also cause cellular pH to drop dramatically. This activates the H^+/Na^+ exchanger, and the Na^+ that enters the cell is then itself exchanged for Ca^{2+}. These changes may trigger necrosis in affected cells, generally within around a half an hour of onset of ischemia. In addition, an area of apoptosis may develop around the periphery of the damaged area, perhaps in response to factors released by necrotic cells.

Myocardial infarction typically produces radiating chest pain and other discomfort, but may be almost asymptomatic. One reliable measure of cardiac damage is the presence of enzymes in the bloodstream (some forms of *lactate dehydrogenase* (LDH) or *creatine kinase* (CK)) that normally occur only in cardiac cells. Another recently validated marker of myocardial damage is *B-type natriuretic peptide (BNP)*, which serves as a marker of congestive heart failure. Following a heart attack, there may be some proliferation of surviving cells, as well as fibrosis in the damaged area.

Although the proven clinical response to myocardial infarction is to re-establish blood flow to the blocked area as soon as possible, this *reperfusion* may itself produce what has been termed *ischemia-reperfusion injury*. Reperfusion may produce elevated levels of reactive oxygen species, and, in fact, inhibition of cytochrome P450 (one of the generators of reactive oxygen species [ROS]) may reduce the tissue damage that occurs during a heart attack.

Interestingly enough, brief exposure to anoxia may protect myocytes from later anoxic exposures. The mechanism of this protection may be related to production of adenosine, which, in binding to its receptors, triggers the opening of ion channels in the mitochondrial membrane. This may protect the mitochondria from depolarization by producing a temporary hyperpolarization.

Reactive Oxygen Species

See also:
> *Effects of Toxicants on Lipids*
> *Chapter 4*

> *Free Radical-Induced Damage*
> *Chapter 8*

THE VASCULAR SYSTEM

The heart pumps blood through a network of vessels known as the *vascular system* (Figure 9.3). There are two primary circulation systems in the body. In the *pulmonary circuit*, blood leaves the right ventricle of the heart through the pulmonary arteries, which then carry the blood to the lungs for oxygenation. Oxygenated blood returns from the lungs through the pulmonary veins and enters the left atrium. In

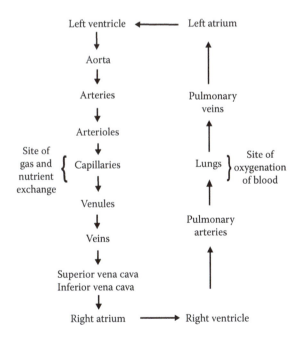

FIGURE 9.3 Circulation of blood, showing the systemic circulation on the left-hand side of the figure and the pulmonary circulation on the right-hand side of the figure.

the *systemic circuit*, blood from the left ventricle leaves the heart through the aorta and is distributed throughout the body through vessels called arteries. Blood returns from the tissues to the heart through the veins.

Arteries, the vessels that carry blood away from the heart, are generally large elastic vessels. They consist of an inner layer of epithelial cells and connective tissue (containing many elastic fibers) called the *endothelium*, a middle layer of smooth muscle, and an outer connective tissue covering. As the distance from the heart increases, arteries branch out and decrease in diameter. Other changes occur also, with the relative amount of elastic fibers decreasing and the relative amount of smooth muscle increasing.

Among the major arteries in the body are the *carotid arteries*, which supply the head and neck region (including the brain); the *coronary arteries*, which supply the heart itself; the *subclavian arteries*, which supply the chest, shoulders, and arms; the *celiac* and *mesenteric arteries*, which supply the gastrointestinal organs; the *renal arteries*, which supply the kidneys; and the *iliac arteries*, which supply the pelvic region and legs.

Eventually, arteries become *arterioles*, which are much smaller vessels consisting only of the endothelial layer and a few smooth muscle cells. Arterioles then branch into *capillaries*, which consist of simply an endothelial layer. The endothelium of capillaries may be continuous, or it may have pores or gaps. Capillaries are where gas and material exchange between the blood and the tissues occurs.

At the junction between arterioles and capillaries, there is a band of smooth muscle. This *sphincter* regulates blood flow in the capillary by contracting (shutting off blood flow) and relaxing (allowing blood flow). Additional methods of regulation of blood flow include contraction of smooth muscle in arterioles and routing of blood through vessels called *anastomoses* that supply a direct connection between arterioles and *venules* and bypass capillaries. Also, not all capillaries are open at any one time, allowing both up- and downregulation of blood flow to tissues based on oxygen demand and other factors.

To return blood to the heart, capillaries merge to form *venules*, which in turn merge to form *veins*. Veins have much thinner, less muscular walls than arteries. Veins also have valves to prevent backflow of blood. These are necessary because the blood pressure that keeps blood moving in arteries drops very low in veins. The *jugular vein* (which drains the head and neck region), as well as other veins from the upper part of the body, merge to form the *superior vena cava*. The *hepatic* (from the liver), *renal* (from the kidneys), and other veins from the lower part of the body merge to form the *inferior vena cava*. The superior vena cava and inferior vena cava then flow into the right atrium.

The lymphatic system, which returns excess fluids and other constituents from the tissues to the circulatory system (through a connection located at the vena cava), will be discussed in a later chapter.

EFFECTS OF TOXICANTS ON THE VASCULAR SYSTEM

HEMORRHAGE
Perhaps the most direct way for toxicants to affect the vascular system is to impact the integrity of the capillaries and produce hemorrhage, or uncontrolled bleeding.

The toxicants best known for producing this effect are the snake venoms. Specifically, most venoms contain proteins called *metalloproteinases*, which damage capillaries, often very rapidly following venom exposure. Pathological changes which are seen include separation of endothelial cells from the

Snake Venom

See also:
    Reptiles
                                                    Chapter 20

basement membrane, and erosion of the basement membrane, as well as damage to the endothelial cells themselves. Bacterial endotoxins, fungal mycotoxins, and other compounds can also directly damage endothelial cells, leading to loss of integrity of capillaries.

VASCULAR SPASMS AND BLOOD PRESSURE

Regulation of blood flow occurs both at the systemic and at the local levels. The sympathetic nervous system governs the state of relaxation or contraction of vascular smooth muscle (an effect mediated primarily through α-*adrenergic receptors*), although some tissues (skeletal muscle, for example) may also respond to cholinergic input. Hormones can also play a role in regulation of blood flow at the systemic level. One of the most important hormone systems is the *renin-angiotensin system*, which, along with the hormone *aldosterone*, regulates blood pressure through effects on both vasoconstriction and blood volume and composition. Locally, oxygen, nitric oxide, and other metabolic products can also affect blood flow.

Thus, through either systemic or local mechanisms, substances can produce changes in vascular smooth muscle tone and thus change blood flow to an area of tissue. Endogenous compounds such as catecholamines, for example, interact with the *alpha adrenergic* receptor system on vascular smooth muscle to produce vasoconstriction. Widespread vasoconstriction increases the resistance to blood flow and thus blood pressure generally must rise in order to maintain adequate flow. This is one of the mechanisms by which cocaine produces its cardiovascular toxicity, since one effect of cocaine is to increase release of catecholamines. The resulting vasoconstriction can lead to elevated blood pressure, heart attack, and stroke. Conversely, many drugs that are used to treat high blood pressure block catecholamine action by blocking alpha adrenergic receptors.

Another example of a toxicant that affects vascular smooth muscle is a class of compounds called *nitrates*. Nitrates and related compounds, probably through formation of nitric oxide, can activate an enzyme called *guanylate cyclase* that interacts with other enzymes to produce relaxation of the smooth muscle. This vasodilation is one of the ways in which the drug *nitroglycerin* reduces heart pain (angina) that is caused by reduced blood flow to cardiac tissue. Exposure to nitrates may occur in the explosives or pharmaceutical industries and can produce headache (caused by dilation of cerebral blood vessels) or dizziness (caused by reduced blood pressure). After a period of time, however, a tolerance to the nitrates may develop and symptoms may disappear. At this point, however, cessation of exposure may trigger reflexive vasospasms (in other words, vasoconstriction) and sudden death from myocardial infarction may even occur.

Hypertension, or high blood pressure, is a complex condition that is not well understood, however evidence has accumulated that chronic exposure to toxicants may play a role in some cases. *Cadmium*, for example, has produced hypertension in rats at levels of 5 ppm in drinking water, and exposure to *lead* may also be a risk factor for hypertension. There is evidence that exposure to mercury, also, may be associated with development of hypertension. Residents of Minamata, Japan, for example, who were exposed to high levels of methylmercury in the 1960s, demonstrated a significantly higher incidence of hypertension than residents who were exposed to lower levels. On the opposite side of the spectrum, one can find the condition *orthostatic hypotension* (also called postural hypotension). This condition, characterized by dizziness or fainting when rising to a standing position from either a prone or sitting position, typically occurs as a result of pooling of blood in the extremities. It can be produced by drugs or toxicants which block alpha adrenergic receptors in the vasculature, preventing the compensatory vasoconstriction which would normally return blood to the central core. Classes of drugs that can produce this include tricyclic antidepressants and antipsychotic drugs.

ATHEROSCLEROSIS

Atherosclerosis is a condition characterized by accumulation of plaques, lipid-containing masses that can form in the lumen of blood vessels and can severely narrow them. This, of course, restricts blood flow, and if a blood clot forms or becomes lodged at a plaque, blood flow may be completely blocked. This is particularly damaging if it occurs in arteries that supply tissues that depend heavily on a constant supply of oxygen. For example, blockage in cerebral vessels leads to death of brain tissue and is termed a *cerebrovascular accident* (or *stroke*). Blockage of coronary vessels leads to death of cardiac tissue and, as we have already discussed, is termed a myocardial infarction (or heart attack).

The process by which plaques form is not completely understood, but probably involves damage to endothelial cells, proliferation of smooth muscle, invasion of the area by immune system cells, adhesion of platelets, and accumulation of lipids by the cells involved. Risk factors for development of atherosclerosis include high levels of cholesterol and *trans*-fatty acids in blood (which results from a combination of genetic factors and diet), high blood pressure, age (it is more common in older than in younger individuals), and sex (it is more common in men than in women, but the risk factor for women rises at menopause).

Toxicant exposure has also been implicated in some cases of atherosclerosis. Exposure to *carbon disulfide* has been reported in both laboratory and epidemiological studies to produce a significant increase in incidence of atherosclerosis. Carbon disulfide may initiate or accelerate the atherosclerotic process by direct injury to endothelial cells, by alterations in metabolism that increase cholesterol levels, or by a combination of these mechanisms. Chronic exposure to *carbon monoxide* also appears to accelerate the production of atherosclerotic plaques. It is unclear whether this is a direct effect on the vessels or a by-product of CO-induced hypoxia (lack of sufficient oxygen). In either case, the carbon monoxide found in cigarette smoke may be one factor behind the observation that smokers are at higher risk for atherosclerosis than nonsmokers.

Most recently, an additional environmental risk factor for atherosclerosis has been found: particulate air pollution. Studies have long indicated that exposure to air pollution poses an increase in risk for cardiovascular disease, and now laboratory studies have shown that exposure to particulate matter (particularly small particles) does, in fact, promote development of atherosclerotic lesions. The mechanism of action is not yet clear, but may involve the ability of these particles to trigger oxidative stress, which in turn may lead to inflammation.

Carbon Disulfide

See also:
   Other Cytotoxic Compounds
   Chapter 10

   Carbon Disulfide
   Appendix

Carbon Monoxide

See also:
   Effects of Toxicants on
   Transport Proteins
   Chapter 4

   Effects of Toxicants on
   Hemoglobin
   Chapter 9

   Carbon Oxides
   Chapter 17

   Carbon Monoxide
   Appendix

Angiogenesis

Finally, the process of angiogenesis (the formation of new blood vessels) is one that plays a major role both in tissue repair and in disease progression (cancer, for example). This process can be initiated by endothelial cells themselves, or by circulating progenitor cells which appear to originate in the bone marrow. Toxicants have been shown to affect angiogenesis in assays such as the CAM (chick chorioallantoic membrane) assay, which measures development of capillaries in chick embryos. Using this model, for example, both mainstream and sidestream cigarette smoke have been shown to significantly disrupt angiogenesis.

THE BLOOD

Blood consists of a liquid called plasma and a variety of cells, including *red blood cells*, *white blood cells*, and cell fragments called *platelets*. Plasma is mostly water, but also contains dissolved salts, nutrients, gases, and plasma proteins. These include *albumins*, which are transport proteins; *globulins*, which have roles in transport and immune function; and *fibrinogen*, a soluble protein that is converted into the insoluble *fibrin* during the blood-clotting process.

Red blood cells, or *erythrocytes*, are biconcave discs with no nuclei. Proteins on the surface of red blood cells are what determine a person's blood type. Red blood cells contain the oxygen-carrying molecule *hemoglobin* (Figure 9.4). Hemoglobin is a protein composed of four subunits, each of which contains a *heme molecule* (a porphyrin ring containing an iron atom). Each heme molecule is capable of combining with one molecule of oxygen (O_2). A number of factors influence the binding of oxygen to hemoglobin. When partial pressure of oxygen is low, cells produce a molecule called 2,3-diphosphoglycerate that interacts with hemoglobin to encourage release of oxygen. Other factors that enhance oxygen release include low blood pH (a reflection of higher carbon dioxide levels) and higher temperatures. Higher blood pH (a reflection of lower carbon dioxide levels) and lower temperatures, on the other hand, enhance binding of oxygen to hemoglobin. A phenomenon called *cooperativity* also exists. The four heme subunits cooperate together in that the release of one oxygen molecule alters the conformation of the hemoglobin molecule and facilitates the release of the other oxygen molecules. Likewise, binding of one oxygen molecule facilitates binding of others.

Red blood cells have a lifetime of about 120 days. They are produced in the bone marrow in a process stimulated by a hormone called *erythropoietin* (made by the kidneys) that is released in response to oxygen deficiency. Production also requires sufficient quantities of vitamin B_{12}, folic acid, and *iron*, which is partially recycled from red cells that have been destroyed. Erythrocyte cell membranes (ghosts) possess

FIGURE 9.4   The structure of hemoglobin: a heme unit.

acetylcholinesterase activity similar to that of neurosynapses; butyrylcholinesterase activity is present in the blood serum. Activities of these enzymes in drawn blood are sometimes monitored as signs of possible exposure to acetylcholinesterase inhibitors such as the organophosphorus insecticides.

There are five types of white blood cells, or *leukocytes*, normally found in the blood. *Neutrophils, eosinophils*, and *basophils* are characterized under the microscope by the presence of visible granules in their cytoplasm. Neutrophils carry out phagocytosis, eosinophils help regulate and control allergic reactions, and basophils release histamine and other mediators of allergic reactions. *Monocytes* and *lymphocytes*, on the other hand, lack granules. Monocytes leave the bloodstream and upon entering tissues become macrophages—cells that are also important in phagocytosis. Lymphocytes are involved in the production of specific immune responses, responses that are directed against a specific invader (more about this in Chapter 12).

Platelets are cell fragments that are involved in the process of *hemostasis* (cessation of blood loss). When a blood vessel is damaged, smooth muscle fibers near the injury contract (slowing blood loss from the damaged area), platelets adhere to the damaged endothelium, and proteins called *clotting factors* initiate the conversion of the soluble protein fibrinogen into the insoluble fibrin. More platelets stick to the strands of fibrin, and the clot draws the edges of the damaged area together. After repair occurs, the clot dissolves.

EFFECTS OF TOXICANTS ON THE BLOOD

ANEMIAS, HEMOLYSIS, AND RELATED DISORDERS

One major site at which chemicals can interfere with functioning of the blood is the bone marrow. Damage to bone marrow can lead to *pancytopenia*, a decrease in the numbers of red and white cells and platelets. Severe damage or outright destruction of bone marrow prevents stem cells from producing any new cells, a condition called *aplastic anemia*. Toxicants that can cause aplastic anemia include drugs such as *chloramphenicol*, an antibiotic; *lindane*, an insecticide that is a chlorinated cyclohexane derived from benzene; and *benzene* itself. Chronic exposure to benzene levels of 100 ppm or higher can produce either a reversible pancytopenia or the more severe aplastic anemia. Benzene exposure has also been linked to development of *acute myelogenous leukemia*. It is probable that a metabolite of benzene, perhaps *benzoquinone*, is responsible for these effects. Toxicants that damage dividing cells (such as *radiation* or some anticancer drugs) can also produce pancytopenia.

Disorders of the blood, including anemia, can be diagnosed through measures such as red blood cell count (RBC) and *hematocrit*, or packed cell volume (PCV). Hematocrit is defined as the percentage by volume of red blood cells in blood and is normally in the range of 40%–45%.

Rather than producing broad effects on bone marrow, some toxicants may affect one or more blood cells specifically. Several drugs produce decreases in

Benzene

See also:
   Toxicant-Induced
    Immunosuppression
             Chapter 13

Benzene
                     Appendix

platelet numbers, while others inhibit production of various classes of white blood cells. Red blood cell production may be altered by availability of iron, as well as by chemicals that interfere in the synthesis of heme. The reduction of red blood cells due to blockade of heme synthesis is called *sideroblastic anemia* and can be identified by accumulation of iron in cells of the bone marrow (which show a characteristic staining pattern when treated with the stain Prussian Blue). *Lead* is a well-known producer of sideroblastic anemia through its inhibition of the enzyme *ALA-D* and other enzymes important to heme production. Genetic variations of the enzyme ALA-D between individuals may lead to differences in individual susceptibility to lead. Inhibitors of DNA synthesis produce a condition called *megaloblastic anemia*. Megaloblastic anemia may be a side effect of drugs used to treat cancer (such as methotrexate or cytosine arabinoside) which directly or indirectly interfere with DNA synthesis.

Lead

See also:
  Interference with Cell
    Division
                          Chapter 7

  Myelinopathies
  Other neurotoxicants
                          Chapter 10

  Toxicant-Induced
    Immunosuppression
                          Chapter 13

  Metals
                          Chapter 17

  Lead
                          Appendix

Red blood cell levels are affected not only by the actions of toxicants on bone marrow, but also by the actions of toxicants on circulating cells. A decrease in the numbers of red blood cells resulting from the destruction of circulating cells is called *hemolytic anemia*, and may be identified by a rise in the number of reticulocytes, which are immature red blood cells. In one type of hemolytic anemia, oxidants such as *phenylhydrazine* or *aniline* produce reactive peroxides that are detoxified through reactions involving the oxidation of *glutathione*. The oxidized glutathione is then reduced by an enzyme, *glutathione reductase*, a step that requires NADPH (which is generated in the red blood cell during glycolysis in a series of steps called the hexosemonophosphate shunt). If activity of the oxidants outstrips the ability of the red cell to produce NADPH, then glutathione cannot be reduced, peroxides may accumulate, and oxidative damage to hemoglobin may occur. The decreased solubility of the damaged hemoglobin causes it to precipitate and form visible deposits called *Heinz bodies*. Heinz bodies distort the shape of the red blood cell, often leading to its destruction through *hemolysis*. Some individuals suffer a genetic deficiency in an enzyme, glucose-6-phosphate dehydrogenase (G6PD), which is necessary for NADPH generation. These individuals would be even more susceptible to hemolysis by oxidants than unaffected individuals.

Other toxicants, such as the heavy metals lead and mercury, can cause red blood cell hemolysis through other mechanisms. Lead, for example, increases hemolysis, probably through damage to the cell membrane.

EFFECTS OF TOXICANTS ON HEMOGLOBIN

Some toxicants produce their effects by interfering with the binding of oxygen molecules to hemoglobin. *Carbon monoxide*, for example, binds at the same site on the hemoglobin molecule as does oxygen, but with an affinity 245 times higher. Thus, low levels of carbon monoxide are able to produce significant binding to hemoglobin and resulting displacement of oxygen. Exposure to an atmosphere containing 0.1% carbon monoxide can lead to symptoms such as headache, nausea, tachycardia, and even death from oxygen deprivation in a matter of hours. Opportunities for exposure to carbon monoxide are common, because it is produced during the process of combustion of fossil fuels. Smokers, in fact, may have up to 10% of their hemoglobin saturated with carbon monoxide (as compared to less than 1% in nonsmokers).

Carbon Monoxide

See also:
  Effects of Toxicants on
    Transport Proteins
                Chapter 4

  Atherosclerosis
                Chapter 9

  Carbon Oxides
                Chapter 17

  Carbon Monoxide
                Appendix

Recently, however, new evidence has raised questions as to whether the hypoxic aspect of carbon monoxide toxicity is its only mechanism of action. Some scientists have argued that compensatory increases in blood flow and oxygen delivery to the brain during carbon monoxide intoxication may provide adequate oxygen levels for normal functioning. If that is the case, another explanation for the observed toxicity is necessary. Hypotheses that have been raised include binding of carbon monoxide to mitochondrial cytochromes, activation of nitric oxide-producing immune cells, and effects on neurotransmitter systems (carbon monoxide has itself, of course, now been shown to be a neurotransmitter). Of course, it may be that all of these mechanisms, including hypoxia, interact to produce the observed toxic effects. Perhaps additional research will clarify this question.

Another series of toxicants that can interact with hemoglobin are the *nitrites, nitrates, aromatic amines*, and other nitrogen-containing compounds. These compounds (or in some cases, their metabolites) can oxidize the heme molecules, converting the iron atom from a ferrous to a ferric state. Ferric iron cannot combine with oxygen and thus the hemoglobin molecule (now called *methemoglobin*) cannot function normally. Red blood cells have a system (the *diaphorase I system*) containing an enzyme, *methemoglobin reductase*, which is capable of reducing methemoglobin. The enzyme requires NADH (which is supplied by glycolysis). A second system for reducing methemoglobin also exists (the *diaphorase II system*), and it can be activated by administration of the compound *methylene blue* (a dye). This second system requires NADPH (supplied by the pentose phosphate shunt) (Figure 9.5).

Nitrates

See also:
  Effects of Toxicants on
    Transport Proteins
                Chapter 4

  Vascular Spasms and Blood
    Pressure
                Chapter 9

  Phosphorus and Nitrogen
                Chapter 17

  Preservatives and
    Antioxidants
                Chapter 19

  Nitrates
                Appendix

Due to the presence of methemoglobin reductase and due to the fact that levels of methemoglobin must reach around 10%–20% to produce clinical symptoms, and 70% to produce fatalities, methemoglobinemia is

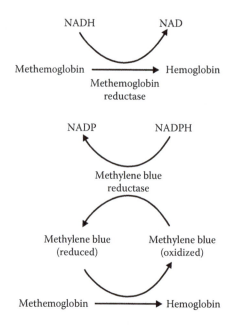

FIGURE 9.5 Reduction systems for methemoglobin. At top, the endogenous system involving the enzyme methemoglobin reductase; at bottom, the mechanism through which methylene blue can be used to reduce methemoglobin.

not a common problem. One exception is in infants, who have lower levels of methemoglobin reductase than adults, and who may be exposed to nitrates in drinking water (particularly in rural areas). There are also a few local anesthetics (*lidocaine*, for example) and other drugs that can induce methemoglobinemia.

EFFECTS OF TOXICANTS ON PLATELETS AND COAGULATION

Finally, there are several mechanisms through which toxicants can impact platelets and the coagulation process (aside from the already mentioned condition of pancytopenia, which includes reduction in the platelet number). For example, binding of drugs or toxicants to platelets can trigger platelet destruction by the immune system, in some cases. Nonsteroidal anti-inflammatory agents (NSAIDS) can also impact platelet function through inhibition of cyclooxygenases. Of course, agents used to suppress coagulation (such as warfarin), while clinically useful in the prevention of heart attacks and strokes, can produce problems in their own right if dosages are not carefully titrated. These drugs interfere with the synthesis of *coagulation factors*, proteins necessary for coagulation, and may themselves be affected by other drugs as well as dietary components.

BIBLIOGRAPHY

Araujo, J.A., Barajas, B., Kleinman, M., Wang, X., Bennett, B.J., Gong, K.W., Navab, M., Harkema, J., Sioutas, C., Lusis, A.J., and Nel, A.E., Ambient particulate pollutants in the ultrafine range promote early atherosclerosis and system oxidative stress, *Circ. Res.*, 102, 589, 2008.

Bloom, J.C. and Brandt, J.T., Toxic responses of the blood, in *Casarett and Doull's Toxicology*, Klaassen, C.D., Ed., McGraw-Hill, New York, 2008, chap. 11.

Budas, G.R., Disatnik, M.-H., and Mochly-Rosen, D., Aldehyde dehydrogenase 2 in cardiac protection: A new therapeutic target? *Trends Cardiovas. Med.*, 19, 158, 2009.

Combs, A.B., Ramos, K., and Acosta, D., Cardiovascular toxicity, in *Introduction to Biochemical Toxicology*, Hodgson, E. and Smart, R.C., Eds., Elsevier, New York, 2001, chap. 26.

Ejaz, S. and Lim, C.W., Toxicological overview of cigarette smoking on angiogenesis, *Environ. Toxicol. Pharmacol.*, 20, 335, 2005.

Escalante, T., Rucavado, A., Fox, J.W., and Gutierrez, J.M., Key events in microvascular damage induced by snake venom hemorrhagic metalloproteinases, *J. Proteomics*, 74, 1781, 2011.

Gorman, D., Drewry, A., Huang, Y.L., and Sames, C., The clinical toxicology of carbon monoxide, *Toxicology*, 187, 25, 2003.

James, R.C., Hematotoxicity: Toxic effects in the blood, in *Industrial Toxicology*, Williams, P.L. and Burson, J.L., Eds., Van Nostrand Reinhold, New York, 1985, chap. 4.

Kang, Y.J., Toxic responses of the heart and vascular system, in *Casarett and Doull's Toxicology*, Klaassen, C.D., Ed., McGraw-Hill, New York, 2008, chap. 18.

Lang, C.H., Kimball, S.R., Frost, R.A., and Vary, T.C., Alcohol myopathy: Impairment of protein synthesis and translation initiation, *Int. J. Biochem. Cell Biol.*, 33, 457, 2003.

Lewis, W., Cardiomyopathy, nucleoside reverse transcriptase inhibitors and mitochondria are linked through AIDS and its therapy, *Mitochondrion*, 4, 141, 2004.

Logue, S.E., Gustafsson, A.B., Samali, A., and Gottleib, R.A., Ischemia/reperfusion injury at the intersection with cell death, *J. Mol. Cell. Cardiol.*, 38, 21, 2005.

Ramos, K.S., Melchert, R.B., Chacon, E., and Acosta, D., Jr., Toxic responses of the heart and vascular systems, in *Casarett and Doull's Toxicology*, Klaassen, C.D., Ed., McGraw-Hill, New York, 2001, chap. 18.

Smith, T.L., Koman, L.A., Mosberg, A.T., and Hayes, A.W., Cardiovascular physiology and methods for toxicology, in *Principles and Methods of Toxicology*, 4th ed., Hayes, A.W., Ed., Taylor & Francis, Philadelphia, 2001, chap. 27.

Van Stee, E.W., Cardiovascular toxicology: Foundations and scope, in *Cardiovascular Toxicology*, VanStee, E.W., Ed., Raven Press, New York, 1982.

Wallace, K.B., Doxorubicin-induced cardiac mitochondrionopathy, *Pharmacol. Toxicol.*, 93, 105, 2003.

Yorifuji, T., Tsuda, T., Kashima, S., Takao, S., and Harada M., Long-term exposure to methylmercury and its effects on hypertension in Minamata, *Environ. Res.*, 110, 40, 2010.

Zhang, Y. and Ren, J., ALDH2 in alcoholic heart diseases: Molecular mechanisms and clinical implications, *Pharmacol. Ther.*, 132, 86, 2011.

10 Neurotoxicology

FUNCTION OF THE NERVOUS SYSTEM

In general, the nervous system has three functions. First of all, specialized cells detect *sensory* information from the environment and then relay that information to other parts of the nervous system (such as the brain, for example). A second segment of the system directs the *motor* functions of the body, often in direct response to sensory input. Finally, part of the nervous system is involved in processing of information. These *integrative* functions include such processes as thought, consciousness, learning, and memory. All of these functions are potentially vulnerable to the actions of toxicants.

ANATOMY AND PHYSIOLOGY OF THE NERVOUS SYSTEM

The nervous system consists of two fundamental anatomical divisions: the *central nervous system* (CNS) and the *peripheral nervous system* (PNS). The CNS includes the brain and spinal cord, while the PNS consists of all other nervous tissue that lies outside the CNS.

The CNS is structurally quite complex, but can be divided into four major areas based on the process of neural development. Within the brain, there is a forebrain region that consists of the *cerebrum* (*cerebral cortex* and *basal ganglia*), the *thalamus*, and the *hypothalamus*. There is also a small *midbrain* region, and a *hindbrain* consisting of the *medulla*, *pons*, and *cerebellum*. The fourth major area is the *spinal cord*. The location of these areas is diagrammed in Figure 10.1.

The PNS includes both *afferent nerves* that relay sensory information from specialized receptors to the CNS and *efferent nerves* that relay motor information from the CNS to various muscles and glands. Efferent nerves that carry motor information to skeletal muscles make up the *somatic* or voluntary nervous system, while efferent nerves that carry motor information to smooth muscles, cardiac muscle, and various glands are part of the *autonomic* or involuntary nervous system.

On the histological level, there are two types of cells found within the *nervous* system: neurons and *glial cells*. Neurons are the cells directly responsible for transmission of information. An individual neuron consists of a *cell body* (also called a *soma* or *perikaryon*) and processes called *dendrites* and *axons* (Figure 10.2). Each part of the neuron has a specific function. The cell body contains a nucleus, mitochondria, endoplasmic reticulum, and other organelles and is where most cellular metabolism occurs (including virtually all of the protein synthesis). Dendrites are branching extensions of the cell body, specialized for reception of incoming information. Axons transmit information to other neurons. A cell may have many dendrites, but generally has only one axon. Many neurons bundled together form what is called a *nerve*.

FIGURE 10.1 The anatomy of the central nervous system showing the major anatomical divisions of the brain. (Modified with permission from http://www.shutterstock.com/pic-113340007/stock-vector-brain-sections-vector.html?src=&ws=1. Accessed on June 22, 2014.)

Glial cells function as supporting cells. *Astrocytes* are a type of glial cell that provide structural support and also play a role in metabolism and support of neurotransmission. In fact, recent discoveries indicate that these glial cells express neurotransmitter receptors and transporter systems and play a major role in the regulation of neurotransmitter levels for glutamate, GABA, and other neurotransmitter systems. *Microglia* are phagocytic, engulfing and digesting dead material and debris, thus removing it from the CNS. *Ependymal cells* line the ventricles (fluid-filled cavities) of the brain. The remaining two types of glial cells are involved in the formation of *myelin*, a lipid-rich substance that covers many axons and aids in efficient conduction of information. These are the *oligodendroglial cells* (which produce myelin in the CNS) and the *Schwann cells* (which produce myelin in the PNS).

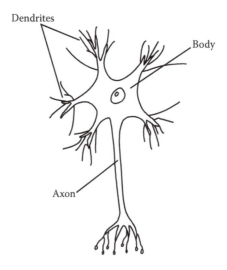

FIGURE 10.2 The structure of an individual neuron.

Within the nervous system, information is passed along the nets of interconnected neurons by chemical and electrical signals. Within each neuron, the signal is electrical in nature, with each electrical impulse being initiated at the dendrites then traveling through the cell body and down the axon. Communication between neurons, on the other hand, is primarily chemical in nature. Neurons do not physically contact each other—there is a small gap between the axon of one neuron and the dendrite of another. This junction, consisting of the axon, dendrite, and gap between them, is called a *synapse*.

When an electrical impulse reaches an axon, it triggers the release of small molecules called *neurotransmitters*, which then migrate across the *synaptic gap* and bind to *receptors*, often with an integrated ion channel, on the dendrites of the next neuron. This chemical binding, and the conformational change it induces in the receptor and ion channel, then triggers the start of an electrical impulse in that next neuron. In this manner, information is relayed throughout the nervous system. Some motor neurons pass their impulses along not to other neurons, but to voluntary muscles. This nerve–muscle connection is known as the *neuromuscular junction*.

EFFECTS OF TOXICANTS ON THE NERVOUS SYSTEM: GENERAL PRINCIPLES

The nervous system is a vulnerable target for toxicants due to the critical voltages that must be maintained in cells and the all-or-nothing responses when voltages reach threshold levels. In addition, the nervous system is a very active tissue metabolically and is thus vulnerable to toxicants that interfere with energy metabolism. The architecture of the nervous system also contributes to its vulnerability, as it consists of a complex network of cells spread throughout the body which nonetheless are required to function as a rapid and reliable communications system.

The role of the nervous system in directing many critical physiological operations also means that any damage may well have widespread and significant functional consequences. And finally, damage to neurons is sustained more permanently than other cells due to the relative lack of regeneration, particularly in the central nervous system.

THE BLOOD–BRAIN BARRIER

The central nervous system does, however, have one critical feature for protection against injury by toxic chemicals—the *blood–brain barrier* (Figure 10.3). Anatomically, the blood–brain barrier consists of modifications to the endothelial cells that line capillaries in the brain. These changes, such as the specialized *tight junctions* between the cells, distinguish CNS endothelial cells from endothelial cells found in other parts of the body. Tight junctions have been found to contain an array of unique proteins such as *occludin* and a family of proteins known as the *claudins*. However, the role of most of these proteins is not yet completely understood. Some glial cells are most likely a part of the blood–brain barrier also, since long processes from astrocytes are found wrapped around capillaries in many parts of the brain.

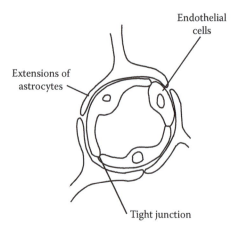

FIGURE 10.3   The blood–brain barrier. A combination of tight junctions between endothelial cells and the surrounding of capillaries by extensions of astrocytes prevents easy passage of many molecules from blood into brain tissue.

The functional result of the blood–brain barrier is to keep many blood-borne molecules from entering the central nervous system. Therefore, most toxicants that affect the central nervous system tend to be small, highly lipid-soluble, nonpolar molecules—if they were not, they could not diffuse through the barrier. There are, however, ways to pass the blood–brain barrier that do not depend on diffusion. Several specific transport systems are associated with the blood–brain barrier and allow the transport of essential nutrients (such as glucose and amino acids) and ions into the brain, as well as the transport of other molecules out of the brain back into the blood. These generally protective pump systems would include the *P-glycoprotein transporter* (which plays a role in resistance to cancer chemotherapy), as well as other ATP-dependent pumps (often called *ABC* (for "ATP-binding cassette") *transporters*. Expression of some of these pumps are inducible, allowing the barrier to respond to varying physiological conditions.

P-Glycoprotein

See also:
 Absorption

                                    Chapter 2

Some evidence points to the existence of a metabolic blood–brain barrier as well as an anatomical one. Although tissues of the CNS are low in levels of detoxifying enzymes such as the cytochrome P450 system, there are other enzymes, such as *monoamine oxidase* or *catechol-O-methyl transferases* that can metabolically change molecules as soon as they cross the endothelium, thus modifying their potential toxic effects.

While most of the central nervous system is afforded the protection of the blood–brain barrier, there are a few areas in the CNS, such as the pituitary and the hypothalamus, in which the blood–brain barrier is reduced or lacking. This is, in fact, undoubtedly critical in allowing the brain to respond to changes in blood levels of hormones and other circulating molecules. The peripheral nervous system is also not protected. The blood–brain barrier is also not well developed at birth and thus provides less protection in infants.

Any damage to the blood–brain barrier by either toxicants (such as lead) or disease states produces a complex array of consequences. On the one hand, a damaged barrier may potentially allow a greater number and concentration of toxicants into the brain. On the other hand, since the blood–brain barrier provides a barrier not only to toxicants, but also to therapeutic compounds, a damaged barrier may actually enhance delivery of drugs to the CNS. A damaged blood–brain barrier can be detected by watching for the appearance of CNS-specific proteins such as *S100* (which is normally found in astrocytes) in the general circulation.

The differential protection that the blood–brain barrier affords to the CNS as opposed to the PNS can often help explain the different action of compounds on the two parts of the system. For example, an antidote for organophosphate poisoning, *pralidoxime* (2-PAM), is effective in the peripheral but not the central nervous system, due to its inability to cross the blood–brain barrier. Fortunately, the complementary antidote, atropine, a natural product from *Atropa belladona*, is highly efficacious in the CNS.

EFFECTS OF TOXICANTS ON THE NERVOUS SYSTEM: GENERAL CATEGORIES

The effects of toxicants on the nervous system can be grouped into several functional categories. These categories will be introduced here and then explored in more detail in later sections.

Some toxicants affect the passage of electrical impulses down the axon. These toxicants interfere with the passage of sensory, motor, and also integrative impulses, leading to effects such as *paresthesias* (abnormal sensations such as tingling or hot or cold sensations), *numbness, weakness,* and *paralysis*. A second category of toxicants affects synaptic transmission between neurons, leading to either under- or overstimulation of a part of the nervous system. There are also toxicants that affect myelin, as well as toxicants that damage axons. Exposure to other toxicants can lead to neuronal cell death (producing a variety of physiological effects), and the effects of a final category of toxicants are produced through unknown mechanisms.

EFFECTS OF TOXICANTS ON ELECTRICAL CONDUCTION

In order to carry out its functions, the nervous system needs to be able to efficiently conduct information from one part of the body to another. As discussed briefly in the introductory sections, information in the nervous system is passed along from one neuron to another in the form of impulses having both electrical and chemical components. We will now look at this process in more detail and describe how it can be affected by various toxicants.

At rest, a neuron has a *resting membrane potential* of approximately –70 mV, indicating that negatively charged ions outnumber positively charged ions inside the cell (in the intracellular fluid) but not outside the cell (in the extracellular fluid). This uneven distribution of charge is caused in part by the fact that although the membrane is quite permeable to some ions, it is not permeable to many of the large, negatively charged proteins found within the neuron. Thus, these molecules are trapped inside the neuron, contributing to the excess negative charge within.

Ion Channels

See also:

Effects of Toxicants on Receptors and Ion Channels

Chapter 4

The membrane potential is also due to the uneven distribution of various ions across the neuronal membrane. This membrane contains several *ion channels* that allow passive flow of various ions through the membrane. These channels are typically constructed of proteins and are specific for a given ion. For example, there are channels specific for sodium ions and channels for potassium ions, among others.

The opening of these ion channels allows ions to pass across the membrane (down their concentration gradient) and thus affect the membrane potential. Conversely, changes in the membrane potential can also affect the opening and closing of the channels (probably by changing the shape and arrangement of the molecules that form the channel). In fact, because of their reaction to local changes in membrane potential, these channels are often called *voltage-gated channels.*

The opening and closing of a single sodium ion channel can be observed in a *patch clamp* experiment, in which current is measured as the test potential is changed. Sodium ion channels, for example, are huge proteins composed of four domains, each with six subdomains traversing the neuronal membrane. Opening of one channel in response to changing voltage is called activation, and allows the passive flow of sodium ions. Activation of the channel is thought to result from a twisting of one transmembrane helix in each domain, moving charged amino acids so that ions can flow across. Fast inactivation can occur from the final open state when a loop folds over the intracellular mouth of the channel like a flap (as described by Hille and Catterall, 1999).

Another component found in the neuronal membrane is an active transport system called the *sodium–potassium pump.* This energy-dependent system simultaneously transports three molecules of sodium (a positively charged ion) out of the cell and two molecules of potassium (also positively charged) into the cell. Some of the potassium that is pumped into the cell will leak back out because the potassium channels in a resting neuron are usually at least partially open. However, the action of the pump plus the attractive force of the negatively charged molecules trapped within the cell still manage to maintain a higher concentration of potassium inside the cell than outside. The sodium ion channels, unlike the potassium ion channels, are nearly all closed in the resting neuron. Because of this, almost all of the sodium that was pumped out of the cell remains outside the cell.

Action Potential

See also:

Effects of Toxicants on Voltage-Activated Ion Channels

Chapter 4

Chloride ions are also unevenly distributed across the neuronal membrane. Chloride is found in much higher concentrations outside than inside the neuron, due to a combination of factors. First of all, the negatively charged chloride ions are repulsed by the other negatively charged ions within the cell. Chloride ions may also be actively transported outward by a chloride pump. The uneven distribution of chloride, as well as that of sodium and potassium, also contributes to the formation of the negative membrane potential (Figure 10.4).

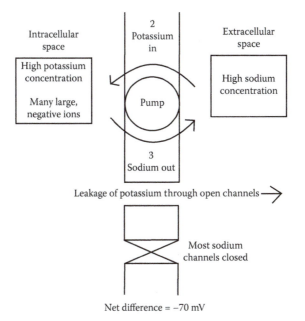

FIGURE 10.4 Resting membrane potential. Differential permeability of the neuronal membrane and the Na^+/K^+ pump create a potential difference or voltage across the membrane.

The resting membrane potential of a neuron changes during the propagation of an electrical impulse, or *action potential* (Figure 10.5). When a neurotransmitter binds to a receptor on a dendrite, it triggers the opening of receptor cation channels (more details on how this happens later on), making the membrane potential in that region less negative. This change in membrane potential then causes the opening of sodium channels (which are, as you remember, voltage-gated) in a chain reaction.

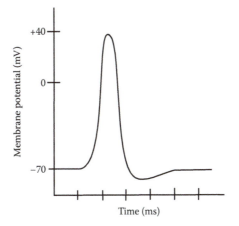

FIGURE 10.5 The change in resting membrane potential that occurs during the action potential.

The action potential occurs when this reaction builds to the point that it becomes self-sustaining. At that point, the *threshold*, the membrane potential rapidly changes from −70 to +30 mV (a process called *depolarization*). This change in potential then spreads rapidly across the entire neuronal membrane.

Almost as soon as a part of the membrane is depolarized, though, the process of *repolarization* begins. Sodium channels close and potassium channels open, allowing positive potassium ions to leave the cell and thus restore the negative membrane potential. The sodium–potassium pump will then restore intra- and extracellular levels of sodium and potassium to their original levels. During the process of repolarization, the neuron is said to be in a *refractory state*, during which it cannot conduct another action potential.

Tetrodotoxin, Saxitoxin

See also:
  Effects of Toxicants on
    Voltage-Activated Ion
    Channels
                    Chapter 4

  Cyanobacterial Toxins
  Dinoflagellates and
    Paralytic Shellfish Toxins
  Mollusks
  Fish, Birds, and Mammals
                   Chapter 20

  TTX, STX
                    Appendix

A number of neurotoxicants can interfere with the propagation of electrical impulses. *Tetrodotoxin* (TTX) is a toxicant of biological origin, found in a number of frogs, fish, and other species, including the blue-ringed octopus of Australia and the puffer fish, which is a popular food in Japan. Tetrodotoxin blocks the generation of an action potential by binding to a site on the outside of the neuronal membrane and blocking the sodium channels. Effects of tetrodotoxin include motor weakness and paresthesias. Higher doses may cause paralysis not only of skeletal (voluntary) muscle, but also of smooth muscle in blood vessels, which may then lead to severe hypotension and circulatory failure. Most tetrodotoxin poisonings occur in people who have eaten improperly prepared puffer fish. Although tetrodotoxin concentration in the fish muscle is relatively low, levels may be high (up to 10 mg/g) in other organs such as the liver. The LD_{50} of tetrodotoxin in mice is around 300 µg/kg.

Another biological toxin that blocks sodium channels is *saxitoxin* (STX). Saxitoxin is produced by organisms called *dinoflagellates* (of the genus *Gonyaulax*, among others). These organisms serve as a food source for various shellfish and fish. Although many fish are killed by exposure to this toxin, many shellfish are not susceptible and can accumulate a milligram or more of the toxin in their tissues. Then, the shellfish may be eaten by unsuspecting humans who may then become ill or perhaps even die. To prevent this, shellfish harvesting may be prohibited in affected areas during periods of dinoflagellate *blooms* (large increases in population).

A third biological toxin is *batrachotoxin* (BTX), found in South American frogs and used as an arrow poison. Although batrachotoxin (like tetrodotoxin and saxitoxin) prevents the passage of nerve impulses, its mechanism of action is somewhat different. Batrachotoxin increases the permeability of the resting neuronal membrane to sodium by preventing closing of the sodium channels. Thus, the membrane potential (which depends in part on uneven distribution of sodium across the membrane) cannot properly develop, and the action potential cannot be created or propagate. Batrachotoxin can act from either the inside or outside of the membrane, which is probably more a reflection of its high degree of lipid solubility than an indication of where on the channel it is binding. Batrachotoxin also modifies the selectivity of the channel (its ability to exclude ions other than sodium). It is extremely toxic: less than 200 mg may be fatal to a human.

Several scorpion and sea anemone toxins have a similar action to batrachotoxin, delaying the closing of sodium channels. These peptides act from the outside of the membrane and can prolong action potential duration from several milliseconds to several seconds by binding strongly to open sodium channels.

Organochlorine Pesticides

See also:
   Material Cycling in
   Ecosystems
                               Chapter 14

   Pesticides
                               Chapter 17

   Organochlorine Pesticides
                               Appendix

The sodium ion channel is also the axonic target of DDT and synthetic pyrethroid insecticides. Resistance to these insecticides was observed in various populations of house flies with several independent genes conferring the resistance. Certain resistant populations were knocked down by exposure to DDT, but recovered; however, other resistant populations were not knocked down due to a single gene on chromosome 3 named *knockdown resistance*, or *kdr*. This type of resistance was traced to a point mutation in the sodium ion channel that changed amino acid 1014 from leucine to phenylalanine. Later, resistance also evolved in the tobacco budworm moth, and it was also traced to the homologous amino acid 1029, but in this case leucine was changed to histidine. Other populations of this moth possessed a mutation of valine to methionine at amino acid 421; depolarization in these moths was less sensitive to permethrin (a pyrethroid) exposure, and voltage gating was altered so that peak activation was delayed until depolarization was reduced 13 mV more than in susceptible moths.

The plant alkaloids *aconitine* and *veratridine* also delay closing of sodium channels. These compounds probably act at different sites, though, than either batrachotoxin or the scorpion or sea anemone toxins.

EFFECTS OF TOXICANTS ON SYNAPTIC FUNCTION

There are two types of synapses in the nervous system: the synapse between two nerve cells and the synapse between a nerve and a muscle cell or gland. Both, however, operate on similar principles.

Where two neurons come together (Figure 10.6), the axonal membrane is termed the *presynaptic membrane*, the dendritic membrane is termed the *postsynaptic membrane*, and the gap which forms between them is termed the synapse. The dendrites and cell body of any one neuron may receive inputs from many axons, and one axon may connect (through multiple branching axon terminals) to many dendrites and cell bodies.

At some synapses, the electrical impulse in the presynaptic neuron is communicated to the postsynaptic membrane directly through electrical current. In most cases, though, communication between the pre- and postsynaptic neurons is chemical in nature. When an electrical impulse reaches the end of the presynaptic axon, the change in membrane potential triggers the release of the chemical messenger, which is the neurotransmitter. The neurotransmitter then diffuses across the synaptic gap and, acting as a *ligand*, binds to the receptor molecules on the postsynaptic

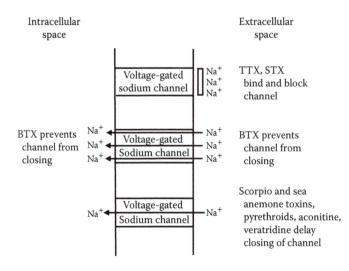

FIGURE 10.6 Actions of toxicants on ion channels in the neuronal membrane.

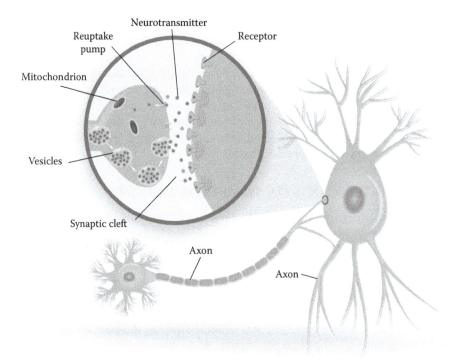

FIGURE 10.7 The synapse, showing release of a neurotransmitter from the presynaptic neuron and binding of the neurotransmitter to receptors on the postsynaptic neuron. (Modified with permission from http://www.shutterstock.com/pic-163236482/stock-vector-structure-of-a-typical-chemical-synapse-neurotransmitter-release-mechanisms-neurotransmitters-are.html?src=&ws=1. Accessed on June 22, 2014.)

membrane (Figure 10.7). This binding affects the membrane potential of the postsynaptic neuron. Neurotransmitters and *neurohormones* affect only neurons containing the corresponding selective receptor protein and thus, they do not excite all neurons.

There are two general types of neuroreceptors: *excitatory* and *inhibitory*. Each is activated by specific neurotransmitter ligands. Binding of neurotransmitters to excitatory neuroreceptors opens ion channels, allowing cations (positive ions such as sodium) to enter the cell, and thus making the membrane potential a little less negative. This change lasts for only a few milliseconds, after which the potential returns to its original level. If enough neurotransmitter molecules are released simultaneously, however, many cation channels will be opened and the effects will add up in a process called *summation*. As more and more cation channels open, the membrane potential becomes less and less negative, until finally the threshold is reached and an action potential may be generated.

Other neurotransmitters bind to inhibitory neuroreceptors. Binding of these neurotransmitters to their receptors opens potassium and chloride channels. This allows more potassium to leave the cell and chloride to enter, making the membrane potential even more negative, and thus preventing the initiation of an action potential.

There are many different neurotransmitter substances in the nervous system. At one time, it was thought that each neuron manufactures and releases only one single type of neurotransmitter; however, this has since been shown to be untrue. A single neuron may also receive inputs from several different types of neurons, each releasing a different neurotransmitter. The results may be either excitatory or inhibitory, but the response is ultimately determined by the numbers and types of specific receptors in the postsynaptic membrane.

Four of the major types of neurotransmitters are *acetylcholine*, the *biogenic amines*, the *amino acids*, and the *neuropeptides*. Lately, evidence has shown that gases, too, can act as neurotransmitters. *Nitric oxide*, for example, was shown to be a neurotransmitter in the mid-1990s, and *carbon monoxide* was identified as one soon after.

Nitric Oxide

See also:
Excitotoxicity
Chapter 10

Since these substances are gases, they cannot be stored in the neuron, but must be manufactured immediately prior to release. Nitric oxide is synthesized by the enzyme *nitric oxide synthase* (of which there are three forms) and has a half-life of only a few seconds (one reason why it was so hard to detect as a neurotransmitter). Carbon monoxide is synthesized by *heme oxygenase*. Also, these neurotransmitters do not bind to receptors in the traditional fashion, but may act by covalently binding to and modifying the activity of target proteins.

Another unusual neurotransmitter is *D-serine*, an amino acid that interacts with the NMDA receptor system. This neurotransmitter is not only unusual in that it has a D-dextrorotary rather than an L-levorotary structure, but it is actually released by glial cells, which were long thought to play little or no role in neurotransmission.

ACETYLCHOLINE

Acetylcholine (ACh) is an important neurotransmitter in the autonomic nervous system (the system that controls involuntary muscle movement). The autonomic nervous system has two branches: the sympathetic and the parasympathetic. The sympathetic and parasympathetic branches control many of the same muscles and glands, but the effects of each branch on those muscles and glands differ greatly. The physiological state of the body reflects the balance between the two influences. Stimulation of the sympathetic branch leads to what is often called the fight-or-flight response: tachycardia (increase in the heart rate), dilation of bronchioles, dilation of the pupil, constriction of peripheral blood vessels, and decrease in digestive activity. Parasympathetic stimulation leads to bradycardia (decrease in the heart rate), constriction of bronchioles, constriction of the pupils, increase in peristalsis (activity of digestive smooth muscle), and increase in secretions.

In both branches, the connection between the central nervous system and the muscle or gland that is to be controlled consists of two neurons. The first neuron (the *preganglionic*) connects the CNS with a group of nerve cells called a *ganglion*, and the second (*postganglionic*) neuron originates in the ganglion and connects with the muscle or gland it controls. Acetylcholine is released by the preganglionic neurons of both branches and by the postganglionic neurons of the parasympathetic branch. Acetylcholine is also released by the neurons that control voluntary muscle movement and is released by neurons in areas of the central nervous system as well (Figure 10.8).

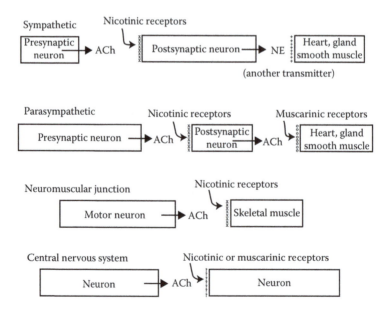

FIGURE 10.8 The neurotransmitter acetylcholine (ACh) and its locations in the nervous system. Types of cholinergic receptors are also shown.

Acetylcholine is synthesized within the neuron from the molecules *acetyl–CoA* and *choline*. The rate of synthesis is dependent on the supply of choline, which is taken up from outside the neuron by a sodium-dependent transport mechanism. A molecule called *hemicholinium* (HC-3) is able to block this transport process, leading to a reduction in acetylcholine levels (Figure 10.9). Acetylcholine is stored in the neuron, perhaps in membrane-bound droplets called vesicles or perhaps elsewhere in the cytoplasm. Release of packets of acetylcholine molecules from the axon is a calcium-dependent process that occurs regularly even when the neuron is resting. A much larger, synchronized release of the neurotransmitter is triggered by the rapid influx of calcium that accompanies action potential-induced changes in the membrane potential.

One of the most deadly toxicants known, *botulinum toxin* (with an LD_{50} in some animals as low as 10 ng/kg), binds to the nerve axon and interferes with the release of acetylcholine (Figure 10.9). For more on toxic and therapeutic effects of botulinum toxin, see the case at the end of this chapter.

Once released from the axon terminal, acetylcholine molecules migrate across the synaptic gap and bind to the acetylcholine receptors found on the postsynaptic membrane in cholinergic synapses. There are two different types of receptors that respond to acetylcholine: *nicotinic receptors* (named on the basis of their response to *nicotine*) and *muscarinic receptors* (named on the basis of their response to the drug *muscarine*). Nicotinic receptors are found on neurons in autonomic ganglia of both branches and on skeletal muscle. Muscarinic receptors are found where neurons of the parasympathetic system connect to smooth muscle and glands (Figure 10.8).

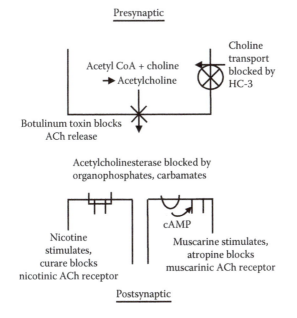

FIGURE 10.9 Sites of action of various toxicants on the cholinergic system. Acetylcholinesterase is tethered to the postsynaptic membrane where it degrades ACh to terminate each chemical signal.

When acetylcholine binds to a nicotinic receptor, it directly opens a cation channel, immediately allowing an influx of cations and making the membrane potential less negative. Action at a muscarinic receptor is somewhat slower. When acetylcholine binds to one of these receptors, a group of proteins called G proteins are activated, initiating a series of biochemical changes (involving second messenger compounds such as cyclic AMP), which then ultimately regulate the opening and closing of calcium and potassium channels.

Nicotine

See also:
Tobacco
Appendix

Compounds such as nicotine and muscarine are cholinergic *agonists* (Figure 10.9). This means that they bind to the nicotinic or muscarinic receptor and mimic the effects of acetylcholine. Because nicotinic receptors are found in both the sympathetic and parasympathetic systems, poisoning with nicotine leads to a mix of sympathetic and parasympathetic symptoms, including increased heart rate, increased blood pressure, and increased gastrointestinal motility and secretion. Nicotinic receptors are also found within the central nervous system and it is the binding of nicotine to these receptors which produces the psychoactive effects (mild euphoria, relaxation) related to nicotine exposure. Human toxicity due to acute exposure to high levels of nicotine is most often seen in children who ingest tobacco products, although it may also occur in workers in the tobacco industry.

Other compounds also bind to the nicotinic or muscarinic receptor, but with different results. Instead of mimicking acetylcholine, they compete with acetylcholine for binding and block the receptor, preventing the initiation of an action potential in the postsynaptic neuron (Figure 10.9). One example of such a nicotinic *antagonist* or *blocker* is *curare*. The term *curare* actually includes a number of different arrow poisons used by South American Indians. These poisons contain a number of alkaloids such as D-*tubocurarine* that bind to and block the nicotinic receptor.

Effects of blocking the nicotinic receptors at the neuromuscular junction include motor weakness and paralysis. Duration is brief, and reversible, but death may occur from paralysis of the diaphragm muscle (which is essential to breathing). Effects that result from blocking of receptors at the ganglia include decreases in blood pressure and heart rate. Because of their effects, these compounds are sometimes used in conjunction with general anesthetics to induce muscle relaxation during surgery.

Acetylcholinesterase

See also:
Serine Hydrolases
Chapter 3

Effects of Toxicants on Enzymes
Chapter 4

Organophosphates
Appendix

Atropine, a muscarinic blocker, is found in the plants *Atropa belladonna* ("Deadly nightshade") and *Datura stramonium* ("Jimson weed"). Known since ancient times, *Atropa* is named for Atropos, the Fate who cuts thread of life; *belladonna* means "beautiful lady," a reflection of the plant's cosmetic use to widen the pupil of the eye. Another muscarinic blocker, *scopolamine*, is found in the plant *Hyoscamus niger* ("Henbane"). Blockade of muscarinic receptors blocks the effect of the parasympathetic neurons on muscles and glands and leads to an autonomic system imbalance in favor of the sympathetic branch. As little as 2.0 mg of atropine in a human can produce tachycardia, dilation of the pupils, dilation of bronchioles, decrease in peristalsis, and decrease in secretions such as saliva. Atropine also has some effects on the central nervous system, producing general excitation with small doses and depression with larger doses. Both atropine and scopolamine can be hallucinogenic. Much larger doses are necessary to affect the CNS than to produce autonomic effects, probably due to the exclusion of the compounds by the blood–brain barrier.

After acetylcholine has been released and binding to the receptors has occurred, there must be a way to inactivate the neurotransmitter. Otherwise, receptors would continue to be stimulated and action potentials would continue to be produced long after the original impulse had passed. In the case of acetylcholine, the remaining molecules are hydrolyzed in a reaction catalyzed by an enzyme called *acetylcholinesterase* (AChase). Hydrolysis of acetylcholine occurs in three steps: reversible binding to the enzyme, acetylation of a serine residue at the enzyme active site (yielding choline), and deacetylation on attack by a hydroxyl ion (yielding acetate). *Organophosphates* (OPs) and *carbamates* are two classes of pesticides that can bind to and inhibit acetylcholinesterase (Figure 10.9). Because inhibition is reversible in the case of carbamates, their action is of short duration. Inhibition by OPs, however, is a different

Organophosphates

See also:
  Serine Hydrolases
                        Chapter 3

  Effects of Toxicants on
    Enzymes
                        Chapter 4

  Acetylcholine
  Axonopathies
                        Chapter 10

  Pesticides
                        Chapter 17

  Organophosphates
                        Appendix

matter. The bond formed between acetylcholinesterase and most OPs is quite stable and only slowly reversible. In fact, recovery from poisoning by some OPs (such as military nerve agents) may depend on synthesis of new acetylcholinesterase.

Effects of OP and carbamate poisoning reflect overstimulation of the parasympathetic nervous system and include slowing of heart rate, constriction of the pupils, bronchoconstriction, and increase in secretions (four classic symptoms are salivation, lacrimation, urination, and defecation). Overstimulation at the neuromuscular junction produces twitching and cramps; central effects include anxiety, restlessness, and confusion that can lead to coma. Death is usually by respiratory failure brought on by paralysis of respiratory muscles and inhibition of the central nervous system centers that control respiration.

The oral LD_{50} for organophosphates ranges from one to several thousand milligrams per kilogram. Antidotal treatment is with *atropine*, which blocks muscarinic receptors. In addition, pralidoxime (2-PAM) helps accelerate the reversal of acetylcholinesterase inhibition.

BIOGENIC AMINES

A second major group of neurotransmitters and neurohormones is the biogenic amines, which includes the neurotransmitters *norepinephrine, epinephrine, dopamine, serotonin,* and *histamine* (Figure 10.10). Norepinephrine is the neurotransmitter

FIGURE 10.10   The biogenic amine neurotransmitters and their locations in the nervous system.

released by the postganglionic neurons of the sympathetic nervous system. It is also released by some neurons of the central nervous system. Many of these norepineph-rine-releasing neurons (as well as many neurons that release epinephrine, many that release dopamine, and many that release serotonin) originate in the medulla, pons, or midbrain (a grouping of areas that is commonly called the brain stem) and lead to many other areas of the brain. Histamine-releasing neurons are found in highest concentrations in the hypothalamus.

Norepinephrine, epinephrine, and dopamine together are known as the *catechol-amines*. Toxicants can interfere with this group of neurotransmitters through many of the mechanisms we have already discussed (Figure 10.11). The drug *reserpine*, for example, interferes with the storage of biogenic amines in the axon, leading to a shortage of these neurotransmitters. This results in a decrease in sympathetic activ-ity (producing slowing of the heart rate), an increase in digestive activity, and an increase in secretion. *Amphetamine*, on the other hand, exerts part of its stimulatory effects through promoting increased release of norepinephrine.

As is the case with acetylcholine, there are two major categories of receptors that respond to the catecholamines: *alpha and beta adrenergic receptors*. They can be differentiated on the basis of their relative sensitivities to norepinephrine and epinephrine. Alpha and beta receptors can be further subdivided into alpha one and alpha two, beta one and beta two populations. Alpha one receptors are found on

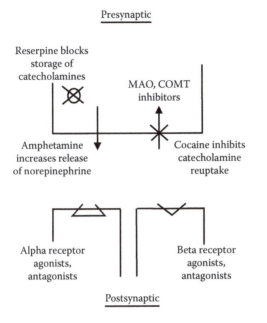

FIGURE 10.11   Effects of various toxicants on the biogenic amines.

smooth muscles and glands, while alpha two receptors are thought to be involved in feedback inhibition of various neurons throughout the nervous system. Beta one receptors are found in heart tissue; beta two receptors (like alpha one receptors) are found on smooth muscle and glands.

Many compounds interact with catecholamine receptors, some as agonists (stimulating the sympathetic nervous system) and some as antagonists (depressing the sympathetic nervous system) (Figure 10.11). Many nasal decongestants are alpha agonists that work by constricting blood vessels of the nose. Alpha agonists can also be used to treat hypotension (low blood pressure), such as might accompany shock. Many beta two agonists are extremely useful in widening bronchial airways constricted by asthma; because these compounds are specific for beta two and do not stimulate beta one receptors, there are no cardiac side effects, such as increased heart rate. Amphetamine stimulates both alpha and beta receptors and, of course, is a CNS stimulant as well.

Several drugs act as alpha blockers, producing decreases in blood pressure, increases in activity of gastrointestinal muscles, and nasal stuffiness due to dilation of blood vessels. One group of compounds that interacts with alpha receptors in a complex manner are the *ergot alkaloids*. These compounds are produced by the fungus *Claviceps purpurea*, which grows on rye and other grains. Its effects have been known for centuries, and epidemics of ergot poisoning were fairly common during the Middle Ages. Some ergot alkaloids act as partial agonists and others as antagonists of alpha receptors. Ergot alkaloids stimulate smooth muscle contraction, and one of the predominant symptoms of ergotism was gangrene of the extremities caused by constriction of the smooth muscle of blood vessels in the arms and legs.

Spontaneous abortion due to stimulation of uterine muscle was also common. These drugs are still used in the treatment of migraine (a condition which may result in part from increases in blood flow in cranial arteries).

Beta receptor blocking drugs such as *propranolol* are widely used to manage cardiovascular disorders due to their ability to decrease both heart rate and blood pressure. Other drugs have been developed recently that are specific for beta one receptors, thus eliminating the side effect of bronchoconstriction, which would occur if beta two receptors are also blocked.

Cocaine

See also:
Major Categories of Illegal Drugs: Neuroactive Drugs
Chapter 16

Deactivation of the catecholamines following their release from the axon is achieved through a different mechanism than inactivation of acetylcholine. Instead of being enzymatically broken down, catecholamines are returned to the axon through a reuptake mechanism. Catecholamine reuptake requires sodium and potassium and is energy-dependent. Inhibition of reuptake (by *cocaine*, for example) may lead to overstimulation of the postsynaptic neuron. Cocaine also stimulates the release of dopamine, and the two mechanisms together combine to produce the euphoria and other psychoactive effects associated with cocaine exposure. Cocaine, of course, also has toxic effects on the cardiovascular system.

Monoamine Oxidase

See also:
Other Enzymes Carrying Out Oxidation
Chapter 3

Within the axon, catecholamines can be broken down by the two enzymes monoamine oxidase (MAO) and catechol-*O*-methyltransferase (COMT) (Figure 10.11). Inhibitors of these enzymes (for example, the antidepressant drug *chlorpromazine*) can increase levels of catecholamines in the brain. There are two forms of MAO—MAO-A and MAO-B—which differ in their substrate specificity and distribution in the body. MAO-A acts on serotonin, epinephrine, and norepinephrine, whereas MAO-B acts on dopamine as well as other compounds.

Nonselective MAO inhibitors (inhibitors that block both MAO-A and MAO-B), as well as selective MAO-A inhibitors, have a potentially dangerous side effect. MAO in the gastrointestinal system is responsible for the metabolism of *tyramine*, an amine found in foods such as cheese, wine, cured meats, and chocolate. Tyramine interacts with the sympathetic nervous system to produce a sharp increase in blood pressure, and elevated levels of tyramine can lead to symptoms ranging from headache to a potentially lethal hypertensive crisis. Since tyramine is predominantly metabolized by MAO-A, selective MAO-B inhibitors are less likely to cause this problem, which is often termed the "cheese reaction."

As for effects of toxicants on the remaining biogenic amines, histamine and serotonin, little is known. There are specific receptors for both serotonin and histamine in the brain, and there is conflicting evidence as to whether some hallucinogenic drugs such as LSD may interact with serotonin receptors. Like catecholamines, serotonin is inactivated by reuptake, but through a separate reuptake mechanism. No reuptake process for histamine has been found.

Amino Acid Neurotransmitters

Probably the most significant amino acid neurotransmitter is *gamma-aminobutyric acid* (GABA). GABA is an inhibitory transmitter, produced by neurons throughout the nervous system and acting only at the inhibitory GABA receptor and chloride ion channel. It is found in particularly high concentrations within the cerebellum and spinal cord, and in cells originating in the hippocampus and leading to the midbrain. Like most neurotransmitters, GABA is manufactured and stored in the presynaptic neuron and, following release, binds to a postsynaptic receptor. GABA binding triggers an increase in chloride permeability, allowing chloride to enter the neuron and making the membrane potential more negative. Reuptake of GABA occurs in both the presynaptic neuron and nearby glial cells.

GABAergic pathways seem to be important in control of emotions. *Benzodiazepines* (antianxiety drugs better known by trade names such as Valium® and Librium®) interact with GABA through a mechanism that is not yet clear. Although they seem to mimic the effects of GABA, they are ineffective if GABA itself is not present. *Picrotoxin*, a powerful stimulant derived from the seeds of an East Indian plant, antagonizes the effects of GABA and also blocks the action of benzodiazepines. The inhibitory effects of GABA may also be important in motor control. Loss of GABAergic neurons occurs in the genetic neurodegenerative disorder *Huntington's disease* and may be responsible for the involuntary movements (*chorea*) characteristic of the disease. The cyclodiene insecticides *dieldrin* and *chlordane* also block the GABA receptor.

Another inhibitory transmitter is the amino acid *glycine*. Glycine acts primarily in the brain stem and spinal cord. *Tetanus toxin* binds to presynaptic membranes and prevents release of glycine, while *strychnine* (lethal dose in humans of 1 mg/kg), a powerful convulsant, binds to and blocks the postsynaptic glycine receptor. Antagonism of glycine's inhibitory effect leads to the sustained muscle contraction characteristic of both toxicants.

Glutamate and *aspartate* are excitatory amino acids. *Kainic acid* is a glutamate agonist that kills neurons, as can glutamate itself in large concentrations. Mechanisms of neuronal death due to *excitotoxic* effects of glutamine and related amino acids will be discussed later in this chapter.

Neuroactive Peptides

The neuropeptides differ in several ways from the other transmitters in the nervous system. Neuropeptides act at much lower concentrations, and their actions last longer. In addition, neuropeptides are generally made in the neuronal cell body rather than in the axon. Like other neurotransmitters, neuropeptides may affect membrane potential, or they may be released along with a neurotransmitter and alter its release or binding.

There are probably 50–100 neuropeptides (Table 10.1); one of the better known groups is the *opioid* peptides. This category includes *enkephalins* and *endorphins*. Several different opioid receptors occur throughout the central nervous system and may have inhibitory effects on pathways involved in the transmission of pain impulses. In at least one type of receptor, binding of the peptide to the receptor is

TABLE 10.1

Some Types of Neuroactive Peptides, Categorized by Localization in the Body

Gastrointestinal-related peptides (most found in the brain and in neurons innervating the gastrointestinal tract)	VIP, CCK, substance P, enkephalins, gastrin, neurotensin, insulin, glucagons, bombesin
Hypothalamic-releasing hormones	TRH, LHRH, GH, GHRH, somatostatin
Pituitary peptides	ACTH, endorphin, alpha MSH, prolactin, LH, GH, thyrotropin
Others	Angiotensin II, bradykinin, oxytocin, vasopressin

linked to the opening of potassium channels through the mediation of a second messenger, cAMP. Other receptors may act presynaptically, controlling calcium channels to decrease release of a neurotransmitter.

The opioid peptides are named for *opium*, a drug derived from the juice of the opium poppy. Opium itself consists of many different compounds, including *morphine* and *codeine*, both of which, although they are not peptides, interact with opioid receptors. *Heroin* is a chemical derivative of morphine. Although a few milligrams of morphine or related opioid agonists produce the clinically useful effects of drowsiness and pain relief, they also produce euphoria and have an equally significant history of recreational usage. Tolerance to these drugs develops with continued usage: in other words, progressively higher doses are necessary to produce the same physiological effects. *Naloxone* is an opioid receptor antagonist that can block effects of both endogenous peptides and opioid drugs.

AXONOPATHIES

Another potentially vulnerable part of the neuron is the axon. The axon projects from the cell body and has a *proximal* section (nearest the cell body) and a *distal* section (containing the end of the axon, or *axon terminal*). These terms are relative: there is no distinct point of division between the two regions.

Axonopathies, damages to the axon, are most common in the peripheral nervous system, and the resulting sensory and motor dysfunction is often referred to as a neuropathy. Axonopathies are generally categorized as either proximal or distal. Figure 10.12 shows the potential targets for the production of axonopathies. Typically, the part of an axon that is distal to axonal damage eventually "dies back," in a process called Wallerian degeneration. Peripheral axonopathies may be, at least in part, reversible; conditions in the CNS are not conducive to regeneration of damaged axons. The inability of CNS neurons to regenerate is, of course, an important area of research, and evidence indicates that a combination of inhibitors of neural growth and key differences between CNS and PNS glial cells play a role.

Axon Transport Systems

Unlike the cell body, the axon has limited metabolic capabilities. Most of the molecules that are needed in the axon must be made in the cell body, which therefore needs a high synthetic capacity. These molecules then must be transported down the

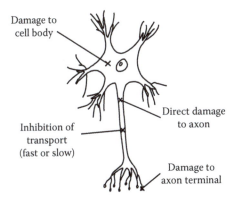

Damage to
cell body

Direct damage
to axon

Inhibition of
transport
(fast or slow)

Damage to
axon terminal

FIGURE 10.12   Potential targets for production of axonopathies.

axon, often traveling considerable distances (several feet for some motor neurons of the spinal cord). Transport originating in the cell body and moving down the axon is called *anterograde transport*. Other materials may be returned to the cell body from the axon, in a movement called *retrograde transport*.

There are two types of systems involved in axonal transport. The first system is called *fast axonal transport*. Fast transport moves substances at the rapid rate of about 400 mm/day and can be either anterograde or retrograde in direction. Fast transport distributes membrane components, mitochondria, and other cellular structures from the cell body to the axon and also returns used membrane components to the cell body for recycling. Fast transport can also move substances absorbed at the axon terminal up to the cell body. Some viruses, including herpes viruses and toxins such as *tetanus toxin*, are thought to enter the cell body in this manner.

Studies of the molecular mechanisms of the fast transport processes are often carried out by radiolabeling molecules and monitoring their progress down the axon. Researchers have also used techniques such as transfecting cells with a gene that produces neurofilament subunits that are tagged with *green fluorescent protein* (GFP). These studies show that the mechanism of fast transport involves some or all of the structural elements of the neuron: the microtubules, microfilaments, and neurofilaments. According to one current theory, microtubules form a track along the axon. Materials to be moved in an anterograde direction are attached to and pulled along the track by molecules called *kinesins*; retrograde transport appears to involve the protein *dynein*. The molecular details of how materials to be transported actually interact with the kinesin and dynein motors, as well as the details of how the motors interact with the tracks, remain to be completely explained.

The second type of transport is *slow transport* or *axoplasmic flow*. Slow transport seems to have two components, one of which moves at the rate of 1 mm/day and the second at 2–4 mm/day. Slow transport is strictly anterograde in direction, with no retrograde motion. Most axonal proteins are moved by slow transport, including many enzymes as well as structural proteins such as *microtubules, microfilaments*, and *neurofilaments*. The mechanism of slow transport is not well understood: new material seems to displace that material in front in a process that has been compared to toothpaste moving through the tube. Evidence does indicate that slow transport depends on phosphorylation of nucleosides, with kinases and phosphatases playing a role.

A third, intermediate transport process also exists, with a rate of around 55 mm/day. This transport process is responsible for transport of organelles such as mitochondria.

Evidence is mounting that various disease states, as well as toxicant-induced injuries, may, in fact, impact axon transport systems. Mutations in genes for proteins involved in transport systems produce conditions such as *Charcot–Marie–Tooth disease* and *hereditary spastic paraplegia*, both of which are characterized by sensory and motor dysfunction. Recently, some of the proteins apparently involved in *Alzheimer's disease* (including *amyloid precursor protein*, the *ApoE4* protein, and *tau protein*) have been shown to interact with proteins of the transport systems, and swelling and abnormal accumulation of transport-associated proteins have been described in mouse models as well as in affected humans. *Huntingtin*, the protein that is mutated in Huntington's disease, has also been shown to be transported, and deletion of the gene for huntingtin in *Drosophila* produced deficits in axonal transport in affected flies.

PROXIMAL AXONOPATHIES

Proximal axonopathies are characterized by a swelling of the proximal axon (called a *giant axonal swelling*). A synthetic aminonitrile compound, *IDPN*, is one of only a few chemicals known to produce a proximal axonopathy. Exposure to IDPN leads to an accumulation of neurofilaments and results in the giant axonal swelling. This effect may be produced by a blockade of slow transport, but the molecular mechanisms are not clear. Following the development of the swelling, the distal portions of the axon, deprived of necessary structural proteins, then atrophy. Breakdown of myelin (which will be discussed later in this chapter) may also occur around the swelling, probably as a result of the axonal disruption. Giant axonal swellings are also associated with the neurological disease *amyotrophic lateral sclerosis* (ALS). The cause of this disease is uncertain, but some evidence has indicated that environmental factors may be involved.

DISTAL AXONOPATHIES

Distal axonopathies involve pathological changes in the distal portions of axons. This damage varies with the toxicant producing the change, but may include swelling, damage to mitochondria, and accumulation of neurofilaments. Like proximal axonopathies, distal axonopathies are also often accompanied by disintegration of myelin. Because damage often appears in the axon terminal first, these axonopathies are also sometimes called *dying-back neuropathies*. Distal axonopathies may follow a single exposure to some toxicants or may be a result of chronic exposure. Even following a single acute exposure, though, the actual onset of symptoms is unlikely to occur sooner than a week following the exposure.

There are several hypotheses concerning the cause of distal axonopathies. One possibility is damage to the cell body of the neuron. If synthesis of cell products such as structural proteins is affected, any shortfall in availability of these products is likely to be found first in the more distal portions of the axon (the end of the line for axonal transport). However, cell body damage has been observed in only a few studies and may well be the result and not the cause of the axonopathy. In addition, peripheral neurons may regenerate following axonopathy, an event that would be unlikely if the neuronal cell body itself was damaged. An alternative hypothesis is that axonopathy is due to damage to the axon itself, perhaps initiating at the axon terminal. Finally, effects on transport mechanism(s) may be the cause. It is also possible that there is no one single cause, but rather different toxicants may act through different mechanisms.

Among the toxicants that produce distal axonopathies are a group of compounds already discussed, the *organophosphates*. In the early 1920s and 1930s, neuropathies were reported in tuberculosis patients treated with the organophosphate compound *phosphocreosote*. Also during that period, cases of *Ginger Jake paralysis* were traced to ingestion of ginger extract still containing traces of the organophosphates used to make the extract. The neuropathy may occur following either acute exposure to high levels or chronic exposure to lower levels. Symptoms begin 1–2 weeks after acute exposure and typically include weakness and perhaps even paralysis of the lower limbs. Recovery is slow and seldom complete. Not all species are sensitive to organophosphate-induced neuropathies; those that are sensitive include humans, cats, sheep, and many birds (much experimental work is done using chickens as models).

Organophosphates

See also:
  Serine Hydrolases
                              Chapter 3

  Effects of Toxicants on Enzymes
                              Chapter 4

  Acetylcholine
                              Chapter 10

  Pesticides
                              Chapter 17

  Organophosphates
                              Appendix

The mechanism of action of organophosphate axonopathy has been clarified in recent years. Since axonopathy is produced by some (but not all) of the acetylcholinesterase-inhibiting organophosphates, many researchers believed that inhibition of a related enzyme (a different esterase) might be involved, and a proposed target molecule was identified. This esterase, termed *neuropathy target esterase* (NTE), seems to play a role in neural development, as mice that are NTE–/– (in other words, that have been genetically engineered to lack NTE expression) do not survive embryonic development. The mechanism by which inhibition of NTE may produce axonopathy, however, is still not understood. However, it appears that "aging" (dealkylation of the phosphoryl group of the enzyme-bound organophosphate) may play a role in inhibition of NTE in a similar manner as it does with acetylcholinesterase. Thus, binding to NTE in and of itself may not be sufficient to produce neuropathy. In fact, administration of non-aging OPs prior to (but not after) administration of an OP which undergoes aging can be protective against the neuropathy.

Another compound that produces distal axonopathy is *acrylamide*. Acrylamide is a monomer that can be polymerized to form the polyacrylamide gels commonly used in the laboratory for separation of proteins (it is nontoxic in its polymeric form). It appears to act primarily by inhibiting fast axonal transport. The gamma diketone *2,5-hexanedione* (a metabolite of the solvents *n-hexane and methyl n-butyl ketone*) also produces distal axonopathy. Chronic exposure to these solvents, used in glues and cleaning fluids, leads to accumulation of neurofilaments in the distal portions of the axon (similar to giant axonal swelling, but in more distal portions of the axon), followed by disruption of myelin. The likely mechanism of action is binding of these compounds with amine groups on neurofilament proteins to form *pyrroles*; oxidation of the pyrroles then leads to cross-linking and neurofilament accumulation. The sensory disturbances and motor weakness that accompany this are sometimes called *glue sniffer's neuropathy*. If the source of exposure is removed, recovery is generally complete in mild cases. Some impairment may remain in severe cases, however, perhaps due to involvement of central nervous system neurons. *Carbon disulfide* produces a similar neuropathy, most likely also through cross-linking of neurofilaments. Carbon disulfide is also involved in the production of various neurobehavioral effects, something that will be discussed later in this chapter.

MYELINOPATHIES

The axons of many neurons in both the central and peripheral nervous systems are covered by an insulating substance called *myelin*. In the central nervous system, myelin is formed when an oligodendroglial cell sends out a process that wraps tightly around a segment of the neuronal axon (Figure 10.13). The myelin-forming cells in the peripheral nervous system are called Schwann cells. Unlike oligodendroglial cells, which can only wrap one axon, one Schwann cell may send out several processes and contribute segments of myelin to many different axons.

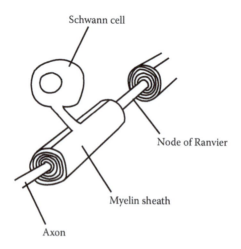

FIGURE 10.13 A myelinated axon in the peripheral nervous system.

In both the central and peripheral nervous systems, there are gaps between myelinated segments of an axon. These gaps are called *nodes of Ranvier*. Electrical conduction in myelinated axons is fundamentally the same as in unmyelinated axons, except that changes in membrane potential occur only at the nodes. Thus, the impulse jumps from one node to the next in a process called *saltatory conduction* (Figure 10.14). Saltatory conduction is both faster and more efficient than the continuous conduction that occurs in unmyelinated neurons.

Not surprisingly, the composition of the myelin sheath is similar to that of a typical cell membrane. Myelin contains proteins such as *myelin basic protein* and *proteolipid protein* in the CNS and *P1 protein* in the PNS, but consists mostly of lipids, including cholesterol, phospholipids, and galactosphingolipid.

Damage to myelin interferes with normal function of the nervous system in much the same ways as damage to axons or interference with electrical conduction. Damage to myelin may block conduction completely or may delay or reduce the amplitude of action potentials (perhaps through transition from saltatory to continuous conduction or through increase in the length of the refractory period). Symptoms of this include numbness, weakness, and paralysis. In addition, action potentials may arise spontaneously in demyelinated neurons, causing paresthesias. Remyelination of demyelinated axons, however, can and does occur in many situations, particularly in the PNS (and to a limited extent in the CNS).

Some toxicants appear to produce myelinopathy through direct effects on myelin rather than damage to axons or oligodendroglial and Schwann cells. More than a thousand people were exposed to one such compound, *triethyltin* (TET), in France in 1954, through contamination of a supposed antibacterial preparation. Triethyltin is highly lipid-soluble and also binds directly to sites within the myelin. Dosages as low as 6 mg/kg lead to splitting of the myelin sheath and production of large fluid-filled vacuoles within the myelin. (This accumulation of fluids, you may recall, is called edema.)

Another highly lipid-soluble compound with strikingly similar effects to triethyltin is *hexachlorophene* (HCP). Hexachlorophene is an excellent antibacterial agent and was at one time widely used until studies on premature infants washed with hexachlorophene showed that some had suffered damage to myelin in both the central and peripheral nervous systems. Hexachlorophene partitions into the myelin, probably disrupting ion distribution, and perhaps interfering with energy management through uncoupling oxidative phosphorylation. Effects of hexachlorophene are generally reversible, as are effects of triethyltin.

Action potential ⟶

FIGURE 10.14   Saltatory conduction in a myelinated axon.

Although *lead* causes demyelination, its effects are limited to the peripheral nervous system. In addition, the mechanism of the demyelination is probably quite different than with hexachlorophene or triethyltin.

Swelling and other morphological changes are seen in the Schwann cells along with damage to myelin, indicating that lead is probably damaging the Schwann cell itself.

A human disease of unknown etiology that affects myelin is *multiple sclerosis* (MS). This disease usually strikes young adults between 20 and 40 years of age, with symptoms of sensory disturbances and motor weakness. Pathological changes occur within the central nervous system and include degeneration of myelin and its replacement by plaques of astrocytes (another type of glial cell). The disease is cyclic, with relapses and improvements probably coinciding with periods of demyelination and remyelination.

Evidence has indicated that MS is an autoimmune disease and that damage to myelin is secondary to activation of immune system cells and release of cytokines that trigger inflammation. Attempts have been made to identify an underlying cause for the disease, and hypotheses have centered on early viral infections that may initiate autoimmune reactions many years down the road. Most recently a retrovirus, *MSRV*, has been isolated from cells and plasma of MS patients. However, the virus is also found in patients with other inflammatory diseases of the nervous system, so it is most likely not the sole causative factor of the disease. There is also evidence for a genetic component involved in predisposition toward acquiring the disease.

EFFECTS OF TOXICANTS DIRECTLY ON NEURONS AND GLIAL CELLS

Finally, some neurotoxicants exert their effects directly on the cell bodies of neurons and glial cells. Neurons that have been damaged by toxicants often show structural changes such as swelling and breakdown of organelles like rough endoplasmic reticulum and mitochondria or damage to synaptic membranes. Chronic exposure to some neurotoxicants results in accumulations of filaments that are then called neurofibrillary tangles. These tangles also result from some diseases such as Alzheimer's. Functional changes in the damaged neuron can include decreases in protein synthesis and oxidative metabolism. These changes may then affect the ability of the neuron to transmit impulses and may ultimately lead to cell death.

Excitotoxicity

One cytotoxic neurotoxicant is the excitatory neurotransmitter *glutamate*. Although glutamate is a normally occurring endogenous neurotransmitter, overstimulation of the glutamate system seems to be associated with a number of toxic and disease states. Glutamate from external sources can have effects also. Low levels of *monosodium glutamate* (MSG), a popular food additive, may produce in some people a group of symptoms often called Chinese restaurant syndrome. These sensitive individuals may react to as little as 1–2 g of MSG with burning or tingling sensations in the upper body, and occasionally even chest discomfort. These symptoms are possibly produced by interaction of glutamate with the peripheral nervous system.

However, it is exposure to high levels of endogenously produced glutamate that has the potential to produce severe central nervous system effects. The damage resulting from overactivation of the glutamate system has been termed excitotoxicity, and the mechanism behind it has been a major area of research for the last several years.

The actual mechanism by which glutamate damages cells and induces cell death is not completely understood. There are a variety of receptors in the CNS that respond to glutamate, but the main type of receptor that seems to be involved with excitotoxicity is the *NMDA receptor* (named for the fact that the receptors respond to *N*-methyl-D-aspartate, an amino acid used in the laboratory). Studies indicate that there are most likely several subtypes of NMDA receptors and that they probably are important in both development and learning and memory. NMDA blockers can prevent development of excitotoxicity in laboratory animals, but have proved to be toxic themselves in humans.

The NMDA receptor acts as an ion channel for Na^+, K^+, and Ca^{2+} ions, and activation requires simultaneous binding of glutamate and another neurotransmitter, glycine. Calcium enters the cell following activation and binds to the regulatory protein *calmodulin*. Calmodulin then activates the enzyme nitric oxide synthase, leading to the production of nitric oxide. The nitric oxide diffuses into nearby neurons, binding to and activating the enzyme *guanylyl cyclase*, which catalyzes the formation of the second messenger *cyclic GMP*. The entry of calcium appears to be important in the mechanism of excitotoxicity, since reduction in extracellular Ca^+ levels is somewhat protective. Likewise, nitric oxide synthase inhibitors offer partial protection against excitotoxicity, supporting a role for nitric oxide in the process.

Nitric Oxide

See also:
Effects of Toxicants on
Synaptic Function
Chapter 10

There is evidence that whether a cell ultimately undergoes apoptosis or necrosis following excitotoxicity may depend on the time course of the event—glutamate overload over a brief period may rapidly deplete ATP and send the cell into necrosis, while a more gradual overload might allow the cell to maintain energy integrity long enough to execute apoptosis.

There are other compounds that also produce excitotoxicity. Kainic acid, a glutamate agonist isolated originally from seaweed, probably acts in a manner similar to that of NMDA, but binds to a different subset of glutamate receptors, the *kainate receptor*. A third type of glutamate receptor, the *AMPA receptor*, may also be important in neurotoxicity.

Exposure to the neurotoxin and animal procarcinogen *cycasin* and to *b-N-methylamino-l-alanine* (BMAA), an excitatory amino acid found in cycad seed, may be associated with development of the neurodegenerative disease *amyotrophic lateral sclerosis–Parkinsonism dementia* (ALS/PD) on the island of Guam. Known as lytico-bodig, the disease is common only among the Chamorro people and has only been seen in individuals who were born prior to 1935. Affected individuals may suffer from Parkinson-like symptoms (rigidity, tremor, and sometimes an Alzheimer-like dementia), ALS-like symptoms (gradual loss of motor function), or some combination of the two.

In the absence of any evidence for a genetic link, exposure to environmental factors such as BMAA has been suggested as the cause of lytico-bodig. The seed of the cycad plant in which BMAA is found has long been used by the Chamorro people to make flour. This practice was particularly common during the Japanese occupation of Guam in World War II, when other sources of flour were in short supply. However, although BMAA has been clearly proven to be neurotoxic in laboratory animals, concentrations of BMAA in cycad flour as prepared on Guam have been measured and have been found to be far short of the dose necessary to produce that neurotoxicity.

Recently, a hypothesis was suggested that might explain this apparent discrepancy. Flying foxes were another food source during World War II, and it is possible that BMAA *bioaccumulated* in the animal (which commonly fed on cycad seeds) and was then passed on to the humans who ate it. This would also explain the disappearance of lytico-bodig, as flying foxes were hunted to extinction and are no longer part of the Chamorro diet. Recent studies have supported this hypothesis, demonstrating that museum specimens of flying foxes do, in fact, contain sufficiently high levels of BMAA to produce neurotoxicity if eaten.

Finally, excitotoxicity such as is produced by glutamate and other compounds may also be involved in the development of damage produced by strokes and trauma. These conditions can lead to an increased accumulation of glutamate in the extracellular space, perhaps through calcium-related release from neurons, or through impact on the glutamate reuptake transporter. This excess glutamate can then interact with receptors to trigger the excitotoxic response.

OTHER CYTOTOXIC COMPOUNDS

Some neurotoxicants damage neurons in very specific areas of the brain. One example is the organometal *trimethyltin* (TMT), which kills neurons in the hippocampus (a region of the cerebrum) and surrounding areas. The hippocampus plays a role in acquisition of memories and is also part of the limbic system, which is important in emotional response. Animals treated with trimethyltin show behavioral changes consistent with disruption of these functions: they are quite aggressive, and there is evidence that memory-related processes may also be affected.

Another specific neurotoxicant is *MPTP*, a contaminant of a synthetic heroin-like drug of abuse called MPPP. MPTP neurotoxicity in humans was first reported by a group of doctors in 1982 who traced the cause of puzzling Parkinson's disease-like symptoms in young adult patients to their exposure to MPTP. In Parkinson's disease, normally a disease of the elderly, neurons are gradually lost from an area of the mid-brain called the *substantia nigra*. These dopamine-producing neurons release their neurotransmitter onto neurons in the basal ganglia (an area of the brain important in movement), and the dopamine deficiency that accompanies their loss results in symptoms including tremor, slow movement, and rigidity.

MPTP not only produces symptoms that are virtually identical to Parkinson's disease, but also appears to act in the same manner, killing cells in the substantia nigra. Because of its lipid solubility, MPTP enters the brain easily, where it accumulates in the affected neurons. MPTP itself is not neurotoxic, but it is metabolized by the enzyme MAO-B (which, as you may recall, breaks down catecholamines) to form a toxic derivative, MPP$^+$, which is taken up by dopaminergic neurons. MPP$^+$ is an inhibitor of oxidative phosphorylation, blocking the electron transport chain, and may also contribute to cell damage through production of free radicals. Further research on MPTP may help the search for causes and treatment of Parkinson's disease.

Oxidative Phosphorylation

See also:
 Effects of Toxicants on
 Enzymes
                                                         Chapter 4

There are some candidate genes that may be involved in what seem to be familial forms of Parkinsonism, however most hypotheses concerning the cause of the disease have centered on environmental agents. MPTP is structurally quite similar to the herbicide *paraquat*, and at least one intriguing study has found a higher incidence of Parkinson's disease among those with high pesticide exposures. Other hypotheses have targeted solvent exposure as a risk factor. And recently, not only paraquat, but also *rotenone, maneb*, and other inhibitors of oxidative phosphorylation have been shown to produce aspects of Parkinson's disease in laboratory animals. Interestingly, nicotine and caffeine have proved to be protective against development of the condition both in animal models and epidemiological studies.

Other toxicants such as *carbon disulfide* also act in part through direct destruction of CNS neurons.

GLIAL CELLS

Effects of toxicants on glial cells, which were at one time discounted in terms of relevance, are now being studied much more thoroughly as the importance of these cells in neurological functioning has become more clearer. For example, ammonia accumulation (generally secondary to liver damage) not only acts on neurons, but can also lead to swelling of astrocytes, thus affecting glutamate levels (metabolism of glutamate takes place in these cells). Dinitrobenzene is another chemical that targets astrocytes, possibly through mitochondrial effects. The list of compounds targeting astrocytes and other glial cells is growing, and it is certain that many more examples of glial toxicity will be elucidated in the future.

OTHER NEUROTOXICANTS

There are many other neurotoxicants, several with mechanisms of action that are not fully understood. Heavy metals such as *organic* and *inorganic mercury* compounds are neurotoxic. Occupational exposure to inorganic mercury in the hat industry several hundred years ago gave rise to the phrase "mad as a hatter," as well as serving as inspiration for the character of the Mad Hatter in Lewis Carroll's *Alice's Adventures in Wonderland*. Symptoms of mercury poisoning may include depression, moodiness, insomnia, and confusion, as well as tremors. Exposure to organic mercury compounds such as *methylmercury* produces tremors, motor dysfunction, and sensory disturbances. A major case of human exposure to methylmercury occurred in Minamata, Japan, and is discussed elsewhere in this book.

Another neurotoxic metal is lead. Aside from the peripheral effects already discussed, lead also has effects on the central nervous system. These effects are collectively termed *lead encephalopathy*. Lead produces excitation of the central nervous system, leading to insomnia, restlessness, irritability, and convulsions. Children are more susceptible than adults, and recovery is in general not complete. Despite decades of research on the metal, the mechanisms behind these effects are still not clear.

Another example of a class of neurotoxicants is the *halogenated hydrocarbon solvents*. Because of their lipid solubility, they enter the central nervous system easily. Exposure to these solvents typically produces disorientation, euphoria, and confusion—symptoms that are reversible upon termination of exposure. Higher concentrations can lead to death, usually from depression of brain areas that stimulate respiration.

Other solvents, including aromatic solvents (*benzene, toluene*), and *alcohols* have similar effects. Several of these toxicants pose significant hazards as drugs of abuse. Toluene, for example, is used as a solvent in paints, glues, and other household products, and may be abused by glue sniffers. Ethanol is perhaps the most widely used drug in this country. Ethanol is a central nervous system depressant, producing depression of inhibitions and mild euphoria at low levels (blood alcohol levels of 0.1%), but leading to impairment of reflexes, decreased sensory function, and loss of consciousness, coma, and even death at high levels (0.4%–0.5%). The mechanism by which alcohol and other solvents produce their effects is not well understood, but probably involves actions on membranes and membrane fluidity. Effects of membrane changes on ion channels have also been implicated.

EFFECTS ON SPECIAL SENSORY ORGANS

Many toxicants that affect peripheral neurons can indirectly affect sensory function, but there are some toxicants that affect specialized sensory organs, such as the eye or ear, directly. *Methanol*, for example, produces edema specifically in the optic nerve, leading to blindness (this is the origin of the phrase "blind drunk"). Some excitatory amino acid neurotransmitters, in addition to their other effects, damage cells in the retina. Finally, a number of compounds, including *2,4-DNP*, *corticosteroids*, and *naphthalene*, can produce reductions in transparency of the lens, a condition commonly known as a cataract.

A few chemicals directly affect the ear. High doses of the antibiotic *streptomycin* can produce dizziness and hearing loss as a result of damage to the vestibular apparatus (which regulates balance) and the cochlea (the organ in the inner ear that responds to pressure waves and sends sensory signals to the brain). (Excessive exposure to noise, of course, can also damage the cochlea and lead to significant hearing loss.) *Aspirin* may produce a temporary hearing impairment characterized by tinnitus (ringing of the ears).

Halogenated Hydrocarbon Solvents

See also:
Primary Biotransformation (Phase I Reactions): Reductions
Chapter 3

Arrhythmias
Chapter 9

Liver Cell Death: Necrosis and Apoptosis
Fibrosis and Cirrhosis
Chapter 11

Damage to the Proximal Tubule
Chapter 12

Halogenated Hydrocarbons
Appendix

DEVELOPMENTAL EFFECTS

The period during which the nervous system develops is quite long, extending from a few days postconception to well into the postnatal (after birth) period. Exposure to toxicants during this developmental period is likely to have quite different effects on an organism than exposure to the same toxicant after growth and development is complete. Developmental neurotoxicants and their effects are discussed in Chapter 7.

METHODS IN NEUROTOXICOLOGY

There are many different methods that can be used to study effects of neurotoxicants. These techniques include both *in vitro* techniques involving isolated tissues or chemicals and *in vivo* techniques involving the whole animal. Although it is impossible to produce here a comprehensive list of all neurotoxicological techniques, some representative approaches will be discussed.

One major technique used in neurotoxicology is the study of *behavior*. If toxicant-treated animals respond differently in behavioral testing than control animals do, it can be an indication that the toxicant has nervous system effects. Effects could be sensory (inability to sense a stimulus), motor (inability to press a bar), or integrative (effects on learning or memory). The ability to acquire the behavior can be studied, as can the effect of varying *reinforcement* (reward) *schedules* on acquisition. Also, the *extinction* of the behavior (how long it takes for the behavior to disappear in the absence of reinforcement) can be studied.

One basic type of behavioral testing is called *operant conditioning*. In this type of testing, reinforcement (which may be either positive or negative) is used to modify a voluntary response. For example, an animal may be trained (using an apparatus called a *Skinner box*) to press a lever in order to obtain a reward. In *classical* or *respondent conditioning*, a stimulus called the *conditioning stimulus* is paired with another stimulus that evokes a response from an animal. In the classic example, Pavlov paired the ringing of a bell with the presentation of food, which evoked the response of salivation in a dog. Eventually, presentation of the conditioning stimulus alone was sufficient to evoke the response.

Some behavioral tests are specifically designed to detect sensory dysfunction. For example, the *acoustic startle chamber* is a box with speakers and a special pressure-sensitive platform. Sounds of different frequencies and intensities are played through the speakers, and the startle reflex of the animal is measured. A variation of this is the *air puff startle*, where instead of sound the stimulus is a puff of air. Sensitivity to vibration can also be measured.

Motor function, too, can be assessed. The activity of an animal in a *maze* or on a flat, open platform called an *open field* can be measured. The motor activity measured by these methods is called *exploratory behavior*. Other tests of motor ability and coordination include descent down a rope, ability to stay on a rotating rod, and tests of walking ability (which is sometimes measured by dipping the animal's feet in ink and examining the footprints it leaves).

Other tests are designed to measure learning. One commonly used test involves an apparatus called a *radial arm maze*. This maze has a center area with eight arms radiating out from it. Each arm has a station at the end where reinforcement (typically in the form of a food pellet) can be delivered. In a typical test, pellets are placed in each arm, and the rapidity with which the animal is able to collect all of the pellets is measured. Pellets can also be placed in any combination of the arms, and the ability of the animal to learn new patterns can be tested.

Finally, general observation of various behaviors can also be an important testing method. Neurotoxicants can cause alterations in feeding behavior, mating behavior, reproductive behavior, and other social behaviors.

On a physiological level, some studies focus on the measurement of electrical activity of neurons. An *electroencephalogram* (EEG) measures electrical activity of the brain by measuring the potential difference (voltage) between electrodes that are placed on the scalp in humans and sometimes directly onto the surface of the brain in experimental animals. *Sensory-evoked potentials*, the changes in EEG resulting from sensory stimulation, can be measured. Seizures can also be chemically or electrically induced in experimental animals, and the EEG monitored. Toxicants may alter evoked potentials or affect the induction and course of seizures.

Electrical conduction can also be studied on the biochemical level. One common technique for studying effects of toxicants on ion channels in neurons is the *voltage clamp*. The voltage clamp uses an external energy source to maintain membrane potential at a desired value. Other biochemical techniques focus on synaptic functions. Receptors can be isolated and characterized, *and receptor binding* by various molecules studied. Many other biochemical studies are possible, including studies of neurotransmitter levels, enzyme activities, axonal transport, etc.

Finally, recent advances in imaging techniques have allowed researchers to get a glimpse of the living, functioning brain. *Computerized tomography* (CT scan) *equipment* rotates an x-ray source around the head, shooting multiple narrow beams of x-rays through the brain, and measuring the degree to which x-rays are absorbed at each point and ultimately reconstructing (by computer) an image of a brain slice. Even more exciting is the *positron emission tomography* (PET) scan. Molecules such as glucose analogs (compounds that are structurally similar to glucose) or neurotransmitters are radioactively labeled with isotopes of oxygen, carbon, and nitrogen with short half-lives. As these isotopes decay, the positrons that are emitted combine almost immediately with electrons, thus releasing detectable gamma radiation. Detectors measure the radiation and construct an image showing the location of the labeled molecules in the brain. Changes in PET images over time can indicate changes in brain activity.

CASE STUDY   Botulinum Toxin

Botulinum toxin is actually a group of at least seven different toxic proteins produced by seven different *serotypes* (identifiable subtypes) of the bacterium *Clostridium botulinum*. This organism can be found throughout the world, with different serotypes occurring in different geographical areas. In the United States, for example, type A predominates in the western part of the country, type B in the eastern part of the country, and type E is found near the Great Lakes and in Alaska. Most cases of human poisoning involve serotypes/toxins A, B, and E.

The term *botulism* actually comes from the Latin word for sausage, *botulus*, since the illness was first noted in the early 1800s as occurring among individuals who had eaten improperly prepared sausage. Because *C. botulinum* is a strict anaerobe and can form spores that are highly resistant to heat, it can grow quite well in sealed containers. Thus, the majority of cases of botulism poisoning in the United States were at one time related to consumption of home-canned foods such as corn, carrots, and beans. Currently, however, botulism is more likely to occur in commercially prepared foods, including baked potatoes, cheese sauces, and stews.

Food poisoning is not the only way to acquire botulism, though. A second type of botulism is termed *infant botulism* and occurs when the gastrointestinal tract of infants becomes infected with *C. botulinum* spores. Infants are particularly susceptible to this infection, as their gastrointestinal tract is both less acidic and less populated with benign flora than adult tracts (although adults with GI disease may also be at risk). Because honey frequently contains *Clostridium* spores, pediatricians recommend that children not be given honey until they are past 1 year of age. *Clostridium* infections can also occur through wounds and through needle puncture sites in intravenous drug users.

Botulinum toxins are among the most toxic substances known, with an estimated LD_{50} in humans of 1 ng/kg. The molecules are produced by the bacterium as an inactive polypeptide of 150 kDa mw and are cleaved by a protease to form two separate chains that are then linked by a disulfide bond.

The molecular mechanism of action of botulinum toxins is reasonably well established. Although the toxin is widely distributed in the bloodstream, cholinergic neurons appear to be a preferential target due to the existence of high-affinity binding sites for the toxin on the nerve terminals. Binding of the toxin to these high-affinity sites allows entry into the neuron via the process of receptor-mediated endocytosis. Inside the neuron, the target for toxins A and E is the molecule *synaptosome-associated protein* 25 kDa (SNAP-25). This protein belongs to a class of proteins called *SNAREs*, which are critical in the process by which synaptic vesicles fuse with target membranes to release neurotransmitters. By cleaving SNAP-25, the botulinum toxins block the release of the acetylcholine from the neuron. The molecular target for toxin B, *VAMP/synaptobrevin*, is another SNARE protein.

The physical symptoms of botulism poisoning reflect the interruption of impulse transmission by neurons that use acetylcholine and include muscle weakness and paralysis (neuromuscular junction effects), blurred vision (autonomic effects), and other effects such as nausea and diarrhea. Botulism can only be definitively diagnosed, however, by the presence of the toxin in blood or other tissues. Treatment involves supportive therapy (such as ventilation of an affected patient if paralysis of the diaphragm has occurred) and administration of an antitoxin (which is commonly available for toxin types A, B, and E). Antibiotics may also be given.

There is another side to botulinum toxin, though, as the medical community has developed a number of useful clinical applications for this powerful, biologically active substance. The toxin has long been used therapeutically to treat muscle spasms, and now several studies have indicated that it may also be useful in the treatment of migraine headache. The mechanism involved in headache relief is not clear, but may involve blockade of release of neurotransmitters, including some of those involved in the pain pathways. The toxin may also prove helpful in blocking excessive sweating (which may be related to autonomic dysfunction).

Of course, injections of type A botulinum toxin (under the trade name BOTOX®) are also used to paralyze those facial muscles responsible for wrinkles, temporarily rendering the face smoother and presumably younger looking. In this procedure, small amounts of the toxin are injected directly into the muscle, minimizing any systemic effects. The resulting paralysis seems to last for around 3 months. During that time, recovery at the molecular level occurs, first through sprouting of new accessory structures at the affected nerve terminal, then through apparent recovery in the original affected region. Interestingly enough, although immunity to botulinum toxin does not commonly develop in individuals who develop botulism (probably because the amount of the toxin in the body is too small to trigger an effective immune response), it can develop in up to 20% of patients treated clinically.

Unfortunately, botulinum toxin also has potential as a biological warfare agent, and in fact has already been deployed. During World War II, botulinum toxin was part of both Japan and Germany's weapons programs, and in the 1990s botulinum toxin was produced and loaded into weapons by Iraq. Also in the 1990s, a terrorist group in Japan tried but failed on more than one occasion to disperse *C. botulinum* in downtown Tokyo. Although the toxin can be delivered in food or water, the greatest concern with terrorism is inhalational delivery. A vaccine is currently available for individuals at high risk for encountering the toxin (laboratory workers, for example), and other vaccines are currently in development.

BIBLIOGRAPHY

Aarts, M.M. and Tymianski, M., Novel treatment of excitotoxicity: Targeted disruption of intracellular signaling from glutamate receptors, *Biochem. Pharmacol.*, 66, 877, 2003.

Anthony, D.C., Montine, T.J., Valentine, W.M., and Graham, D.G., Toxic responses of the nervous system, in *Casarett and Doull's Toxicology*, Klaassen, C.D., Ed., McGraw-Hill, New York, 2001, chap. 16.

Arnon, S.S., Schechter, R., Inglesby, T.V., Henderson, D.A., Bartlett, J.G., Ascher, M.S., Eitzen, E. et al. Botulinum toxin as a biological weapon, *JAMA*, 285, 8, 2001.

Arundine, M. and Tymianski, M., Molecular mechanisms of glutamate-dependent neurodegeneration in ischemia and traumatic brain injury, *Cell. Mol. Life Sci.*, 61, 657, 2004.

Asbury, A.K. and Brown, M.J., The evolution of structural changes in distal axonopathies, in *Experimental and Clinical Neurotoxicology*, Spencer, P.S. and Schaumburg, H.H., Eds., Williams & Wilkins, Baltimore, 1980, chap. 12.

Banack, S.A. and Cox, P.A., Biomagnification of cycad neurotoxins in flying foxes: Implications for ALS-PDC in Guam, *Neurology*, 61, 387, 2003.

Blumenfeld, A., Botulinum toxin type A for the treatment of headache: Pro, *Headache*, 44, 825, 2004.

Brown, A.W., Aldridge, W.N., Street, B.W., and Verschoyle, R.D., The behavioral and neuropathologic sequelae of intoxication by trimethyltin compounds in the rat, *Am. J. Pathol.*, 97, 59, 1979.

Caya, J.G., Agni, R., and Miller, J.E., *Clostridium botulinum* and the clinical laboratorian, *Arch. Pathol. Lab. Med.*, 128, 653, 2004.

Comella, C.L. and Pullman, S.L., Botulinum toxins in neurological disease, *Muscle Nerve*, 29, 628, 2004.

Engelhardt, B., Development of the blood–brain barrier, *Cell Tissue Res.*, 314, 119, 2003.

Eulenburg, V. and Gomeza, J., Neurotransmitter transporters expressed in glial cells as regulators of synapse function, *Brain Res. Rev.*, 63, 103, 2010.

Griffin, J.W. and Price, D.L., Proximal axonopathies induced by toxic chemicals, in *Experimental and Clinical Neurotoxicology*, Spencer, P.S. and Schaumburg, H.H., Eds., Williams & Wilkins, Baltimore, 1980, chap. 11.

Gunawardena, S.L. and Goldstein, L.S.B., Cargo-carrying motor vehicles on the neuronal highway: Transport pathways and neurodegenerative disease, *J. Neurobiol.*, 58, 258, 2004.

Hille, B. and Catterall, W.A., Electrical excitability and ion channels, in *Basic Neurochemistry*, 6th ed., Siegel, G.J., Agranoff, B.W., Fisher, S.K., Albers, R.W., and Uhler, M.D., Eds., Lippincott, Williams, & Wilkins, Philadelphia, 1999, chap. 6.

Hinckelmann, M.-V., Zala, D., and Saudou, F., Releasing the brake: Restoring fast axonal transport in neurodegenerative disorders, *Trends Cell Biol.*, 23, 634, 2013.

Lee, D., Park, Y., Brown, T.M., and Adams, M.E., Altered properties of neuronal sodium channels associated with genetic resistance to pyrethroids, *Mol. Pharm.*, 55, 584, 1999.

Lewin, R., Parkinson's disease: An environmental cause? *Science*, 229, 257, 1985.

Li, S.-H. and Li, X.-J., Huntingtin–protein interactions and the pathogenesis of Huntington's disease, *Trends Genet.*, 20, 146, 2004.

Lock, E.A., Zhang, J., and Checkoway, H., Solvents and Parkinson disease: A systematic review of toxicological and epidemiological evidence, *Toxicol. Appl. Pharmacol.* 266, 345, 2013.

Lowndes, H.E. and Baker, T., Toxic site of action in distal axonopathies, in *Experimental and Clinical Neurotoxicology*, Spencer, P.S. and Schaumburg, H.H., Eds., Williams & Wilkins, Baltimore, 1980, chap. 13.

Marchi, N., Cavaglia, M., Fazio, V., Bhudia, S., Hallene, K., and Janigro, D., Peripheral markers of blood–brain barrier damage, *Clin. Chim. Acta*, 342, 1, 2004.

Miller, D.S., Regulation of P-glycoprotein and other ABC drug transporters at the blood–brain barrier, *Trends Pharmacol. Sci.*, 31, 246, 2010.

Miller, C.C.J., Ackerley, S., Brownlees, J., Grierson, A.J., Jacobsen, N.J.O., and Thornhill, P., Axonal transport of neurofilaments in normal and disease states, *Cell. Mol. Life Sci.*, 59, 323, 2002.

Moser, V.C., Aschner, M., Richardson, R.J., and Philbert, M.A., Toxic responses of the nervous system, in *Casarett and Doull's Toxicology*, Klaassen, C.D., Ed., McGraw-Hill, New York, 2013, chap. 16.

Nelson, R.M., Lambert, D.G., Green, A.R., and Hainsworth, A.H., Pharmacology of ischemia-induced glutamate efflux from rat cerebral cortex *in vitro*, *Brain Res.*, 964, 1, 2003.

Nicotera, P., Molecular switches deciding the death of injured neurons, *Toxicol. Sci.*, 74, 4, 2003.

Pope, C., Karanth, S., and Liu, J., Pharmacology and toxicology of cholinesterase inhibitors: uses and misuses of a common mechanism of action, *Environ. Toxicol. Pharmacol.*, 19, 433, 2005.

Schober, A., Classic toxin-induced animal models of Parkinson's disease: 6-OHDA and MPTP, *Cell Tissue Res.*, 318, 215, 2004.

Snyder, S.H., Opiate receptors and beyond: 30 years of neural signaling research, *Neuropharmacology*, 47, 274, 2004.

Sotgiu, S., Pugliatti, M., Fois, J.L., Arru, G., Sanna, A., Sotgiu, M.A., and Rosati, G., Genes, environment, and susceptibility to multiple sclerosis, *Neurobiol. Dis.*, 17, 131, 2004.

Stokin, G.B., Lillo, C., Falzone, T.L., Brusch, R.G., Rockenstein, E., Mount, S.L., Raman, R., Davies, P., Masliah, E., Williams, D.S., and Goldstein, L.S.B., Axonopathy and transport deficits early in the pathogenesis of Alzheimer's disease, *Science*, 307, 1282, 2005.

Weiss, B. and Cory-Slechta, D.A., Assessment of behavioral toxicity, in *Principles and Methods of Toxicology*, 4th ed., Hayes, A.W., Ed., Taylor & Francis, Philadelphia, 2001.

Williamson, M.S., Martinez-Torres, D., Hick, C.A., Castells, N., and Devonshire, A.L., Analysis of sodium channel gene sequences in pyrethroid-resistant houseflies, in *Molecular Genetics and Evolution of Pesticide Resistance*, ACS Symposium Series, Vol. 645, Brown, T.M., Ed., American Chemical Society Books, Washington, D.C., 1996.

Winrow, C.J., Hemming, M.L., Allen, D.M., Quistad, G.B., Casica, J.E., and Barlow, C., Loss of neuropathy target esterase in mice links organophosphate exposure to hyperactivity, *Nat. Genet.*, 33, 477, 2003.

Yamada, M. and Hasuhara, H., Clinical pharmacology of MAO inhibitors: safety and future, *Neurotoxicology*, 25, 215, 2004.

11 Hepatic Toxicology

ANATOMY AND PHYSIOLOGY OF THE LIVER

LIVER STRUCTURE

The liver is a large organ located in the upper abdomen and in humans is separated into two major and two minor lobes (Figure 11.1). Blood enters the liver via two sources: the *hepatic artery*, which brings blood from the systemic circulation, and the *portal vein*, which brings blood directly from the gastrointestinal tract. Blood exits the liver via the *hepatic vein*, and a manufactured substance called *bile* passes out through the hepatic duct and then either through the common bile duct to the small intestine or through the cystic duct to the gall bladder for storage.

The cells in the liver are arranged in distinctive hexagonal patterns that have been called *lobules* (Figure 11.2). Within a lobule, columns of epithelial cells called *hepatocytes* appear to radiate outward from a *central vein*, which is actually a branch of the hepatic vein. These columns of hepatocytes have between them channels called *sinusoids*, which are lined with highly permeable endothelial cells and which also contain phagocytic cells called *Kupffer cells*. *Hepatic stellate cells* store fat and are located between the endothelial cells and hepatocytes. Three other vessels are found at each of the outer corners of the hexagonal lobule: a branch of the portal vein, a branch of the hepatic artery, and a bile duct. These three vessels together are sometimes called the *portal triad*.

Blood flow in the lobule follows a regular pattern. Blood flows in through the branches of the hepatic artery and portal vein, passes through the sinusoids, and flows out through the central vein. Bile, which is manufactured in the hepatocytes, flows out through *bile canaliculi* (located between adjacent hepatocytes) to the bile ducts.

Studies have shown, though, that lobules are not self-contained functional units. Each hepatic artery–portal vein pair supplies blood not just to one lobule, but to a region of cells that overlaps two or more lobules. This area has been termed an *acinus* (Figure 11.3). The more current and functionally accurate picture of the liver focuses on acini rather than lobules. The characteristics of hepatocytes vary with their location in the acinus. Three *acinar zones* have been defined, based on the distance from the blood-supplying vessels. Cells in *zone 1* show a high activity of enzymes involved in respiration; zone 1 also seems to be the site where regeneration and replacement of liver cells begins (the new cells then migrate outward through zones 2 and 3). Cells in *zone 3*, on the other hand, show high cytochrome P450 activity. An alternative descriptive scheme (based on the earlier lobule model) describes cells as being *centrilobular* (near the central vein, roughly corresponding to zone 3), *periportal* (near the portal area, corresponding to parts of zones 1, 2, and 3),

225

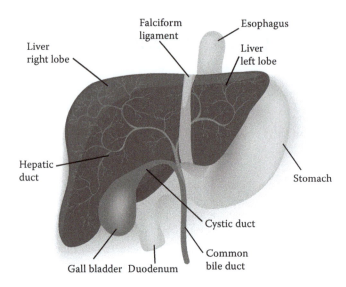

FIGURE 11.1 The anatomy of the liver. (http://www.shutterstock.com/pic-165558941/stock-vector-human-liver-anatomy-liver-gallbladder-esophagus-stomach-and-duodenum-vector-diagram.html?src=&ws=1. Accessed on June 22, 2014.)

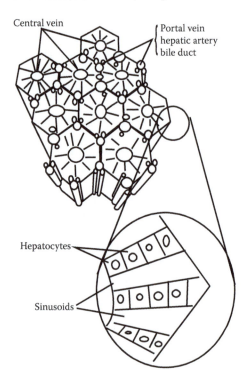

FIGURE 11.2 The hepatic lobule, showing the arrangement of central vein, portal triads, hepatocytes, and sinusoids.

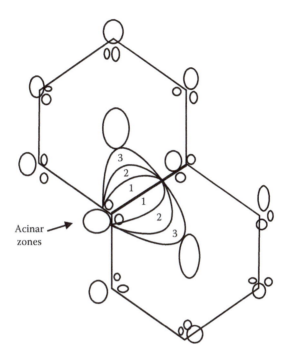

FIGURE 11.3 The hepatic acinus and surrounding zones.

or *midzonal* (along the edge between lobules, corresponding to the remainder of zone 1).

FUNCTION OF THE LIVER

The liver is an organ with an important role in many metabolic processes, as well as being a critical organ in toxicology. One main function of the liver is, of course, to assist in the absorption, metabolism, and storage of nutrients. Nutrient-containing blood from the gastrointestinal tract travels first to the liver (via the portal vein), where carbohydrates, lipids, and vitamins are removed. When blood glucose levels rise (following a meal, for example), hepatocytes are capable of converting sugars, fats, and amino acids into glucose and then storing glucose in the form of the polysaccharide glycogen. Alternatively, excess glucose can also be converted to *triglycerides*. Conversely, when blood glucose levels fall, hepatocytes break down glycogen to release glucose into the bloodstream.

Hepatocytes also play a critical role in protein metabolism by modifying and breaking down amino acids, converting the amine group into ammonia and then into urea for elimination. The amine group can also be used to synthesize other amino acids and used in protein synthesis (such as in making the protein component of lipoproteins).

The liver is particularly important in lipid metabolism. Hepatocytes synthesize and secrete a substance called *bile*, which contains water, ions, cholesterol derivatives known as *bile acids* or *bile salts*, and *bile pigments* such as *bilirubin*, which are released when hemoglobin is broken down (the liver is important in maintenance of proper blood volume and composition, storing blood, and phagocytizing damaged red blood cells). Bile is stored in the gall bladder and secreted into the small intestine, where it plays an important role in aiding in the digestion and absorption of lipids (about 80% of the bile acids are reabsorbed by the small intestine), as well as in the excretion of bile pigments and other wastes (which are not reabsorbed). Hepatocytes can break down, synthesize, and store fats, and can package lipids together with proteins to form the lipoproteins that transport lipids through the bloodstream to other cells.

Lipid Peroxidation

See also:
Effects of Toxicants on Lipids
Chapter 4

Liver Cell Death: Necrosis
and Apoptosis
Chapter 11

Excess levels of bilirubin may arise in the case of either increased production (such as what might happen with large-scale destruction of blood cells) or impaired excretion (due to hepatic dysfunction). This will cause *jaundice*, a yellow discoloration of the skin. Jaundice is particularly common in newborns, due to the rate of blood cell turnover that is much higher than in adults. Newborn jaundice is generally mild and can be treated by *phototherapy*, which consists of exposing the infant to bright lights (in the blue wavelengths), which will convert bilirubin into a form that is more easily excreted. Severe cases can be treated with transfusion. Left untreated, high levels of bilirubin can be neurotoxic to infants, resulting in *kernicterus*, which is characterized by brain damage to the basal ganglia. Recent evidence, however, has also argued for a protective effect of low levels of bilirubin, which can act as an antioxidant to protect cells against free radicals.

Xenobiotic Metabolism

See also:
Biotransformation
Chapter 3

The role of the liver in xenobiotic metabolism and excretion of toxicants is discussed in detail in Chapter 3, so only the basics will be reviewed here. Many hepatocytes contain enzyme systems capable of chemically altering toxic compounds, usually to a less toxic form. There are two basic types of metabolic alterations that can occur, and these are usually referred to as *phase I* and *phase II* reactions. Phase I reactions involve oxidation or hydrolysis (or sometimes even reduction) of the compound and are carried out by an enzyme system called the *cytochrome P450 system* or by various *hydrolases*. The cytochrome P450 system involves a number of different enzymes, including multiple forms of cytochrome P450 itself. Some of these forms are involved in metabolism of steroids and other endogenous compounds, whereas others metabolize xenobiotics. Two of the major groups of P450 enzymes involved in xenobiotic metabolism are a group of enzymes that is inducible by phenobarbital and a group that is commonly referred to as cytochrome P448 and that is inducible by polycyclic aromatic hydrocarbons (PAHs).

Phase II reactions involve conjugation of the toxicant (or often a metabolite resulting from a phase I reaction) with some other molecule. Generally, this action increases the size and water solubility of the toxicant, leading to enhanced excretion.

TYPES OF TOXICANT-INDUCED LIVER INJURY

The liver is vulnerable to toxicant-induced injury on several counts. As a site where significant xenobiotic metabolism occurs, liver cells are at risk for exposure to the toxic bioactivated metabolites that result from the metabolism of some toxicants. The direct routing of blood to the liver from the gastrointestinal tract from which ingested xenobiotics are absorbed, as well as the tendency for some compounds to undergo *enterohepatic cycling* (repeated reabsorption from bile and return to the liver), also increases the vulnerability of liver cells to assault from toxicants. Finally, the multiple functional roles of the liver offer multiple potential targets for toxicants. This chapter will focus on the various ways in which toxicants can interact with the liver to produce injury.

FATTY LIVER

Since the liver is the site of synthesis, storage, and release of lipids, it stands to reason that interference with these processes could lead to an accumulation of fats in the liver itself. Acute exposure to compounds such as *carbon tetrachloride, ethionine,* and *tetracycline* (an antibiotic) or chronic exposure to *ethanol* can block the secretion of a type of lipid called *triglycerides,* leading to the development of *hepatic steatosis,* or what is most commonly called a *fatty liver.* In this condition, which affects around 30 million individuals in the United States alone, anywhere from 5% to 50% of the liver's weight is fat. These cases can be divided into two categories: alcoholic fatty liver disease and nonalcoholic fatty liver disease (NAFLD), with a strong correlation existing between development of NAFLD and development of the metabolic condition diabetes mellitus.

The mechanisms by which fatty liver is produced are not completely clear, but most likely involve interference with the normal regulation of lipoprotein synthesis. In this process, the liver takes free fatty acids and, by combining them with glycerol, synthesizes triglycerides. (The chemical reaction of an acid with an alcohol like glycerol to produce an ester plus water is called *esterification.*) These triglycerides are then combined with phospholipids, cholesterol, and proteins to form *very low-density lipoproteins* (VLDLs). VLDLs then enter the bloodstream, carrying triglycerides to other cells.

There are a number of ways in which toxicants may produce alterations in this process. For example, there is some experimental evidence that exposure to ethanol leads to an increase in triglyceride synthesis. One possible mechanism for this effect would be through interaction with regulatory proteins such as *sterol regulatory*

Carbon Tetrachloride

See also:
   Primary Biotransformation (Phase I Reactions): Reductions
                  Chapter 3

   Arrhythmias
                  Chapter 9

   Other Neurotoxicants
                  Chapter 10

   Liver Cell Death: Necrosis and Apoptosis
   Fibrosis and Cirrhosis
                  Chapter 11

   Damage to the Proximal Tubule
                  Chapter 12

   Halogenated Hydrocarbons
                  Appendix

Ethanol

See also:
  **Other Enzymes Carrying
  Out Oxidation**
                Chapter 3

  Examples of Teratogens
                Chapter 7

  **Cardiomyopathies and Other
  Effects on Cardiac Muscle**
                Chapter 9

  Other Neurotoxicants
                Chapter 10

  **Liver Cell Death: Necrosis
  and Apoptosis
  Fibrosis and Cirrhosis**
                Chapter 11

  **Forensic Toxicology and
  Alcohol Use**
                Chapter 16

  Ethanol
                Appendix

Knockout Mice

See also:
  **Model Organisms and
  Comparative Genomics**
                Chapter 5

Cytochrome P450

See also:
  The Role of Cytochrome P450
                Chapter 3

element-binding protein-1c (SREBP-1c). This protein, along with other transcription factors (such as *carbohydrate response element-binding protein* (ChREBP) and *PPAR-γ*), regulates synthesis of triglycerides by activating genes that code for the enzymes in those synthetic pathways. This same regulatory protein also blocks oxidation of fatty acids by mitochondria, thus potentially increasing lipid levels through two mechanisms. The use of knockout mice to study the role of these proteins in the development of fatty liver has been particularly useful. Interference with protein synthesis or conjugation of triglycerides with proteins is another potential mechanism for producing fatty liver, as is interference with export of VLDLs to the bloodstream.

Induction of one form of cytochrome P450, CYP2E1, has also been implicated in the pathogenesis of fatty liver. CYP2E1 is induced following exposure to ethanol and is also upregulated in obesity and diabetes, perhaps as a result of the increased triglyceride levels that are also associated with those conditions. Increased levels of the CYP2E1 enzyme may lead to an increase in the production of free radicals and other potentially reactive metabolites that might contribute to lipid peroxidation and membrane damage. Damage to endoplasmic reticulum could then lead to inhibition of protein synthesis and thus block VLDL synthesis. Other evidence has indicated that inhibition of release of the lipoprotein may also be a factor.

Finally, inflammation may also play a role in fatty liver. The cytokine *tumor necrosis factor alpha* (TNF-α), which is involved in the inflammatory response, can trigger release of oxygen radicals from mitochondria, as well as promote apoptosis. Evidence indicates that increase in TNF-α activity is associated with fatty liver and that blockade of TNF-α activity may be effective in treating patients with the condition.

Interestingly enough, the presence of the excess fat in the liver does not necessarily affect the functioning of the hepatocytes. Steatosis may, however, in some cases progress to cirrhosis (see later in this chapter) and other serious problems, including liver cancer.

CHOLESTASIS

Stoppage of bile flows is known as *cholestasis*. Cholestasis occurs in humans following administration of drugs such as *steroids, phenothiazines,* and *tricyclic antidepressants.* It is characterized by the development of jaundice, a condition characterized by a yellowish discoloration of the eyes and skin (resulting from the buildup of bile pigments such as bilirubin). The molecular mechanisms behind drug- or toxicant-induced cholestasis may involve interference with the transport systems involved in bile formation, or, in some cases, damage to bile ducts.

LIVER CELL DEATH: NECROSIS AND APOPTOSIS

A number of compounds have been reported to cause hepatic *necrosis*, or cell death. Necrosis of hepatocytes is characterized by accumulation of vacuoles in the cytoplasm, damage to endoplasmic reticulum, swelling of mitochondria, destruction of the nucleus, and disruption of the plasma membrane. Necrosis is often described as being *focal* (confined to a limited area), *zonal* (occurring in a particular zone, usually zone 3), *diffuse* (scattered throughout the liver), or *massive*, and its location is frequently described using the descriptive terms (introduced earlier) centrilobular, midzonal, or periportal. Necrosis not only affects the cells under-going the process, but potentially leads to involve-ment of neighboring cells, as well, often through release of molecules that trigger a secondary inflammatory response in the region.

One possible cause of hepatic necrosis is *lipid peroxidation.* Compounds such as *carbon tetra-chloride, chloroform, bromobenzene,* and other *halogenated hydrocarbons* are metabolized by cytochrome P450 to form free radicals, reactive metabolites that can bind to and damage macromol-ecules (particularly in the centrilobular areas where the highest levels of P450 are found). Unsaturated fatty acids in membranes are particularly vulnera-ble to attack by free radicals. Carbon tetrachloride exposure has been shown to produce damage to hepatocyte membranes, including smooth and rough endoplasmic reticulum, thus reducing xeno-biotic-metabolizing ability as well as reducing pro-tein synthesis. In fact, a small initial dose of carbon tetrachloride protects against injury from a later larger dose, probably by destroying P450 and limit-ing the ability of the liver to bioactivate the later dose. Further evidence that carbon tetrachloride produces lipid peroxidation is found in the increased production of molecules called *conjugated dienes,*

Necrosis

See also:
  Necrosis
                              Chapter 4

Lipid Peroxidation

See also:
  Effects of Toxicants on
   Lipids
                              Chapter 4

  Function of the Liver
                           Chapter 11

Carbon Tetrachloride

See also:
  Primary Biotransformation
   (Phase I Reactions):
   Reductions
                              Chapter 3

  Arrhythmias
                              Chapter 9

  Other Neurotoxicants
                           Chapter 10

  Fatty Liver
  Fibrosis and Cirrhosis
                         Chapter 11

  Damage to the Proximal
   Tubule
                         Chapter 12

  Halogenated Hydrocarbons
                         Appendix

which are frequently used to monitor the occurrence of lipid peroxidation. Administration of *antioxidants* (to reduce or prevent lipid peroxidation) prevents some but not all toxic effects of carbon tetrachloride, indicating that even though lipid peroxidation is a factor, other mechanisms are probably also involved. Endogenous enzymes such as *superoxide dismutase* may also play a role in limiting peroxidation *in vivo*.

Acetaminophen

See also:
  Glutathione Conjugation
                Chapter 3

Acetaminophen
                Appendix

Another potential cause of hepatic necrosis is the production of other types of reactive metabolites. For example, *acetaminophen*, a common over-the-counter analgesic, primarily undergoes conjugation through the sulfate or glucuronide phase II systems. However, an alternative pathway involves metabolism by the CYP2E1 isoform of cytochrome P450 to the active metabolite *N*-acetyl-*p*-benzoquinone imine (NAPQI). This metabolite is highly electrophilic and capable of binding to and damaging cellular macromolecules. At low doses of the drug, the small amount of this reactive metabolite which is produced can be safely conjugated with glutathione and eliminated. At higher doses, however, glutathione stores may become depleted and toxicity may result. Thus, other drugs and compounds that deplete glutathione also have the potential to enhance acetaminophen toxicity, while compounds such as *N*-acetylcysteine, which supplies cysteine for glutathione synthesis, may be protective.

Acetaminophen has also been shown to stimulate formation of *nitric oxide* in hepatocytes. Nitric oxide, which is produced by liver cells in response to inflammatory signals, can actually block apoptosis at low levels by inducing stress proteins, promoting the synthesis of cGMP, and inhibiting caspase activity. However, at higher concentrations, NO may contribute to damage, combining with superoxides to form reactive species such as *peroxynitrite*. Thus, a combination of NAPQI and peroxynitrite production may ultimately be responsible for acetaminophen hepatotoxicity.

Because of the role of ethanol as both a substrate and an inducer of CYP2E1, there is a complicated relationship between co-exposure to ethanol and acetaminophen and the development of liver damage. Low doses of ethanol can actually be protective, competing with acetaminophen for binding to CYP2E1, thus decreasing production of the toxic metabolite NAPQI. Chronic ethanol exposure, however, can induce CYP2E1 levels, leading, on the long term, to increased production of NAPQI. It is likely, however, that ethanol exposure does not have a significant impact on acetaminophen toxicity at therapeutic doses of the drug.

As a reminder, any toxicant (such as acetaminophen) which depends on bioactivation to produce its effects will tend to produce necrosis primarily in the centrilobular region (located in zone 3). This area, as you may recall, is where the greatest P450 activity is located. Also, the toxicity of any of these compounds can be potentiated by compounds that induce cytochrome P450 activity. Ethanol, for example, not only potentiates effects of

Phase Two Reactions

See also:
  Secondary Metabolism
  (Phase II Reactions)
                Chapter 3

acetaminophen, but can also potentiate the effects of carbon tetrachloride and other halogenated hydrocarbons, probably also through its effects on P450.

Also, the ability of the liver to perform phase II reactions is a significant factor in determining the toxicity of not only acetaminophen, but other compounds as well. Many other reactive metabolites produced in phase I also undergo binding to glutathione and other phase II cofactors, thus limiting binding to cellular sites. Therefore, competition of other toxicants for binding to these cofactors, or dietary depletion of cofactors may potentiate the toxicity of these reactive metabolites as well.

Hepatotoxic metabolites can also be produced through enzyme systems other than P450. Ethanol, for example, besides undergoing metabolism via P450, can also be metabolized to an aldehyde by *alcohol dehydrogenase*, and then on to acetate by *aldehyde dehydrogenase*. Some individuals possess a slow aldehyde dehydrogenase isoform, resulting in a buildup of acetaldehyde following ethanol exposure and resulting in acute toxicity. The metabolism of ethanol also leads to depletion of the cofactor NAD, which can have a negative impact on mitochondrial functioning.

To summarize, which of the many effects of these toxicants are actually responsible for the death of the cells in hepatic necrosis is still a subject of considerable debate. It is probably not inhibition of protein synthesis, because toxicants such as ethionine can do this for hours without killing the cell. Damage to mitochondria has also been suggested as a lethal trigger, but as with inhibition of protein synthesis, ATP depletion can be observed without necrosis. Many theories point to effects on calcium homeostasis, since increased intercellular calcium levels frequently accompany cell death, and damage to membranes (such as the endoplasmic reticulum or mitochondrial membrane) could lead to release of sequestered calcium. Entry of external calcium into the cell through a damaged plasma membrane may or may not be involved, as studies have indicated that an external pool of calcium is not necessary to produce necrosis. It is difficult, in addition, to determine whether an observed increase in calcium in necrotic cells has actually caused the death of the cells or if it is a result of it.

Finally, under some circumstances, hepatocytes can also be induced to undergo *apoptosis*, or programmed cell death. For example, cholestasis often results in apoptosis of hepatocytes. The mechanism for the apoptotic effect of cholestasis is not

---

Ethanol

See also:
    Other Enzymes Carrying Out Oxidation
                  Chapter 3

    Examples of Teratogens
                  Chapter 7

    Cardiomyopathies and Other Effects on
     Cardiac Muscle
                  Chapter 9

    Other Neurotoxicants
                  Chapter 10

    Fatty Liver
    Fibrosis and Cirrhosis
                  Chapter 11

    Forensic Toxicology and Alcohol Use
                  Chapter 16

    Ethanol
                  Appendix

Apoptosis

See also:
  Apoptosis
                    Chapter 4

clear, but the trigger is probably the buildup of bile acids in the liver. Bile acids have been shown in cell culture to be hepatotoxic, and experimental evidence indicates that they interact with the *Fas* pathway, one of the major pathways of apoptosis. Other situations that may trigger apoptosis in hepatocytes include treatment with *troglitazone*, a drug that is used to treat type II diabetes, as well as viral infection (*viral hepatitis*). Following apoptosis, cell remnants are typically absorbed by surrounding cells or by the phagocytic Kupffer cells, generally without inflammatory involvement.

FIBROSIS AND CIRRHOSIS

Chronic exposure to hepatotoxicants can lead to *fibrosis,* which is characterized in the liver by excess production and accumulation of collagen as well as other components of the extracellular matrix. This can result in a significant disruption of blood flow in the liver, with the accompanying physiological consequences. The mechanism behind the production of fibrosis involves proliferation of the hepatic stellate cells, which produce and secrete extracellular matrix proteins. These cells can be activated by reactive oxygen species as well as growth factors, cytokines, and other molecules involved in inflammation. Fibrosis was once thought to be irreversible, but recent evidence now indicates that, in at least some cases, both the rate of deposition and the rate of degradation of the extracellular matrix may be able to be modified so as to result in the reversal of a fibrotic condition.

Ethanol

See also:
  Other Enzymes Carrying Out Oxidation
                    Chapter 3

  Examples of Teratogens
                    Chapter 7

  Cardiomyopathies and Other Effects on Cardiac Muscle
                    Chapter 9

  Other Neurotoxicants
                    Chapter 10

  Fatty Liver
  Liver Cell Death: Necrosis and Apoptosis
                    Chapter 11

  Forensic Toxicology and Alcohol Use
                    Chapter 16

  Ethanol
                    Appendix

Cirrhosis is a condition typically resulting from chronic exposures to toxicants and is characterized by widespread fibrosis and development of scar tissue. Chronic exposure to ethanol is a leading cause of cirrhosis in humans, but the mechanism underlying the effect is the subject of considerable debate. Malnutrition frequently accompanies alcoholism, and some investigators hypothesize that it is this factor, rather than the alcohol, that causes the cirrhosis. Evidence has been presented showing that rats that are maintained on an adequate diet can be exposed to ethanol without developing cirrhosis, but other studies have indicated that monkeys develop precirrhotic changes with exposure to ethanol even if no nutritional deficiencies develop. Cirrhosis is generally irreversible.

Many hepatotoxicants, including carbon tetrachloride and chloroform, have also been shown to be hepatic carcinogens in laboratory animals. One group of hepatic carcinogens is the *aflatoxins*. These toxins are produced by a fungus that grows on grain and other foods. Aflatoxin B1, for example, is metabolized by cytochrome P450 to a reactive epoxide, which then can bind to DNA. Some *polychlorinated biphenyls* (PCBs) may also be hepatic carcinogens. The most well-known human hepatic carcinogen is probably *vinyl chloride*, the monomer used in the manufacture of the polymer polyvinyl chloride (PVC). Its carcinogenic potential was discovered when it became clear that workers exposed to vinyl chloride were developing an unusually large number of cases of the relatively rare type of liver cancer known as *angiosarcoma*. Nonalcoholic fatty liver disease is also a risk factor for the development of liver cancer, as is exposure to the hepatitis B virus.

The mechanisms of carcinogenesis in the liver are undoubtedly similar to mechanisms in other tissues. Also, as in other tissues, increased cell turnover as a result of chronic injury and inflammation can produce an environment in which mutations are more likely to be propagated prior to repair.

Carcinogenesis

See also:
   The Mutational Theory of
   Carcinogenesis
                                        Chapter 6

Carbon Tetrachloride

See also:
   Primary Biotransformation
   (Phase I Reactions): Reductions
                                        Chapter 3

   Arrhythmias
                                        Chapter 9

   Other Neurotoxicants
                                        Chapter 10

   Fatty Liver
   Liver Cell Death: Necrosis and
      Apoptosis
                                        Chapter 11

   Damage to the Proximal Tubule
                                        Chapter 12

   Halogenated Hydrocarbons
                                        Appendix

MISCELLANEOUS EFFECTS

Toxicants can also damage sinusoids, enlarging them so that red blood cells can enter and block the lumen. One drug that can do this is acetaminophen. The *pyrrolizidine alkaloids* can also produce sinusoidal damage.

Exposure to other toxicants (such as the anesthetic *halothane*) can cause a condition resembling viral hepatitis, with headache, nausea, vomiting, dizziness, and jaundice. This effect may be caused at least in part by a reaction of the immune system to the drug.

Some individuals may also suffer what can be described as *idiosyncratic liver injury*. These events, while undoubtedly related to individual genetic and environmental factors, are by definition unpredictable and affect only a few of the individuals who are exposed to a drug or toxicant. One example of a drug known to produce idiosyncratic hepatotoxicity is troglitazone, the drug referred to earlier which was approved for the treatment of type II diabetes. Out of somewhere between 1 and 2 million patients who took the drug, a very few developed hepatotoxicity (leading to withdrawal of the drug from the market). Although evidence indicates that the drug can be metabolized to form electrophilic metabolites, it also appears that metabolism is predominantly protective. The drug has also been shown to induce apoptosis, perhaps through direct effects on mitochondria or perhaps through production of cholestasis. Thus, individual genetic differences in either xenobiotic metabolism or energy metabolism, as well as cellular changes associated with disease states (such as diabetes) can render select individuals susceptible to effects which might not be problematic for the rest of the population.

RESPONSE TO LIVER INJURY

In response to liver injury, hepatic tissues are often infiltrated by cells of the immune system such as macrophages and neutrophils. Although these cells help to remove foreign materials and cell debris, they also produce chemicals such as nitric oxide that may be toxic to surrounding healthy cells. Thus, as in many tissues, inflammation may be either helpful or damaging, depending on the degree and circumstances of the response.

Hepatocytes also have a significant regenerative capacity. If a portion of the liver is lost due to physical or chemical injury, the remaining portions will increase in mass until the approximate original mass of the liver is regained. Regeneration of lost or damaged tissue is primarily due to the replication of the remaining hepatocytes; however, there are stem cells in the liver called *oval cells* that can serve as precursors of hepatocytes. Following chemical injury (as opposed to surgical removal of tissue), regeneration is more likely to include proliferation of oval cells. Replication of hepatocytes appears to be triggered by cytokines including TNF-α and growth factors HGF (hepatocyte growth factor) and TGF-α.

EVALUATING LIVER INJURY AND TREATING DISEASE

Several methods, both clinical and experimental, are used to test for injury to the liver. Serum enzyme tests look for activity of enzymes in the blood that are normally found only in the hepatic cells. Increased serum activities of these enzymes may indicate damage to hepatocytes and subsequent leakage of the enzymes. Enzymes that are typically assessed may include *aminotransferases* such as serum glutamic-oxaloacetic transaminase (SGOT) and serum glutamic-pyruvic transaminase (SGPT), serum *alkaline phosphatase* (AP), serum *lactate dehydrogenase* (LDH), and many others. Some of these enzymes are more specific for liver injury than others (which may be elevated when other tissues are also injured). On the other hand, some are specific enough not only to indicate liver injury, but also to actually aid in diagnosing the type of injury.

The damaged liver is of great interest as a model in the development of transgenic cell therapy. This is in part due to the natural regenerative capacity of hepatocytes (vs. neurons, for example). A *transgene* (DNA that is incorporated into the cell from another source) might encode, for example, the normal sequence of a protein missing from the diseased liver. One approach would be to introduce a therapeutic transgene directly using an engineered virus as a vector. Another approach would be to culture cells from a patient, modify the cells using the virus, and then introduce the transgenic cells into the liver.

Prospects for curing liver diseases have also risen due to several fortunate properties observed in the early phases of experimental cell therapy. *Hematopoietic stem cells* are the current workhorses of experimental cell therapy and have been used in attempts to treat lymphoma and other cancers. They are harvested from bone marrow and are capable of differentiation into various types of blood cells. Injected hematopoietic stem cells also migrate into liver and a few other organs in a type of homing response, a property that has been exploited in experimental therapy of liver disease in mice. Interestingly, hematopoietic stem cells have also been observed to differentiate into hepatocytes. As cell therapy technology develops to employ other types of stem cells, diseases of the liver will be among the most promising candidates for treatment.

CASE STUDY   Reye's Syndrome

In 1963, a doctor in Australia named Ralph Reye published his observations of a number of children who developed a condition characterized by a combination of encephalopathy (neurological problems) and hepatic dysfunction. This condition, which came to be known as Reye's (or Reye) syndrome, carries with it a high level of mortality, particularly if it is not recognized and treated promptly.

From the beginning, Reye's syndrome has been puzzling. First of all, the disease occurs almost exclusively in children and young adults under the age of 18. It also tends to occur following a viral illness, most commonly chicken pox or influenza. However, a single common etiologic (causative) infectious agent cannot be identified. Symptoms include vomiting, listlessness, and drowsiness progressing to aggressive behavior, delirium, and coma. Reported pathology includes swelling of the brain, as well as fatty changes in the liver. In fact, diagnosis of Reye's relies not only on the combination of history of viral illness and unexplained vomiting, but also on the elevation of serum liver enzymes (which, as you may recall, indicate hepatic damage). On the histological level, liver cells show proliferation of smooth endoplasmic reticulum and peroxisomes and enlarged mitochondria. This involvement of mitochondria is one of the factors that helps distinguish Reye's from some of the inherited metabolic disorders that may mimic it.

Although it became clear early on that prior viral infection was strongly associated with the development of Reye's syndrome, the fact that most children with viral illnesses do not develop Reye's led researchers to search for additional factors that might be contributory agents. Because Reye's is relatively rare, establishing a link between the disease and causative factors is difficult. However, multiple studies have shown a strong association between development of Reye's and use of aspirin during the viral illnesses that typically precede its onset. In fact, the evidence was strong enough that in 1986 the FDA required a warning on products containing aspirin. This warning states that children and teenagers who have viral illnesses should not be treated with products containing aspirin. There remained some controversy in the medical community about whether aspirin is really a contributing factor, but the facts are that since the publication of this warning, the incidence of Reye's in the United States has dramatically declined. While there were 555 cases reported in children in 1980, this had dropped to fewer than 2 cases per year in the mid- to late-1990s.

Recent research in hepatotoxicity may be able to provide some answers as to the molecular mechanism behind Reye's syndrome, and one hypothesis developed by Trost and Lemasters focuses on mitochondria as the key. Salicylate, the active metabolite of aspirin, has been shown to induce the mitochondrial permeability transition (see Chapter 4) in hepatocytes, leading to mitochondrial uncoupling. This disruption of hepatic metabolism would be expected to lead to metabolic changes such as those seen in Reye's, including hypoglycemia, increase in fatty acid levels, and increase in levels of ammonia. Excess ammonia, in turn, has been shown to lead to brain edema and other neurological effects. Thus, most of the symptoms of Reye's correlate well with salicylate toxicity.

At least one question, though, remains: Why does Reye's almost invariably follow a viral illness? There may be a molecular answer to this. First, there is evidence that viral infections may also act on mitochondria, disrupting calcium metabolism, and resulting in increased calcium levels in hepatocytes. Other studies have then shown that increased calcium levels significantly enhance the ability of salicylate to invoke the MPT. Thus, the viral infection may sensitize hepatocytes to salicylate toxicity, setting the stage for the development of Reye's syndrome.

Still, Reye's syndrome is rare, even in individuals exposed to the combination of the two factors of viral infection and aspirin. This implies that there must be other factors also involved in its development. Perhaps only individuals with a particular genetic predisposition are susceptible or perhaps there are other environmental factors that have not yet been identified. Nonetheless, over the past 40 years, the combination of epidemiological and laboratory research has done a great deal to advance the understanding of this once totally mysterious syndrome.

BIBLIOGRAPHY

Belat, E.D., Bresee, J.S., Holman, R.C., Khan, A.S., Shahriari, A., and Schonberger, L.B., Reye's syndrome in the United States from 1981 through 1997, *N. Engl. J. Med.*, 340, 1377, 1999.

Browning, J.D. and Horton, J.D., Molecular mediators of hepatic steatosis and liver injury, *J. Clin. Invest.*, 114, 147, 2004.

Fausto, N. and Campbell, J.S., The role of hepatocytes and oval cells in liver regeneration and repopulation, *Mechanisms Dev.*, 120, 117, 2003.

Felipo, V. and Butterworth, R.F., Mitochondrial dysfunction in acute hyperammonemia, *Neurochem. Int.*, 40, 487, 2002.

Greenberg, D.A., The jaundice of the cell, *Proc. Natl Acad. Sci. USA*, 99, 15837, 2002.

Ishii, H., Common pathogenic mechanisms in ASH and NASH, *Hepatol. Res.*, 28, 18, 2004.

Jaeschke, H., Toxic responses of the Liver, in *Casarett and Doull's Toxicology*, Klaassen, C.D., Ed., McGraw-Hill, New York, 2008, chap. 13.

Jaeschke, H., Gores, G.J., Cederbaum, A.I., Hinson, J.A., Pessayre, D., and Lemasters, J.J., Mechanisms of hepatotoxicity, *Toxicol. Sci.*, 65, 166, 2002.

Jaeschke, H., Gujral, J.S., and Bajt, M.L., Apoptosis and necrosis in liver disease, *Liver Int.*, 24, 85, 2004.

Kirschstein, R. and Skirboll, L.R., Stem Cells: Scientific Progress and Future Research Directions, NIH, 2001, available at http://stemcells.nih.gov/info/scireport/.

Leung, T.-M. and Nieto, N., CYP2E1 and oxidant stress in alcoholic and non-alcoholic fatty liver disease, *J. Hepatol.*, 58, 395, 2013.

Meyer, S.A. and Kulkarni, A.P., Hepatotoxicity, in *Introduction to Biochemical Toxicology*, Hodgson, E. and Guthrie, F.E., Eds., Wiley Interscience, New York, 2001, chap. 20.

Parkinson, A., Biotransformation of toxicants, in *Casarett and Doull's Toxicology*, Klaassen, C.D., Ed., McGraw-Hill, New York, 2001, chap. 6.

Plaa, G.L. and Charbonneau, M., Detection and evaluation of chemically induced liver injury, in *Principles and Methods of Toxicology*, Hayes, A.W., Ed., Taylor & Francis, Philadelphia, 2001, chap. 24.

Reye, R.D.K., Morgan, G., and Baral, J., Encephalopathy and fatty degeneration of the viscera. A disease entity in childhood, *Lancet*, 2, 749, 1963.

Rockey, D.C. and Shah, V., Nitric oxide biology and the liver: Report of an AASLD research workshop, *Hepatology*, 39, 250, 2004.

Rumack, B.H., Acetaminophen hepatotoxicity: The first 35 years, *Clin. Toxicol.*, 40, 3, 2002.

Smith, M.T., Mechanisms of troglitazone hepatotoxicity, *Chem. Res. Toxicol.*, 16, 679, 2003.

Treinen-Moslen, M., Toxic responses of the liver, in *Casarett and Doull's Toxicology*, Klaassen, C.D., Ed., McGraw-Hill, New York, 2001, chap. 13.

Trost, L.C. and Lemasters, J.J., Role of the mitochondrial permeability transition in salicylate toxicity to cultured rat hepatocytes: Implications for the pathogenesis of Reye's syndrome, *Toxicol. Appl. Pharmacol.*, 147, 431, 1997.

Wallace, A.D. and Meyer, S.A., Hepatotoxicity, in *A Textbook of Modern Toxicology*, 4th Edition, Hodgson, E., Ed., John Wiley and Sons, 2010.

Wang, J.-S. and Groopman, J.D., Hepatic disorders, in *Occupational Health*, Levy, B.S. and Wegman, D.H., Eds., Lippincott, Williams & Wilkins, Baltimore, 2000, chap. 34.

Williams, K.H., Shackel, N.A., Gorrell, M.D., McLennan, S.V., and Twigg, S.M., Diabetes and nonalcoholic fatty liver disease: A pathogenic duo, *Endocrine Rev.*, 34, 84, 2013.

12 Renal Toxicology

FUNCTION OF THE KIDNEYS

In general, the kidneys play a major role in the maintenance of a constant internal environment within the body. This dynamic process, known as *homeostasis*, allows the body to maintain optimal conditions within its cells, even in the face of external changes in the environment.

One specific function of the kidneys is to excrete waste (including soluble xenobiotics and conjugates) from the blood through formation of urine. The kidneys also act to regulate levels of water and salts such as potassium and sodium in the body. In addition, hormones and enzymes produced by the kidney are important in the regulation of blood pressure, the maintenance of stable pH levels in blood and body fluids, the regulation of calcium metabolism, and the production of red blood cells. Therefore, toxicant-induced kidney damage has the potential to effect significant physiological changes that extend well beyond the boundaries of the organ itself.

ANATOMY AND PHYSIOLOGY OF THE KIDNEYS

The paired kidneys are located in the abdominal area, near the posterior wall. The structure of a kidney is defined by several morphological features (Figure 12.1). Each kidney is covered by an outer *capsule*. Underneath the capsule is a layer of tissue called the *cortex* and an inner zone known as the *medulla*. Blood enters the kidney through the *renal artery* and leaves via the *renal vein*. The cortex receives the bulk of the blood flow to the kidney and has a much higher rate of oxygen utilization than the medulla. (As we will see, most of the energy-intensive processes in the kidney occur in the cortex.) The relatively small *papillary region*, although characterized by a low volume of blood flow, may nonetheless experience high concentrations of toxicants as *urine*, the waste-containing fluid formed in the tissues of the kidney, is collected and passed through the renal pelvis and out the *ureter*.

The functional unit of the kidney is the *nephron* (Figure 12.2), with each kidney containing around a million nearly identical nephrons. Each nephron is composed of a *glomerulus*, which is a knot of capillaries, and which is surrounded by a structure called *Bowman's capsule*. Leading out of Bowman's capsule is a tubule consisting of a *proximal portion*, a loop (*loop of Henle*), and a *distal portion*, which empties into a *collecting duct*. The glomerular portion of all nephrons is located in the cortex, but while the tubules of some nephrons are found only in the cortex, the tubules of other nephrons (those located deep in the cortex) extend far down into the medulla.

The kidney basically acts as a biological filter. Fluid from the blood is filtered out of the glomerulus (the glomerular capillaries are quite permeable), driven by

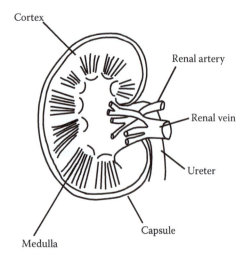

FIGURE 12.1   The anatomy of the kidney.

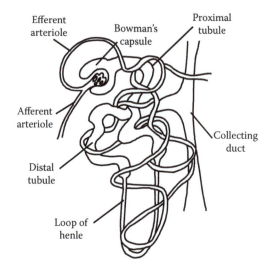

FIGURE 12.2   The nephron, showing Bowman's capsule, the proximal tubule, the loop of Henle, the distal tubule, and the collecting duct, as well as the afferent and efferent arterioles and glomerulus.

a combination of hydrostatic pressure (blood pressure) and osmotic pressure. Approximately 20% of blood volume is filtered out in a single pass through the kidney. Blood cells, however, as well as larger molecules such as albumin do not typically pass out of the glomerulus (in fact, if they do, it is often an indication of kidney dysfunction).

The fluid that leaves the vascular system to enter the kidney tissues at this point is termed the *filtrate* and is routed directly into the proximal tubule. As the filtrate passes through the proximal tubule, many substances in the filtrate are *reabsorbed* by the epithelial cells that line the tubule. In fact, 60%–80% of the water and solutes that make up the filtrate will be reabsorbed by these cells. In addition, some substances that were not originally filtered at the glomerulus are picked up from the surrounding blood vessels and *secreted* into the filtrate by the proximal tubule cells. The filtrate is further concentrated in the loop of Henle, and final adjustments to concentrations of water and solutes are made in the distal tubule and collecting duct. The filtrate that exits the nephron is then eventually routed to the ureter and excreted as urine.

EFFECTS OF TOXICANTS ON THE KIDNEY: GENERAL PRINCIPLES

Toxicant-induced damage to the kidney may be mild or severe, reversible or permanent, depending on the toxic agent and the dose. The kidney is particularly susceptible to the effects of toxicants for several reasons. First, blood flow to the kidneys is high (25% of cardiac output), so blood-borne toxicants will be delivered to the kidneys in large quantities. Second, as the kidney removes salts, water, and other substances from the filtrate through the process of reabsorption, any toxicant that is not reabsorbed may become highly concentrated in the remaining filtrate. In fact, some toxicants may become concentrated enough to actually precipitate out, potentially causing blockages. Finally, if a toxicant is reabsorbed, it may accumulate to high concentrations within the epithelial cells that line the tubule themselves. Thus, kidney tubule cells may be exposed to concentrations of a toxicant that are many times higher than the concentration of that toxicant in the plasma. In addition, many cells in the proximal tubule possess cytochrome P450 activity, so if bioactivation of a toxicant occurs, those cells in particular may be affected.

DAMAGE TO THE GLOMERULUS

One site in the nephron at which nephrotoxicants may act is the glomerulus (shown in Figure 12.3). The glomerulus itself is a network of capillaries arising from an *afferent arteriole*, a branch of the renal artery. The walls of the glomerular capillaries are very porous. Blood enters the glomerulus at relatively high pressure (around 60 mmHg). This pressure, which is regulated in part by specialized cells of the afferent arteriole called *juxtaglomerular cells*, forces blood fluids out of the pores, across a basement membrane, and through filtration slits between the *podocytes* (the epithelial cells that are part of Bowman's capsule). The capillaries reunite upon exiting Bowman's capsule, forming an *efferent* arteriole that then branches into a second network of capillaries that wrap around the rest of the tubule. Efferent arterioles eventually empty into the renal vein.

Thus, the glomerulus acts as a filter, allowing the passage of plasma fluids and small molecules into Bowman's capsule. Not only size (remember that blood cells and most plasma proteins are too large to fit through the filter), but also net electrical charge of a molecule affects filtration, with neutral and cationic molecules more likely to pass through the glomerular membrane (which is itself negatively charged) than anionic molecules. In a normal adult, a total of around 125 mL of fluid per

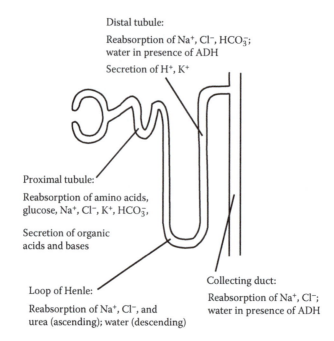

Distal tubule:

Reabsorption of Na^+, Cl^-, HCO_3^-;
water in presence of ADH

Secretion of H^+, K^+

Proximal tubule:

Reabsorption of amino acids,
glucose, Na^+, Cl^-, K^+, HCO_3^-,

Secretion of organic
acids and bases

Loop of Henle:

Reabsorption of Na^+, Cl^-, and
urea (ascending); water (descending)

Collecting duct:

Reabsorption of Na^+, Cl^-;
water in presence of ADH

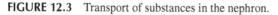

FIGURE 12.3 Transport of substances in the nephron.

minute is filtered by the two kidneys. This number is called the *glomerular filtration rate* (GFR).

There are a number of ways in which toxicants may directly affect the glomerulus. First of all, toxicants may increase glomerular permeability, resulting in *proteinuria*, the leakage of large-molecular-weight proteins into the filtrate and thus into the urine. Other toxicants may damage podocytes, increasing leakage through increasing filtration slit size or by causing detachment of podocytes from the glomerular basement membrane (which lies between epithelial and endothelial cells). One compound that produces this effect is the antibiotic *puromycin*, which may alter podocytes through effects on expression of proteins such as *podocin* and *nephrin* that play a role in slit morphology.

Some toxicants (such as *gentamicin*) can reduce the negative charge of the glomerular membrane, leading to the increased excretion of large anions. Additional toxicants (*amphotericin*, for example) may decrease GFR by causing vasoconstriction of glomerular capillaries. Finally, heavy metals and other chemicals may injure the glomerulus by attracting and interacting with immune system cells (such as macrophages and neutrophils) and stimulating the release of toxic products such as reactive oxygen species. Exposure of glomerular basement membrane proteins through cell damage and death may lead to similar problems through antibody reaction to the exposed proteins.

It should be noted that extra-glomerular factors, as well, can influence GFR. Blockage further on in the tubule, for example, can increase pressure in the tubule and lead to a decrease in filtration, as can obstruction of the bladder or urethra.

Circulatory issues either within (renal vasoconstriction) or outside the kidney (inadequate blood flow secondary to cardiac problems) can also lead to a drop in GFR.

DAMAGE TO THE PROXIMAL TUBULE

As the filtrate traverses the length of the tubule, important processes occur that change its composition (Figure 12.4). The next part of the tubule, the proximal tubule, is perhaps the major site of action for nephrotoxicants. The proximal tubule has two sections: a twisted or *convoluted* section and a straight section or *pars recta* (and a transition area in between the two). The epithelial cells lining the proximal tubule have a *tubular* or *luminal* side facing into the lumen of the tubule (with a convoluted surface called a brush border) and a *peritubular* side facing out toward the efferent capillaries that wrap around the outside of the tubule. These epithelial cells perform the important function of reabsorption of 60%–80% of the filtrate constituents. These constituents are removed from the filtrate by the epithelial cells and are passed back across the endothelial cells of the efferent capillaries and into the bloodstream. The maximum rate of reabsorption varies for different substances and is called the *tubular maximum*, or T_{max}, for that substance. Among the substances that are reabsorbed in the proximal tubule are

- *Electrolytes*: Na$^+$ in the form of NaCl or NaHCO$_3$ is reabsorbed by an active transport mechanism. Na$^+$ diffuses into the proximal tubule cell across the tubular membrane and then is actively pumped out across the peritubular membrane, where it is reabsorbed into the capillaries of the efferent arterioles (Figure 12.5). K$^+$, on the other hand, is actively pumped into the cell across the tubular membrane. Other ions, such as magnesium, calcium, phosphates, and sulfates, are also reabsorbed. Bicarbonate (HCO$_3^-$) is also reabsorbed, but indirectly. The reabsorption of this important buffer is tied to the secretion of H$^+$ and is described later.
- *Glucose*: Glucose is reabsorbed, perhaps through a co-transport mechanism with sodium. Normally, all glucose in the filtrate is reabsorbed. This

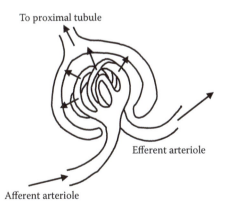

To proximal tubule

Efferent arteriole

Afferent arteriole

FIGURE 12.4 The glomerulus.

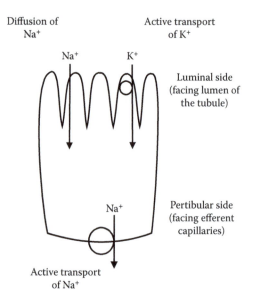

FIGURE 12.5   Transport of sodium and potassium in the proximal tubule cells.

mechanism can be saturated, though, if blood glucose levels are high enough (for example, as a result of diabetes).

- *Amino acids*: Many amino acids are reabsorbed, some more effectively than others. It is probable that several different mechanisms are active in this pH-sensitive process.

Other substances that are reabsorbed include ascorbic acid and, to some extent, urea and uric acid.

Water itself is not actively reabsorbed, but moves passively along its osmotic gradient following the movement of electrolytes. In other words, as solutes are reabsorbed, the concentration of solutes inside the tubule cells becomes higher than the concentration of solutes in the lumen of the tubule, and water will redistribute across the membrane (from the lumen into the cell) to equalize the concentrations. Reabsorption in the proximal tubule is thus isosmotic: although volume of the filtrate decreases, its *osmolality* (a measure of the concentration of dissolved particles in a solution) remains the same.

Another important process also occurs in the proximal tubule: the process of secretion. During this process, substances that remain in the bloodstream and are not filtered are pumped into the tubular lumen by the proximal tubule cells. There appear to be two major secretory transport systems: one for organic anions (substances such as *p-aminohippurate* [PAH] or the antibiotic *penicillin*) and one for organic cations (such as *tetraethylammonium* [TEA] or *N-methylnicotinamide* [NMN]). Some additional electrolytes are secreted as well. H^+, for example, is secreted as part of the process of maintaining the proper pH balance in the body. The secreted H^+ then combines with HCO_3^- in the tubule fluid to form H_2CO_3, which then breaks down to

$CO_2 + H_2O$. The CO_2 is reabsorbed by the proximal tubule cells and combines with water to reform H_2CO_3 within the cell. In this manner, the secretion of H^+ leads to the virtual reabsorption of bicarbonate, even though bicarbonate itself does not actually pass the tubule border.

Here, in the proximal tubule, many toxicants are concentrated through the transport activities of the proximal tubule cells. These toxicants may then act through damaging these epithelial cells. For example, formation of reactive oxygen species such as hydroxyl radicals can lead to membrane damage in proximal tubule epithelial cells, producing decreases in membrane fluidity, effects on membrane-related proteins, or perhaps alterations in calcium homeostasis. If this damage interferes with transport processes, as may be likely, inhibition of reabsorption may result, leading to appearance of glucose or amino acids in the urine (*glycosuria, aminoaciduria*). In addition, inhibition of reabsorption of these and other substances would also diminish the co-absorption of water. This would result in an increase in urine volume, or *polyuria*.

Lipid Peroxidation

See also:
Effects of Toxicants on Lipids
Chapter 4

Eventually, though, severe proximal tubule damage leads to *oliguria* (decrease in urine flow) or *anuria* (stoppage of urine flow). The mechanism by which oliguria and anuria are produced has been questioned. Some have hypothesized that sloughing off of badly damaged proximal tubule cells obstructs the tubular lumen. The sloughing of cells might also be caused by damage to the basement membrane, perhaps mediated through effects on *integrins*, transmembrane proteins that play an important role in cell adhesion. It is also possible that increased leakiness in the proximal tubule may lead to near-complete loss of filtrate or that vascular effects may also be involved, leading to a reduction in GFR.

One class of compounds that acts on the proximal tubule is the heavy metals. In addition to producing the previously mentioned functional indications of proximal cell dysfunction, heavy metals also produce microscopically observable cell damage and necrosis of proximal cells. Metals may act by binding to sulfhydryl groups on membranes and enzymes and disrupting their normal functions. In fact, administration of dithiothreitol, a sulfhydryl-containing mercury chelator, has been shown to protect against mercury-induced renal toxicity. As low a dosage as 1 mg/kg of a *mercuric salt*, for example, has been shown to affect enzymes in the brush border of proximal tubule cells within minutes, with intracellular damage occurring several hours later. It is not clear, however, whether this intracellular damage, including effects on energy metabolism, is the primary cause of toxicity or merely a response to the initial membrane damage. Mercury toxicity is complex and

Mercury

See also:
Vascular Spasms and Blood
Pressure
Chapter 9

Other Cytotoxic Compounds
Chapter 10

Metals
Chapter 17

Mercury
Appendix

may also involve effects on blood flow, with constriction of blood vessels causing a decrease in filtration as well as decreases in oxygen supply. This could then produce hypoxia in renal tissues. The effects of *organomercurials* are similar, perhaps because they are metabolized to inorganic mercury in the kidney tissues.

Although heavy metals may be similar in many of their effects on proximal tubule function, some effects may differ (particularly at low doses). For example, mercury-induced damage is concentrated in that part of the proximal tubule where anion secretion occurs (the S3 region, part of the pars recta or straight part of the tubule). Thus, organic anion secretion is particularly sensitive to mercury, while glucose reabsorption (which occurs in another section, the pars convoluta or convoluted part of the tubule) is affected less. Low doses of *chromium*, on the other hand, produce marked inhibition of glucose reabsorption as a result of damage to the convoluted proximal tubule. In kidney slices *in vitro*, chromium actually stimulates anion secretion at low concentrations (10^{-6} M), although it inhibits secretion at higher concentrations (10^{-4} M).

Cadmium

See also:

    Interference with Cell Division

                        Chapter 7

    Vascular Spasms and Blood
       Pressure

                        Chapter 9

    Metals

                      Chapter 17

    Cadmium

                      Appendix

Halogenated Hydrocarbons

See also:

    Primary Biotransformation
      (Phase I Reactions): Reductions

                      Chapter 3

    Arrhythmias

                      Chapter 9

    Other Neurotoxicants

                      Chapter 10

    Fatty Liver
    Fibrosis and Cirrhosis

                      Chapter 11

    Halogenated Hydrocarbons

                      Appendix

Another metal that acts as a nephrotoxicant is *cadmium*. Cadmium accumulates in the kidney throughout life, with a half-life measured in tens of years in humans. Toxicity occurs when concentrations of cadmium in the kidney reach 200 mg/g kidney weight. Cadmium accumulates in the kidney due to the presence of cadmium- and zinc-binding proteins called *metallothioneins*. These stable cytoplasmic proteins have low molecular weights and contain large amounts of cysteine. Exposure to cadmium, mercury, or other metals (but not zinc) causes an increase in renal metallothionein synthesis. In fact, studies have shown that pretreatment with low doses of cadmium can protect against damage from a later, larger dose. By binding to cadmium, metallothionein may prevent cadmium from binding to and damaging other cellular constituents, particularly in other organs. In the kidney, however, the cadmium–metallothionein complex itself may still damage kidney cells, particularly in later stages of chronic cadmium exposure. This may be happening as a result of the cadmium–metallothionein complex being degraded inside the proximal tubule cell and free cadmium being released as a result.

The parts of the proximal tubule most affected by cadmium are the S1 and S2 regions of the pars convoluta, however the specific mechanism of toxicity is not clear. Cadmium-induced impacts on mitochondrial functioning could result in increased production of reactive oxygen species and thus result in damage or death to cells.

Certain *halogenated hydrocarbons* also affect the proximal tubule. It is likely that these chemicals are metabolically activated by the cytochrome P450 activity found in the proximal tubule, producing free radical metabolites (including the reactive intermediate *phosgene*) that can then damage proximal tubule cell membranes. Covalent binding of metabolites of halogenated hydrocarbons such as *bromobenzene* and *chloroform* to renal proteins occurs in the proximal tubule cells and correlates with tissue damage. In fact, differences in toxicity of these compounds between various species may relate to quantitative and qualitative differences in renal P450 or glutathione concentrations in those species. Additionally, inducers of P450 such as phenobarbital potentiate chloroform toxicity, while SH-containing compounds are protective.

The halogenated hydrocarbon compound *tetrafluoroethylene* is also nephrotoxic. This compound is bound to glutathione by glutathione-*S*-transferases, but the conjugate is then secreted into the bile, where it may be broken down, reabsorbed in the small intestine, and re-enter the bloodstream. It can then gain entry to the cells of the proximal tubule through the organic anion secretion system. Within the proximal tubule cells, the compound is broken down further to yield a reactive compound which may bind to cellular macromolecules in the mitochondria and elsewhere in the cell.

Many antibiotics are nephrotoxic to proximal tubule cells. For example, *aminoglycosides* such as streptomycin, neomycin, and gentamicin produce damage to proximal tubule cells, perhaps through inhibition of phospholipases or through effects on mitochondrial function. The immunosuppressive drug *cyclosporine* produces significant acute nephrotoxicity related to its vasoconstrictive effects, and following chronic exposure may also produce effects related to other mechanisms of action. Some analgesics (such as *acetaminophen*) may also bind to and damage membranes. Species differences in acetaminophen nephrotoxicity, as with halogenated hydrocarbons, may correspond to species-related differences in xenobiotic metabolism.

The herbicide *2,4,5-T,* while not directly nephrotoxic, can inhibit the organic anion secretion system, and at high enough concentrations may also inhibit cation secretion. *Polychlorinated biphenyls* (PCBs), *polybrominated biphenyls* (PBBs), and *TCDD* may indirectly influence nephrotoxicity by increasing renal P450 activity.

2,4,5-T

See also:
  Pesticides
                                                                 Chapter 17

  2,4,5-T
                                                                 Appendix

Finally, some compounds may produce what are called *obstructive uropathies.* *Ethylene glycol*, for example, is metabolized to oxalic acid, which is then deposited in the tubule lumen as calcium oxalate.

Another example of compounds with this type of effect are melamine and cyanuric acid, which can combine to form a melamine–cyanuric acid complex which precipitates out in renal tubules leading to crystal formation and potential renal failure. These chemicals are perhaps best known for being illegally added to pet, livestock, and fish feeds in 2007 and 2008 to boost the nitrogen content, resulting in injury and death in cats and dogs in the United States, Canada, and South Africa. Following this incident, melamine was also found to have been added to infant formula and milk products in China, leading to health complications for almost 300,000 Chinese infants. In this case, melamine was likely forming a complex with uric acid to form the precipitating compound. Needless to say, these incidents have led to worldwide re-assessment of the safety of both animal and human food supplies, as well as the risks associated with exposure to melamine and related chemicals.

THE REMAINDER OF THE TUBULE

After leaving the proximal tubule, the modified filtrate continues into the loop of Henle. In the loop, distal tubule, and collecting duct, the filtrate becomes more concentrated through a process called a *countercurrent mechanism*. Chloride is actively reabsorbed in the ascending arm of the loop and moves into the area between the ascending and descending arms. Because the cells that make up the ascending arm are not very permeable to water, water is pulled from the cells of the descending arm (which are permeable to water) to compensate for the increase in osmolality in the area between the two arms. This creates a gradient, with increasing osmolality (electrolyte concentration) near the tip of the loop. Thus, the filtrate becomes more concentrated as it moves down the loop.

As the filtrate moves back up the ascending arm, the osmolality again begins to decrease. However, the impermeability of the ascending arm to water, along with further reabsorption of sodium and water in the arm as well as the distal tubule, keeps the filtrate at a higher osmolality and at a much lower volume (approximately 5 or 10% of the initial volume) than when it began its trip through the tubule. Because there is significant Na^+/K^+ ATPase activity in the ascending arm of the loop, energy requirements for these cells are quite high. Oxygen levels in this region are thus, conversely, quite low. Therefore, these cells are particularly vulnerable to hypoxia and hypoxic injury.

The actual degree to which urine is concentrated depends, however, on the permeability of the walls of the collecting duct. In the presence of *antidiuretic hormone* (ADH), the cells lining the duct become quite permeable, allowing water to leave the duct and thus concentrating the urine. If no ADH is present, the cells will be impermeable to water and little concentration will occur.

There are a few compounds that seem to exert their effects on the distal segments of the tubule. Many of them are pharmacological agents. Some non-steroidal anti-inflammatory drugs (*aspirin, ibuprofen, phenacetin,* and others) produce damage to the papillary area of the kidney, which is where many of the loops and collecting ducts are located. Chronic exposure to these drugs may result in a condition termed *analgesic nephropathy*. The mechanism behind this damage is not clear, however, it may either be produced through inhibition of cellular prostaglandin activity (which is normally high in the medulla) or may be secondary to the net vasoconstrictive effect of these drugs (which can lead to ischemic damage to renal tissues).

Methoxyflurane, an anesthetic, may block the effects of ADH on the collecting duct (producing polyuria) and can interfere with reabsorption of Na$^+$ and water in the proximal tubules as well. Metabolism may be necessary to produce these effects. *Tetracyclines*, a group of antibiotics, may also produce damage to the medulla.

MEASUREMENT OF KIDNEY FUNCTION *IN VIVO*

Many measurements of kidney function rely on the determination of the *renal clearance* of a chemical compound:

$$C = \frac{UV}{P},$$

where C is the clearance, U the concentration of the compound in the urine, V the urine flow, and P the concentration of the compound in the plasma. While the formula for clearance is simple enough, the physiological meaning of clearance is sometimes a somewhat difficult concept to grasp. *Clearance of a particular substance can be thought of as the volume of plasma that could hypothetically be completely cleared of that substance in 1 min* (Figure 12.6). Of course, because not all blood fluids are filtered, and due to the process of reabsorption, the kidneys do not always completely clear a substance in one pass-through. Thus, a clearance of 30 mL/min for a substance probably does not mean that an actual 30 mL of plasma is being completely cleared of the substance in 1 min, but may instead mean that 120 mL of plasma is being cleared of one quarter of the substance in 1 min. Each drug or chemical has its own clearance rate, which can be determined by experimentally measuring the three variables in the equation, if the compound follows first-order elimination kinetics.

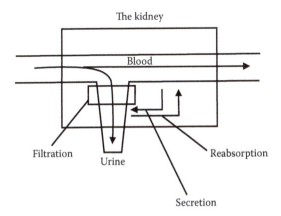

FIGURE 12.6 The concept of clearance. If a substance is completely cleared through secretion, then clearance equals RPF. If a substance is neither secreted nor reabsorbed, clearance is a reflection of GFR.

Some substances, however, through filtration and secretion, are actually completely cleared from the plasma in one pass through the kidney. The clearance of one of these substances is then a measure of total plasma flow through the kidneys. This rate is called *renal plasma flow* (RPF). The substance most commonly used to determine RPF is PAH.

On the other hand, if a substance is neither secreted nor reabsorbed, it will be totally removed only from the plasma filtrate, so its clearance will be a measure of the rate at which plasma is filtered. This is a measure of the already-mentioned GFR. One such substance that is neither secreted nor absorbed is *inulin*, a polymer of fructose.

Along with monitoring RPF or GFR, kidney function can also be assessed by examining urine volume and constituents. Changes in urine volume, osmolality, or pH, as well as the presence of such normally absent substances as protein or glucose (proteinuria or glucosuria), may indicate kidney damage. Increases in levels of *blood urea nitrogen* (BUN) and *plasma creatinine* are also commonly used, and the presence of specific enzymes (maltase, trehalase) in urine normally found only in the brush border of proximal tubule cells may be an early indication of damage to these cells. However, many of these measures are nonspecific in terms of indicating the area of the kidney which is damaged. Newer techniques include the measurement of proteins such as *kidney injury molecule-1 (KIM-1)*, albumin, and *cystatin c* in urine. These molecules can serve as "safety biomarkers" in preclinical drug screening and ultimately may provide more sensitive and specific information on the nature and degree of toxicant-induced renal injury.

One laboratory technique that has provided much useful information on kidney function is the *micropuncture* technique. In micropuncture, fluid can be collected from individual nephrons or capillaries within the kidney of an anesthetized animal (usually a rat or dog). Although this technique allows precise measurements of GFR, RPF, and filtrate volume and composition in a single nephron, the extensive training and experience required to perform the procedure hinder its widespread use.

MEASUREMENT OF KIDNEY FUNCTION *IN VITRO*

Isolated tissues are often used to study renal functions. Slices of renal cortex (which contain both proximal and distal tubules) will accumulate anions and cations and are used to study the secretory process. Besides studying normal function, these studies are quite useful in toxicology. An animal may be dosed with a toxicant prior to preparation of a kidney slice, or the toxicant may be added directly to the slice and its effects on these transport processes studied. Slices will also accumulate glucose, a process that appears to be related to reabsorption *in vivo* (rather than secretion, as in the case of organic anions and cations). It is also possible to test effects of toxicants on either primary renal cell cultures or one of several established renal cell culture lines.

Other techniques used to study renal function *in vitro* include the isolated perfused tubule technique. A segment of a nephron must be dissected out, perfused with fluid, and its function monitored. As with micropuncture, this is a sophisticated and difficult technique and thus not widely used.

In an application of new genomic techniques, attempts are being made to study the impact of renal toxicants on gene expression in the kidney. In one study, puromycin, cisplatin, and gentamicin were administered to rats over a 21-day period. During that time, rats were sacrificed and gene expression analyzed using a cDNA microarray. Patterns of gene expression changes were identified and correlated with traditional pathological and biochemical markers of nephrotoxicity. Techniques such as these show great promise both for diagnostic and for better understanding the link between molecular events and physiological lesions.

Toxicogenomics

See also:
Toxicogenomics
Chapter 5

Metabolomics

See also:
Metabolomics
Chapter 5

Metabolomics is another developing science that can be applied to renal toxicants. Profiles of urinary metabolites are used as indicators of exposure to specific toxicants or to identify genetic populations.

COMPENSATION FOLLOWING RENAL DAMAGE

As with other cells, injury to tubule cells does not necessarily result in cell death, and injured cells may be able to repair themselves sufficiently to regain normal function. When renal cells do undergo necrosis or apoptosis as a result of toxic insult, however, evidence exists for compensation through proliferation and differentiation of undamaged tubule cells. Even loss of whole nephrons can be compensated, at least on the short term, by increases in GFR produced by changes in pressures and flow rates. In fact, because the kidney is able to compensate so well for reductions in function, early identification of renal damage is not always easy. *Chronic renal failure* may result from long-term exposure to a variety of toxicants and may be the result of both direct damage from the toxicants as well as secondary damage due to the cumulative effects of the compensatory mechanisms themselves.

BIBLIOGRAPHY

Amin, R.P., Vickers, A.E., Sistare, F., Thompson, K.L., Roman, R.J., Lawton, M., Kramer, J. et al., Identification of putative gene-based markers of renal toxicity, *Environ. Health Perspect.*, 112, 4, 2004.

Berndt, W.O., Use of the tissue slice technique for evaluation of renal transport processes, *Environ. Health Perspect.*, 15, 73, 1976.

Berndt, W.O., Effects of toxic chemicals on renal transport processes, *Fed. Proc.*, 38, 2226, 1979.

Davis, M.E. and Berndt, W.O., Renal methods for toxicology, in *Principles and Methods in Toxicology*, Hayes, A.W., Ed., Taylor & Francis Group, Philadelphia, 2001, chap. 25.

Dorne, J.L., Doerge, D.R., Vandenbroeck, M., Fink-Gremmels, J., Mennes, W., Knutse, H.K., Vernazza, F., Castle, L., Edler, L., and Benford, D., Recent advances in the risk assessment of melamine and cyanuric acid in animal feed, *Toxicol. Appl. Pharmacol*, 270, 218, 2013.

Gobe, G. and Crane, D., Mitochondria, reactive oxygen species and cadmium toxicity in the kidney, *Toxicol. Lett.*, 198, 49, 2010.

Guan, N., Ding, J., Deng, J., Zhang, J., and Yang, J., Key molecular events in puromycin aminonucleoside nephrosis in rats, *Pathol. Int.*, 54, 703, 2004.

Kohn, S., Fradis, M., Ben-David, J., Zidan, J., and Robinson, E., Nephrotoxicity of combined treatment with cisplatin and gentamicin in the guinea pig: Glomerular injury findings, *Ultrastruct. Pathol.*, 26, 371, 2002.

Middendorf, P.J. and Williams, P.L., Nephrotoxicity: Toxic responses of the kidney, in *Principles of Toxicology: Environmental and Industrial Applications*, Williams, P.L., James, R.C., and Roberts, S.M., Eds., John Wiley & Sons, New York, 2000.

Roch-Ramel, F. and Peters, G., Micropuncture techniques as a tool in renal pharmacology, *Annu. Rev. Pharmacol. Toxicol.*, 19, 323, 1979.

Schnellman, R.G., Toxic responses of the kidney, in *Casarett and Doull's Toxicology*, Klaassen, C.D., Ed., McGraw-Hill, New York, 2008, chap. 14.

Tarloff, J.B. and Wallace, A.D., Nephrotoxicity, in *A Textbook of Modern Toxicology*, Hodgson, E., Ed., John Wiley and Sons, New York, 2010, chap. 14.

Xie, H.-G., Wang, S.-K., Cao, C.-C., and Harpur, E., Qualified kidney biomarkers and their potential significance in drug safety evaluation and prediction, *Pharmacol. Ther.*, 137, 100, 2013.

13 Immunotoxicology

FUNCTION OF THE IMMUNE SYSTEM

The primary job of the immune system is to protect the body from harmful invaders. This involves both distinguishing self from nonself, and then providing nonspecific barriers to invasion as well as customized defenses against specific threats. This is a highly complex undertaking, with tremendous variability between individuals in the responses of the immune system to perceived threats. This, of course, complicates the assessment of effects of toxicants on the system.

The cells that are involved in immune processes are commonly known as the *white blood cells* and include *polymorphonuclear leukocytes* (also called PMNs or *granulocytes*), *lymphocytes*, and *monocytes*. These cells originate and mature in the *lymphatic tissues* (including *bone marrow, thymus, spleen, lymph nodes*, and other tissues spread throughout the body) and travel throughout the lymphatic and circulatory systems. There are also *dendritic cells*, some of which originate in bone marrow, and some of which originate in other tissues. These cells communicate with each other and with other cells of the body through the exchange of chemical messengers called *cytokines*. We will discuss these cells, their functions, and the potential effects of toxicants on the system as a whole.

NONSPECIFIC DEFENSE MECHANISMS

Nonspecific defense mechanisms create barriers against the entry of invaders into the body in general. The protection provided by these mechanisms is sometimes also referred to as *innate immunity*, because it does not depend on prior exposure in order to be effective. This does not mean, however, that innate immunity does not depend on the ability to correctly recognize patterns which might indicate the presence of invaders. *Toll-like receptors*, for example, are found on the surface of some of the cells involved in innate immunity and recognize general patterns (*pathogen-associated molecular patterns*) of proteins which are common to many pathogens. However, neither prior exposure nor recognition of specific (as opposed to general) foreign substances is required to initiate this response.

THE SKIN AND MUCUS MEMBRANES

The first nonspecific barrier against invasion is the skin. The thickness of the epidermis and the keratin coating help prevent entry of foreign substances into the body, and underneath the epithelial layer is a sticky layer of tissue containing hyaluronic acid, which is difficult for invading organisms to penetrate. Also, secretions of oil and sweat glands (which contain lytic enzymes and antibodies) help wash away and destroy any potential invaders. While the epithelial cells that line the respiratory, gastrointestinal, urinary, and reproductive tracts do not provide as complete a barrier, they do secrete protective mucus and have hairs and cilia that help trap particles.

PHAGOCYTOSIS AND OTHER DIRECT ATTACKS

Even if an invader gets past the first line of defense, additional barriers remain. Some cells of the immune system function as *phagocytes*, engulfing and digesting foreign materials and debris from damaged cells. Examples of phagocytic cells include *macrophages* (which develop from monocytes) and *neutrophils* and *eosinophils* (two types of PMN cells). Neutrophils and eosinophils are found in the bloodstream, while macrophages are found in tissues (although they originate as monocytes in the bloodstream). Fixed macrophages are generally immobile and are found fixed in position in various sites in the body. Examples include *Kupffer cells* in the liver and *microglia* in the central nervous system. Free macrophages, on the other hand, can migrate throughout the body. Macrophages are attracted to the chemicals released by damaged cells, bacteria and other microbes, and other cells of the immune system.

Neutrophils are also capable of a second mode of attack upon invaders. This involves the release of hydrogen peroxide, hydroxyl radicals, and other compounds collectively known as *reactive oxygen species* (ROS).

Another type of cell that is important in nonspecific defense is a type of lymphocyte known as a *natural killer* (NK) *cell*. Natural killer cells are able to identify and destroy abnormal cells (cancer cells or cells infected by viruses, for example). They do this by releasing proteins called *perforins*, which literally punch holes in the membrane of the target cell, causing its destruction.

THE COMPLEMENT SYSTEM AND INTERFERONS

There are several groups of proteins found in the bloodstream that contribute to nonspecific defense. The *complement system* is a set of proteins found in plasma that can work together to destroy cell membranes, attract phagocytes, and stimulate activity of various cells of the immune system. Normally in an inactive state, complement proteins are activated by contact with microbes themselves, or with antibodies produced by the specific immune response (more about that shortly).

Another set of proteins important in nonspecific defense are the *interferons* (one type of cytokine). Produced by cells that have been infected by viruses, these chemicals stimulate uninfected neighboring cells to produce *antiviral proteins* (AVPs). AVPs interfere with viral replication, and thus slow the spread of the virus. Interferons can also activate phagocytic cells, ensuring that infected host cells are destroyed.

FEVER

One of the most common physiological responses to infection is *fever*, or elevation of body temperature. This is a response that probably evolved at least 380 million years ago, as it is shared by mammals, birds, and even some lizards and fish. Body temperature is normally regulated in the hypothalamus and reflects a balance between heat produced by cellular metabolism and heat released by various physiological processes (vasodilation, sweating, etc.). Normally, the set point for body temperature is somewhere around 37°C (98.6°F), but during illness the set point may rise to a higher temperature (perhaps around 39°C, which is 102°F).

Molecules that can produce this elevation in temperature are called *pyrogens*, and they include a number of cytokines (interleukins 1 and 6, for example) as well as exogenous molecules such as bacterial endotoxins (some of which may exert their effects directly, but most of which do so through simply increasing levels of the relevant cytokines).

Pyrogenic cytokines seem to increase body temperature through their ability to induce the *cyclooxygenase COX-2* in endothelial cells of vessels within the brain. It is clear that COX-2 plays an important role in fever production, as mice that are deficient in COX-2 are also deficient in the fever response. COX-2 is responsible for production of the prostaglandin *PGE2*, which diffuses from the endothelial cells into neural tissue and binds to receptors on neurons in an area of the hypothalamus known as the preoptic area. This, then, triggers the temperature adjustment, which can be modulated by input from a number of other endogenous compounds. *Antipyretic* drugs like aspirin, acetaminophen, and other *nonsteroidal anti-inflammatory drugs* (NSAIDs) can reduce fever, primarily through their inhibition or downregulation of cyclooxygenases.

At one time thought to be an undesirable side effect to illness, fever is now recognized to be generally beneficial. Many studies have demonstrated that individuals with mild illness whose fevers are not treated actually recover faster than those whose fevers were treated with antipyretics such as acetaminophen. The evidence is less clear-cut in patients with more severe infections, though. How exactly does fever aid in defeating infections? Since most pathogens can carry out proliferation as well at elevated temperatures as at normal body temperature, the additional benefit most likely comes from effects on the host immune system. Some studies have indicated that motility and phagocytosis in some, but not all, white blood cells are enhanced at febrile temperatures. Production, as well as biological activities, of some cytokines may also be enhanced. There is some evidence that heat shock proteins may be induced, too (see Chapter 4 for more on heat shock or stress proteins). However, there is a physiological cost to fever as well. The increased rate of energy metabolism, and thus the additional oxygen demands which are necessary to generate elevated body temperatures can result in additional demands on already stressed systems, such as the respiratory system and cardiovascular system.

THE INFLAMMATORY RESPONSE

When tissues are injured, whether by infection or trauma, they react with a set of responses that have come to be known as the *inflammatory response* (Figure 13.1). First, damage to tissues stimulates a connective tissue cell called a *mast cell* to release a number of different chemicals, including *histamine* and *leukotrienes*. These chemicals produce vasodilation and increased permeability in the blood vessels in the area, bringing increased numbers of macrophages, neutrophils, eosinophils, complement proteins, and other defenders to the area, and aiding in the removal of

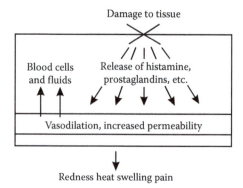

FIGURE 13.1 The inflammatory response.

cellular debris. This increase in blood flow also produces what are often called the *cardinal signs* of inflammation: heat, redness, swelling, and pain. Eventually, fibroblasts are stimulated to lay down collagen, thus producing scar tissue.

While acute inflammation may provide a means for delivery of immune cells to a damaged area, and may thus be productive, chronic inflammation may also play a role in many disease states.

SPECIFIC DEFENSE MECHANISMS

Substances which can activate the body's specific defense mechanisms are called *antigens*. Most antigens are large molecules and they are usually proteins or at least have a protein component (such as glycoproteins or lipoproteins). Larger structures such as cells or viruses that contain these molecules may also be considered antigens. To produce a complete immune response, antigens must be able to both react with *antibodies* (the proteins produced by the specific immune response) and stimulate the production of more antibodies. To do this, antigens must have at least two sites where antibodies can bind. Molecules that have only one of these *antigenic determinant sites* can react with antibodies, but do not stimulate antibody production. These incomplete antigens, or *haptens*, can stimulate a complete response only by binding to another molecule that can supply the necessary second antigenic determinant site. An example of a hapten is the drug *penicillin*, which must bind to proteins in the body before producing the well-known allergic response that some individuals display.

The specific immune response itself is carried out by lymphocytes and consists of two components: a direct attack on the antigen by activated lymphocytes (called *cellular immunity*) and an attack on the antigen by lymphocyte-produced antibodies (called *humoral immunity*).

CELLULAR IMMUNITY

The lymphocytes involved in cellular immunity are called *T cells* (Figure 13.2). There are many different types of T cells circulating in the bloodstream, each of which can be distinguished by the different types of receptors found on their surfaces.

FIGURE 13.2 The process of cellular immunity, showing activation of T cells.

These T cells do not respond directly to free antigen, but only to antigens that have been processed by other cells (*antigen-presenting cells*, a group that includes macrophage and dendritic cells), a step that involves presenting the antigens to the T cells in the proper manner. When macrophages or other phagocytic cells engulf antigens, they break them down and then display the fragments on their cell surface. Infected cells also display proteins from the infecting agent on their surfaces. These antigen fragments are bound to cell surface proteins called *human leukocyte antigen* (HLA) proteins. HLA proteins, produced by a group of genes called the *major histocompatibility complex* (MHC), are found on almost all cells and serve as unique markers that help the immune system to identify self from nonself. It is this combination of HLA protein and foreign antigen that T cells respond to, with the HLA–antigen combination fitting into the T cell surface receptors in a molecular lock-and-key manner.

When T cells with the proper type of receptor (one with the correct molecular fit for that particular antigen) encounter this HLA–antigen combination on a cell surface, they bind to the cell, with the assistance of proteins *CD4* and *CD8*. An exchange of lymphokines (including interleukins 1 and 2) between the T cell and the antigen-presenting cell triggers the T cell to begin to divide and differentiate into one of several activated forms. *Cytotoxic* (also called killer) *T cells* attack and destroy antigens directly by secreting cytotoxic chemicals (such as perforin or tumor necrosis factor), while *memory T cells* remain dormant, but are poised to react swiftly to any later reappearance of that same antigen. Additional types of T cells are also produced: *helper T cells*, which promote further T cell activation, stimulate phagocytic activity, and assist in the humoral immunity process; and *regulatory T cells*, which produce a delayed inhibition of both cellular and humoral responses.

Humoral Immunity

Humoral immunity is mediated by lymphocytes called *B cells* (Figure 13.3). Like T cells, the body has many different types of B cells, which are also differentiated by

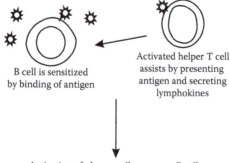

B cell is sensitized
by binding of antigen

Activated helper T cell
assists by presenting
antigen and secreting
lymphokines

Activation of plasma cells, memory B cells

FIGURE 13.3 The process of humoral immunity, showing activation of B cells.

their surface receptors. Among the proteins found on the surface of B cells are *antibodies* (sometimes abbreviated Ab) and *immunoglobulins* (sometimes abbreviated Ig) (Figure 13.4). There are five different classes of antibodies (IgG, IgE, IgD, IgM, and IgA), each of which plays a different role in humoral immunity. All antibodies consist of a pair of light polypeptide chains and a pair of heavy polypeptide chains. Both heavy and light chains have a constant region (which does not vary between antibodies of the same class) and a variable region. The variable region is where antibodies recognize and bind antigens.

When a circulating antigen encounters a B cell with the appropriate antibody displayed on its surface (again, one with the proper molecular shape to bind that antigen), binding occurs and the B cell becomes sensitized. For activation to occur, though, the B cell must also be presented with antigen that is bound to the surface of a helper T cell. The activated cell then divides and differentiates into *plasma cells*, which produce additional antibodies, and *memory B cells*, which (like memory T cells) remain dormant unless a later exposure to the same antigen occurs. The antibodies that are produced can then bind to, immobilize, clump, and mark antigens

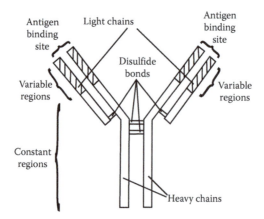

FIGURE 13.4 The general structure of antibodies (immunoglobulins).

for destruction by phagocytes. The process requires approximately 3–5 days for production of functional antibodies.

How does the immune system generate the tremendous diversity of antibodies? One might think of alternative splicing of messenger RNA for a mechanism; however, it was found in the laboratories of Nobel laureate Susumu Tonegawa and associates that the mechanism is rearrangement of DNA (somatic DNA shuffling). During the development of B lymphocytes, the IgG locus of the inherent genome is recombined with about 2.5 million possible outcomes. Nearly every lymphocyte has a uniquely recombined IgG gene, which is then transcribed and translated into an antibody. Possible variants are actually much greater due to a high rate of mutation introduced.

DEVELOPMENT OF IMMUNITY

Immunity is defined as the ability to ward off a specific infection, and it can be developed in two ways. *Active immunity* develops through exposure to the antigen. This exposure may occur naturally or may be deliberate, as in *immunization*, where individuals are exposed to a dead or inactivated pathogen in order to stimulate the development of immunity and thus avoid getting the disease. Immunization works because of the presence of memory cells, which may continue to respond rapidly for years after the initial exposure.

In *passive immunity*, individuals are injected with antibodies against a particular antigen. These antibodies help the individual's own immune system destroy the antigen. Newborn infants are the beneficiaries of a natural type of passive immunization, as antibodies from the maternal circulation can cross the placenta and enter the fetal circulation. Also, antibodies can be passed to the infant through the mother's breast milk (the infant's digestive tract is "leaky" enough to allow absorption of these large molecules).

EFFECTS OF TOXICANTS ON THE IMMUNE SYSTEM

TOXICANT-INDUCED ALLERGIES

Occasionally, the immune system will respond extremely to an inappropriate external stimulus, such as an insect sting, food component, pollen, dust, or even a drug or toxicant. This response is called a *hypersensitivity response* or *allergic response*. There are four different types of allergic responses. The type I response is also known as the *anaphylactic response*, or *immediate hypersensitivity*, and occurs when exposure to an antigen causes IgE antibodies to be produced and bind to sites on basophils and mast cells both in tissues and in the

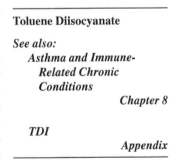

Toluene Diisocyanate

See also:
    Asthma and Immune-
      Related Chronic
      Conditions
                Chapter 8

    TDI
                  Appendix

general circulation. This results in the person becoming *sensitized* to the antigen, and subsequent exposures will result in binding of the antigen to the antibodies on the mast cells. This triggers the release of histamine and other mediators of inflammation, resulting in skin irritation, rhinitis, bronchoconstriction, or in severe cases, rapid systemic vasodilation leading to anaphylactic shock. The reaction usually occurs immediately following exposure to the antigen to which the allergic person has become sensitized.

There are several toxicants that have been observed to produce a type I response in susceptible individuals. *Toluene diisocyanate* (TDI), for example, a chemical used in the manufacture of polyurethane, can act as a hapten, combining with body proteins (most likely the endogenous protein *laminin*) to induce hypersensitivity reactions in exposed individuals. Observations of accidentally exposed workers indicate that the higher the exposure level, the more likely it is that hypersensitivity will develop. Exposure does not necessarily have to be through the respiratory route, as dermal exposure can also produce pulmonary hypersensitivity. These observations have been supported by laboratory studies using the guinea pig, which has proved to be an effective model. Unlike most other allergic reactions, the hypersensitivity induced by TDI may continue after exposure to TDI itself is terminated, perhaps because TDI induces a general increase in reactivity to other irritants.

Another chemical that can elicit a type I response is the antibiotic penicillin (as well as its various derivatives). Allergic reactions to penicillin are the most common drug allergy and are responsible for 75% of the deaths due to anaphylactic shock in the United States. As with TDI, a metabolite of penicillin acts as a hapten, combining with proteins to provoke the immune response. This response may be mild or quite severe, potentially leading to anaphylactic shock. There is a great deal of cross-reactivity involved in allergy to penicillin, and the antibodies produced will recognize not only penicillin, but also other antibiotics with a beta lactam ring structure. Penicillin may also induce type II, III, or IV reactions (as described below).

Type II responses (*antibody-dependent cytotoxic hypersensitivity*) result when IgG or IgM molecules bind to and destroy blood cells or other cells. Exposure to high levels of *trimetallic anhydrides* (TMAs) can trigger this condition. TMAs may also cause a type III response (*immune complex-mediated hypersensitivity*), where antigen–antibody complexes become trapped in vascular tissues and produce inflammation.

Type IV responses (*cell-mediated hypersensitivity*) may take a day or more to develop, and involve the activation and proliferation of T cells. An example of a common type IV response is *allergic contact dermatitis*, also called *sensitization dermatitis*. Individuals may become sensitized to a chemical, often after repeated exposures. By one to three weeks after the sensitizing exposure, further exposure to the chemical can lead to the development of an itchy rash, often characterized by the appearance of grouped blisters and edema.

Formaldehyde

See also:
  Immediate Responses:
    Upper Airway Effects
                          Chapter 8

  Formaldehyde
                          Appendix

Some of the best-known of the agents that cause allergic contact dermatitis are the oils contained in the plants *poison ivy* and *poison oak*. These oils act as haptens, combining with proteins in skin to elicit an immune response. When sensitized persons contact the oil, the skin in the area of contact exhibits the sensitization response. (Contrary to popular opinion, the fluid that forms in the blisters does not contain the oil itself, and thus contact of this fluid with other parts of the body cannot cause the rash to spread.) The rash generally disappears within 1–2 weeks. More serious problems may occur if smoke containing the volatilized oil is inhaled, leading to irritation of the lining of the respiratory tract. Other chemicals that can produce allergic contact dermatitis include *nickel* and *formaldehyde*, as well as some pesticides. *Latex* products may also induce a type IV reaction, an issue of particular concern in the medical and scientific industries, where exposure to gloves and other latex products is common.

Toxicant-Induced Autoimmunity

Normally, the immune system learns to distinguish self from nonself during the process of development. This prevents the immune system from later mounting attacks on normal cells and tissues. When the immune system does inappropriately attack some part of the body, the resulting disease is classified as an *autoimmune disorder.*

Many diseases are now recognized as having a basis in autoimmunity. These include *myasthenia gravis* (caused by attacks on the neuromuscular junction), *multiple sclerosis* (caused by attacks on myelin), *type I diabetes* (caused by attacks on pancreatic beta cells), and *systemic lupus erythematosus* (caused by attacks on various body tissues). The possible role of toxicants in triggering these and other autoimmune disorders is now being investigated. Some autoimmune responses may be triggered by exposure to a toxicant with a molecular structure that is similar to the structure of some normal tissue component. In this case, antibodies produced against the toxicant may also react against the normal tissue. Alternatively, toxicants may damage tissue directly, exposing, in the process, tissue constituents that are normally hidden from immune system surveillance. These previously hidden constituents may not be recognized as self by the immune system and thus may be attacked.

Few specific examples of toxicant-induced autoimmune disorders have been identified; many more have been postulated, but not yet proven, to exist. Isoniazid, a drug used in the treatment of tuberculosis, can produce a lupus-like autoimmune syndrome as a side effect. The mechanism behind this is not clear. Vinyl chloride exposure can also lead to an autoimmune syndrome.

Toxicant-Induced Immunosuppression

Immunosuppression, the decreased responsiveness of some or all of the types of cells of the immune system, can be caused by a number of different factors, ranging from genetic disorders to viral infections to exposure to toxicants. Immunosuppression can even be deliberately induced, as in the case of patients who have undergone organ transplants and hope to avoid rejection of the transplanted tissues, or in the case of patients with autoimmune diseases.

The consequences of immunosuppression depend on the part of the immune system that is affected as well as the degree of suppression. Humoral immunity, cellular immunity, or both may be affected. A number of consequences have been observed (both

Benzene

See also:
   Anemias, Hemolysis, and
    Related Disorders
       Chapter 9

Benzene
       Appendix

in the laboratory and in epidemiological studies) to accompany immunosuppression. These include not only increased susceptibility to various types of infection, but also increased risk of cancer (presumably, immune surveillance against abnormal cells is depressed). Also, immunosuppression can diminish the effectiveness of immunizations in preventing future illnesses.

Because of the complexity of the immune system, there are many possible mechanisms by which drugs and toxicants can produce immunosuppression. *Benzene*, for example, is generally cytotoxic to bone marrow, affecting production of white cells, red cells, and platelets. The mechanism of action of benzene is not completely clear, but it does appear that a metabolite of benzene, and not benzene itself, is responsible for the toxicity. Exposure to benzene has been linked to a decrease in circulating lymphocytes as well as antibodies in humans, and has been shown to lower resistance to infection in rats. *Alkylating agents* (which disrupt DNA replication and thus prevent cell division), *antimetabolites* (which inhibit the synthesis of nucleic acids, again interfering with DNA replication), and *radiation* exposure also produce general immunosuppression through effects on bone marrow (as well as on other lymphoid tissues where white blood cells are proliferating).

Other immunosuppressants affect the action of lymphokines, the molecules through which the various components of the immune system communicate. The drug *cyclosporine A*, for example, inhibits the activation of T cells by inhibiting production of the lymphokine interleukin 2 (IL-2) by helper T cells. Cyclosporine binds to a *cyclophilin* molecule, and the complex binds to and inactivates another molecule, *calcineurin*, which is a phosphatase. This prevents calcineurin from dephosphorylating the DNA-binding protein *NFAT* and thus prevents it from entering the nucleus, where it would act as a transcription factor for the IL-2 gene. *Glucocorticoid* hormones also interfere with lymphokine actions, inhibiting *macrophage migration-inhibitory factor* (MIF), which keeps macrophages from wandering away, and *g-interferon* and interleukin 1 (which stimulate T cells).

Some immunotoxicants act directly on specific lymphoid tissues. Low doses (less than 1 mg/kg) of the compound *2,3,7,8-tetrachlorodibenzo-p-dioxin* (TCDD) produce severe damage to the thymus in guinea pigs, with resulting depression in both antibody production and T cell function. While the

TCDD

See also:
   The Role of Cytochrome P450
                     Chapter 3

   *Stimulation of Cell Growth
   and Division*
                     Chapter 6

   *Endocrine Disrupters
   and Interference with
   Hormonal Controls*
                     Chapter 7

   Pesticides
                     Chapter 17

   TCDD
                     Appendix

PCBs and PBBs

See also:
   *Endocrine Disrupters
   and Interference with
   Hormonal Controls*
                     Chapter 7

   Other Organic Compounds
                     Chapter 17

   *Chemical Contaminants in
   Food*
                     Chapter 19

   PCBs and PBBs
                     Appendix

complete mechanism of action is unclear, the immunosuppressive action of TCDD seems to involve binding to the *aryl hydrocarbon receptor* (AHR), a ligand-activated transcription factor found in the cytoplasm of thymic epithelial cells. This receptor is also found in hepatocytes, where it is involved in induction of one form of cytochrome P450. Some *organotin* compounds also have direct effects on the thymus.

For many toxicants, though, immune system effects and mechanisms of action are much less well defined. Oral or dermal exposure to *polychlorinated biphenyls* (PCBs) lowers circulating antibody levels in mice; however, effects on cellular immunity are not as clear-cut. In a number of experiments, PCBs suppressed T cell functions, but in other experiments T cell functions were enhanced. In humans, PCB exposure has been tentatively associated with decreased antibody levels and increased susceptibility to infection. *Polybrominated biphenyls* (PBBs) have also been shown in the laboratory to suppress antibody production, and at higher exposures to suppress cellular immunity as well. In an epidemiological study, a group of Michigan residents who were inadvertently exposed to PBBs

Lead

See also:
  ***Interference with Cell
    Division***
                    Chapter 7

  ***Anemias, Hemolysis, and
    Related Disorders***
                    Chapter 9

  ***Myelinopathies
  Other Neurotoxicants***
                    Chapter 10

  Metals
                    Chapter 17

  Lead
                    Appendix

through contamination of livestock later displayed a higher percentage of immune system abnormalities than a group of unexposed individuals. However, PCBs and PBBs are known to be contaminated with minute quantities of chlorinated dibenzofurans, compounds with mechanisms of action similar to those of TCDD, which may be responsible for the observed immunosuppressive effects.

Exposure to some metals, including *lead*, has been shown to have adverse effects on immune function. Lead in drinking water appears to increase susceptibility of rats and mice to bacterial and viral infections, and there is epidemiological evidence that people with elevated blood lead levels (such as workers in the lead industry or children exposed to lead in paints) may experience the same effects. Lead affects humoral immunity (perhaps through interference with macrophage function) and may or may not affect cellular immunity (the results from different studies have been contradictory). *Cadmium* and *mercury* have similar patterns of activity.

Compounds such as *polycyclic aromatic hydrocarbons* (PAHs) and pesticides (including *carbamates*, *organochlorines*, and *organophosphates*) have also been suspected of having immunotoxic effects. DDT (an organochlorine), for example, has been shown to have effects on both cellular and humoral immunity, an effect that may be mediated in part through decreases in cytokine production. Both DDT and dieldrin (another organochlorine) may also suppress macrophage function. Mycotoxins such as aflatoxin have also been implicated in immunosuppression. Finally, both synthetic estrogens such as diethylstilbestrol and synthetic androgens have been shown to have effects on immune function. Studies on these and other suspected immunotoxicants continue.

AIDS and Antiviral Drugs

Infection with the sexually transmitted *human immunodeficiency virus* (HIV), a lentivirus with retroviral mechanism of proliferation, leads *to acquired immunodeficiency syndrome* (AIDS). This disease is typified by acute infection followed by clinical latency of several years prior to onset of symptoms. The syndrome results from destruction of CD4+ T cells. Chronic loss of CD4+ T cells can result in susceptibility to various protozoal, bacterial, fungal, and viral infections, several neurological symptoms, and death.

As with other retroviruses, RNA of the infecting virus is processed to DNA, which then enters the nucleus of the host and is integrated with a host chromosome. Just as processing DNA to RNA is called transcription, so the processing of RNA back to DNA is called *reverse transcription* and is catalyzed by the enzyme *reverse transcriptase*. The DNA thus inserted bears a sequence of around 9269 nucleotide bases, including 10 open reading frames (gag, pro, pol, vif, vpr, tat, rev, vpu, env, and nef) producing direct or spliced RNA transcripts.

Treatment of HIV/AIDS was based initially on drugs found to inhibit the function of enzymes encoded among those genes, especially reverse transcriptase and *aspartic proteinase*. A third enzyme, *integrase*, was another potential target of chemical inhibition. Practical therapy of HIV/AIDS now rests on a cocktail of chemical inhibitors of reverse transcriptase, integrase, and aspartic proteinase, as well as drugs designed to block viral entry into cells. This regimen, a combination termed "highly active retroviral therapy," does not eliminate the infection, but does keep viral levels low enough to maintain adequate immune function and thus avoid the opportunistic infections which are so problematic for these patients.

As for the specifics of inhibition of reverse transcriptase, both nucleoside analogs and nonnucleoside compounds have been found that are inhibitory and thus have some efficacy against AIDS. First and most famous is *3′-azido-3′-deoxythymidine* (AZT), a simple azido derivative of the nucleoside thymidine. AZT was originally developed for cancer chemotherapy. Other, similar drugs are derived from deoxycytidine and deoxyguanosine. Nucleoside-type drugs are toxic and poorly tolerated, likely due to nontarget inhibition of DNA polymerases. Nonnucleoside reverse transcriptase inhibitors, such as *thiobenzimidazolone* (TIBO), come in a variety of structures and are better tolerated. Another compound applied against reverse transcriptase is *phosphonoformate*; however, this is a nonselective toxicant. HIV aspartic proteinase inhibitors have also been developed, beginning with derived polypeptide mimics and extending to nonpeptidyl inhibitors such as *indinavir.*

A major hindrance to therapy is that both types of reverse transcriptase inhibitors rapidly select for resistance in the target enzyme due to the high rate of replication of HIV, as well as the apparent presence of either a mixture of viruses or a significant rate of mutation. Resistance was often observed in just a few weeks of administration of a single drug; therefore, combinations of drugs are used to delay resistance based on the theory that multiple resistance will be slower to evolve. Resistance has also arisen in response to aspartic proteinase inhibitors. These drugs were administered to patients, and it was observed that cross-resistance extended from the drug in use to proteinase inhibitors not yet applied in therapy. Highly active retroviral therapy, however, with the strategy of providing multiple barriers to successful viral replication, seems to be the most effective at diminishing the development of resistance in the virus.

METHODS FOR STUDYING IMMUNOTOXICITY

There are several established methods available for studying immune function in the laboratory. Because of the complexity of the system, the National Toxicology Program has recommended a tier-type approach to assessing immunotoxicity which involves several of the tests described below. NTP has also published guidelines to assist in the interpretation of these test results.

The simplest assessments include monitoring white blood cell levels and looking for changes in weight or abnormal histology in lymphoid tissues. Overall function can be assessed by challenging the immune system of the treated animal by exposing it to bacteria or viruses and comparing the results (rate of infection or mortality) to results from control animals.

There are a few animal models for testing specific immunotoxicological conditions. Guinea pigs and sometimes mice are most often used to model asthma and pulmonary-related allergic reactions. Guinea pigs have also been used in assessing allergic contact dermatitis, although mouse models have more recently been developed. In terms of autoimmunity, mouse genetic models exist for diabetes (the nonobese diabetic mouse) and lupus, and there are immunological methods for inducing rheumatoid arthritis in rats.

Other tests focus specifically on assessing cellular immunity. These include the *mixed lymphocyte response* (MLR) assay, which measures the ability of spleen T cells to proliferate when exposed to cells from another individual, and the *cytotoxic T lymphocyte (CTL) assay*, which measures the ability of spleen T cells to proliferate and lyse target cells. The proliferative activity of natural killer (NK) cells can be measured in much the same manner, and the ability of NK cells to lyse target cells can also be quantified. Other assays have been designed to measure phagocytic activity of macrophages.

Humoral immunity can be assessed through quantification of plasma antibody levels, perhaps in response to an antigenic challenge in the form of an injection of a stimulus such as sheep red blood cells (an antibody forming cell assay). Another way to quantify antibody production is to count the antibody-producing cells in the spleen following such an antigenic challenge.

More detailed assessment of the development of both types of immune responses can be undertaken through the use of tools such as the *flow cytometer*, an instrument which can rapidly quantify subpopulations of T cells and B cells based on a number of different cellular characteristics. Most commonly, fluorescent markers are used to identify the different cell surface proteins displayed by different subpopulations of cells, but intracellular staining (of cytokines, for example), can also be used to distinguish one subtype of cell from another.

Finally, measurement of the levels of individual cytokines in the plasma (or in cell cultures), a technique known as *cytokine profiling*, can also be used to diagnose immune response. Cytokine levels can be measured directly through techniques such as ELISA, or levels of cytokine mRNA can be measured using quantitative reverse transcriptase-PCR (RT-PCR).

BIBLIOGRAPHY

Aronoff, D.M. and Neilson, E.G., Antipyretics: Mechanisms of action and clinical use in fever suppression, *Am. J. Med.*, 111, 304, 2001.

Bircher, A.J., Symptoms and danger signs in acute drug hypersensitivity, *Toxicology*, 209, 201, 2005.

Blanca, M., Cornejo-Garcia, J.A., Torres, M.J., and Mayorga, C., Specificities of B-cell reactions to drugs. The penicillin model, *Toxicology*, 209, 181, 2005.

Burns-Naas, L.A., Meade, B.J., and Munson, A.E., Toxic responses of the immune system, in *Casarett and Doull's Toxicology*, Klaassen, C.D., Ed., McGraw-Hill, New York, 2001, chap. 12.

Dean, J.H., House, R.V., and Luster, M.I., Immunotoxicology: Effects of, and response to, drugs and chemicals, in *Principles and Methods of Toxicology*, 4th ed., Hayes, A.W., Ed., Taylor & Francis Group, Philadelphia, 2001, chap. 31.

Diasio, R.B. and LoBuglio, A.F., Immunomodulators: Immunosuppressive agents and immunostimulants, in *Goodman and Gilman's: The Pharmacological Basis for Therapeutics*, 9th ed., Hardman, J.G., Limbird, L.E., Molinoff, P.B., Ruddon, R.W., and Gilman, A.G., Eds., McGraw-Hill, New York, 1996, chap. 52.

Dinarello, C.A., Infection, fever, and exogenous and endogenous pyrogens: Some concepts have changed, *J. Endotoxin Res.*, 10, 4, 2004.

Emini, E.A. and Fan, H.Y., Immunological and pharmacological approaches to the control of retroviral infections, in *Retroviruses*, Coffin, J.M., Hughes, S.H., and Varmus, H.E., Eds., Cold Spring Harbor Laboratory Press, Plainview, NY, 1997, chap. 12.

Hasday, J.D., Fairchild, K.D., and Shanholtz, C., The role of fever in the infected host, *Microbes Infect.*, 2, 1891, 2000.

Holsapple, M.P., Autoimmunity by pesticides: A critical review of the state of the science, *Toxicol. Lett.*, 127, 101, 2002.

Jain, J., McCaffrey, P.G., Miner, Z., Kerppola, T.K., Lambert, J.N., Verdine, G.L., Curran, T., and Rao, A., The T-cell transcription factor NFATp is a substrate for calcineurin and interacts with Fos and Jun, *Nature*, 365, 352, 1993.

Kaplan, B.F.J., Sulentic, C.E.W., Holsapple, M.P., and Kaminski, N.E., Toxic responses of the immune system, in *Casarett and Doull's Toxicology*, Klaassen, C.E., Ed., McGraw-Hill, New York, 2013, chap. 12.

Kimber, I. and Dearman, R.J., Immune responses: Adverse versus non-adverse effects, *Toxicol. Pathol.*, 30, 54, 2002.

National Toxicology Program, Explanation of levels of evidence for immune system toxicity, http://ntp.niehs.nih.gov/files/09-3566_NTP-ITOX-R6.pdf, 2009.

Nossal, G.J.V., Life, death and the immune system, *Sci. Am.*, 269, 52, 1993.

Nossal, G.J.V., The double helix and immunology, *Nature*, 421, 414, 2003.

Selgrade, M.K., Germolec, D.R., Luebke, R.W., Smialowicz, R.J., Ward, M.D., and Sailstad, D.M., Immunotoxicity, in *Introduction to Biochemical Toxicology*, Hodgson, E. and Smart, R.C., Eds., Elsevier, New York, 2001, chap. 23.

14 Ecological Toxicology

INTRODUCTION

A relatively new area within the field of toxicology is *ecological toxicology* or *ecotoxicology*. Although classical toxicology is concerned with assessing effects of toxicants on the molecular, cellular, or physiological levels, ecological toxicology focuses on effects on populations, communities, and ecosystems. This chapter reviews some basic principles of ecology and discusses effects of toxicants on the population, community, and ecosystem levels and how they can be measured. This chapter focuses more on general principles, with specific environmental toxicants discussed in Chapter 17.

EFFECTS OF TOXICANTS AT THE POPULATION LEVEL

POPULATION GENETICS

There are, of course, many different kinds of organisms in the world. Organisms that are structurally and functionally similar and have the ability to produce offspring together are considered to belong to the same *species*. A *population* is a group of organisms of the same species that occupy the same area at the same time.

Some people who study populations focus on *population genetics*, studying changes in the gene pool (the sum of all genes in a population). Normally, each individual has two copies of each of his or her genes (one from each parent). In a population, all copies of a gene may be identical, or there may be two or more variations called *alleles*. For any given gene, individuals within the population may have two identical alleles (in which case they are said to be *homozygous* for that gene) or two different alleles (in which case they are said to be *heterozygous* for that gene). The assortment of alleles that individuals possess is their *genotype*, while the physical characteristics they display is their *phenotype*.

In a population, as individuals mate and produce offspring, alleles for each gene are passed on to the next generation in various combinations. It can be mathematically demonstrated that this reshuffling of alleles from generation to generation should not, however, change the overall frequency of a given allele in the gene pool (a principle called the *Hardy–Weinberg law*; Figure 14.1). The frequencies of the alleles in a population can then be used to calculate the expected frequency of genotypes, which also should not vary from generation to generation (a state called *Hardy–Weinberg equilibrium*).

However, Hardy–Weinberg equilibrium is not typically seen in populations because a number of factors tend to disrupt it. Changes in the gene pool, especially in small populations, can be produced by random fluctuations, disasters that

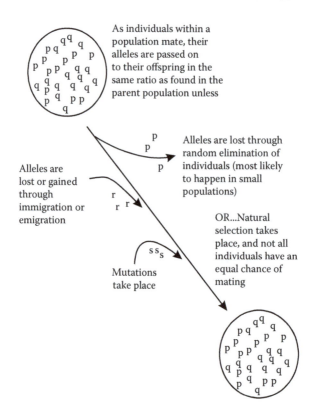

As individuals within a population mate, their alleles are passed on to their offspring in the same ratio as found in the parent population unless

Alleles are lost or gained through immigration or emigration

Alleles are lost through random elimination of individuals (most likely to happen in small populations)

Mutations take place

OR...Natural selection takes place, and not all individuals have an equal chance of mating

FIGURE 14.1   The Hardy–Weinberg principle.

dramatically reduce population size, immigration and emigration, and spontaneous mutations. The most important factor, however, that can alter Hardy–Weinberg equilibrium is *natural selection*.

Natural Selection

The concept of natural selection is based on observations that in any population (1) more individuals are born than will live to reproduce, (2) individuals in a population vary in any number of traits, (3) variations in traits can be inherited, and (4) those individuals whose traits give them an advantage in survival and reproduction (in other words, those individuals who are more fit for survival) are most likely to survive to pass those traits on to the next generation. Genetic fitness is measured by the number of offspring produced; population genetics attempts to estimate the relative fitness of each allele. Natural selection can alter allele frequencies by favoring a particular genotype that confers some fitness advantage. Individuals with this genotype are more likely to survive to reproduce and thus pass on their alleles with a greater frequency than individuals of other genotypes. The result of natural selection is that each succeeding generation of a population should become better adapted to its environment.

Characteristics of an environment that influence natural selection are called *selection pressures*. Many factors can act as selection pressures, including physical characteristics of the environment (such as climate, for example) and biological characteristics such as the presence or absence of other organisms. One special type of selection is *sexual selection*, which operates on those characteristics that influence the likelihood of finding a mate, and can clearly influence reproductive success. Sexual selection can lead to magnificent sexual dimorphism as exemplified in the colorful plumage of some male birds, but also contributes to the general characteristics of both sexes. Some have hypothesized that human intelligence has been greatly influenced by sexual selection.

NATURAL SELECTION, TOXICANTS, AND RESISTANCE

Introduction of toxicants into an environment can also exert strong selection pressures favoring or disfavoring particular characteristics. In the classic example, prior to the Industrial Revolution, the light-colored form of the English peppered moth (a moth that frequents tree trunks and rocks) was much more prevalent than the dark-colored form (which presented an easy target for predators against the light background). Once soot began to blacken the trees and rocks, however, the light-colored moth became more conspicuous, and within a few years almost all peppered moths were of the dark variety.

Another example of selection pressures exerted by toxicants is the effect of pesticides on target species. Random mutations present at extremely low frequencies may be rapidly selected to render a pest species resistant to a particular pesticide. Many cases of resistance occur when a single mutation alters a target protein, making it less sensitive to attack by the pesticide. Recent analysis has shown, however, that some cases of resistance have been due to *amplification* of genes, where many extra copies of a particular gene are found in resistant individuals.

When exposed to a pesticide, the individuals in a population that are most resistant to its effects are those most likely to survive and reproduce. Thus, alleles responsible for pesticide resistance are passed on, and each successive generation becomes more resistant to the pesticide. Development of resistance to a pesticide has been observed in hundreds of species of insects, weeds, fungi, and other organisms. Resistance may develop in response to exposure to pollutants also. The magnitude of these effects is greatly determined by the genetic makeup of the population, dominance of the resistance trait, and the proportion of individuals within the population that are exposed.

An allele for resistance to a pesticide or other environmental toxicant can be selected to fixation (100%) in a closed population. This will happen more rapidly if the resistant allele is recessive, because then only homozygous individuals are selected. When resistance is dominant, both homozygous and heterozygous individuals are selected, resulting in perpetuation of the susceptible allele among heterozygotes. This is an example in which estimating the proportion of heterozygotes can be very important to pest management plans.

Development of resistance also occurs in microbes exposed to antibiotics. The mechanisms by which resistance develops in bacteria have been extensively studied. Resistant bacteria appear to recruit and assemble advantageous alleles on a plasmid or within a transposon in a unit that is called an *integron*. Integrons typically contain both genes for resistance (forming what can be called a *gene cassette*) and genes for *transposase*, *integrase*, and *recombinase* catalytic activities that provide mobility for the entire unit. This phenomenon is battled in hospitals, where integrons have conferred resistance to multiple antibiotics in some strains of *Staphylococcus aureus* and other enterococci. These gene cassettes are not limited to plasmids, as larger assemblies of more than 100 integrons called *superintegrons* occur on a small circular chromosome in *Vibrio cholerae*.

The use of antibiotics as growth promoters in animal feed has raised concerns that antibiotic-resistant strains of bacteria are being selected for in these animals. These antibiotic-resistant strains may have the potential to infect people or may pass resistance-carrying alleles to other bacterial species. There has been at least one well-documented case where an antibiotic-resistant strain of *Salmonella* was apparently transmitted to people through infected beef, causing illnesses and deaths. As a result of these concerns, the FDA submitted a plan in late 2013 under which companies which are engaged in the manufacture of animal pharmaceuticals will voluntarily move some products considered to be "medically important" from over the counter availability to veterinary prescription status.

RECOMBINANT ORGANISMS

In the future, it is also likely that the gene pool will be affected more and more by genetic engineering. *Recombinant organisms* (for bioremediation, for pest resistance in crops, etc.) will be released with greater and greater frequency into the environment. Currently, there are multiple genetically modified plants in common use in agriculture, and their risks and benefits with regard to human health will be addressed in a later chapter.

Micro RNA

See also:
  Effects of Toxicants on Nucleic Acids
                                                    Chapter 4

Aside from human health effects, though, the potential for interbreeding of modified with wild-type organisms has also led to concerns over what has been termed "genetic pollution." *Introgression*, the permanent incorporation of genes from one population into another, has, in fact, been documented between genetically engineered crop plants and wild-type relatives in several species including soybeans, millet, sorghum, and corn. An additional concern has arisen with plants which, instead of being genetically modified to produce a new and different protein (as has been the traditional case), are modified to produce a different range of regulatory RNAs (such as siRNA or miRNA). These double-stranded RNA molecules are relatively stable in the environment, and there is some evidence that organisms feeding on the plant or on plant detritus may take these molecules in, with a resulting impact on gene expression. In fact, some plants have been deliberately engineered to produce siRNAs which are potentially toxic to an insect pest which feeds on the plant. The ecological significance of all of these potential effects has yet to be fully determined, but should prove to be an important area of study in years to come.

POPULATION GROWTH AND DYNAMICS

Some ecologists study the size and characteristics of populations. Populations grow in size due to *natality* (births) and *immigration* and decrease in size due to *mortality* (deaths) and *emigration*. The net effect on population size depends on the balance between these opposing factors. If we ignore immigration and emigration, population growth can be modeled very simply by using the equation

$$\frac{\Delta N}{\Delta T} = rN,$$

where ΔN is the change in the number of individuals in the population and ΔT the time passed. This equation reflects the fact that population growth is proportional to the number of individuals in the population. The factor r is the *intrinsic rate of increase* in the population and is a measure of the balance between rates of natality and mortality. Organisms with high r factors are called *r strategists*. They reproduce early and often, with only a small percentage of offspring surviving to adulthood.

Populations do not, however, grow indefinitely. For any population, there is a *carrying capacity*, K, which is the maximum number of individuals that can be supported by the environment. As the population nears the carrying capacity, the growth rate slows and eventually the population size stabilizes. Population size for r strategists tends to oscillate relatively rapidly from just above to just below carrying capacity. Other organisms, however, tend to maintain a steady population size just at the carrying capacity (Figure 14.2). These *K strategists* are characterized by much lower reproductive rates and longer life spans than r strategists.

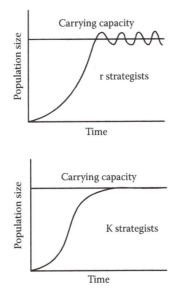

FIGURE 14.2   The relationship between population size and carrying capacity in r strategists and K strategists.

Many factors affect carrying capacity and thus population size. The effects of *density-dependent factors* intensify with increases in population density (the number of individuals per some unit area). One example of a density-dependent factor is *competition* for resources (food, water, shelter, etc.) between members of the same population. The higher the population density, the more intense the competition for finite resources. Another density-dependent factor is *predation*. The higher the population density of a prey population, the more successful predators are likely to be. (We will discuss predator–prey interactions in more detail in the *community* section of this chapter.)

Density-independent factors, on the other hand, affect populations in ways that are not dependent on population density. Drastic weather changes, for example, may kill many individuals in a sensitive population.

Toxicants, in many cases, may act on populations in a density-independent fashion. For example, a chemical spill into an aquatic environment might be expected to kill a percentage of the individuals in a given population, regardless of whether the population is large or small. Toxicants can also interact with other factors in a density-dependent manner. Toxicants that produce sublethal effects may render individuals more susceptible to infection, for example. This may have a greater effect in crowded populations, where pathogens may be more easily transmitted. In general, r strategists can rebound much more quickly from toxicant-induced reductions in population size than can K strategists.

EFFECTS OF TOXICANTS AT THE COMMUNITY LEVEL

All the populations living in the same area at the same time comprise a unit called a *community*. Ecologists who study communities study both community structure as a whole and the interactions between individual species within the community.

One of the most important characteristics of a community is the type and number of species that comprise it. This characteristic is often termed *biodiversity*. It is also important to measure the relative abundance of each species. Species that are particularly abundant or that play particularly important roles in the structure or function of a community are called *dominant* species. The type of community that develops in a given area depends on factors such as climate, soil, and other physical conditions. A relatively stable community that is characteristic of a given region is called the *climax community* for that region.

Community structure is not always static, but may change over time, particularly in response to environmental change. Change in the structure and composition of a community over time is called *succession. Primary succession* refers to the development of a community on a previously uncolonized area; *secondary succession* refers to changes that occur following some perturbation of an existing community. During each of the various stages of succession, organisms dominate that are best able to adapt to the current conditions. These organisms, in turn, affect conditions in such a way as to allow other species to survive and prosper. These changes continue until a stable stage evolves, which is usually the climax community for the region.

Along with physical conditions, the other major factors involved in the determination of community structure are the interactions between different populations within the community. For example, different species may compete for a common resource. In fact, two species that are too similar in resource and environmental requirements (in other words, too similar in their *niche*, or role that each plays in the community) generally cannot coexist in the same community. Other interactions include *predation* (where the predator survives by killing and eating the prey), *parasitism* (where the parasite derives nourishment from the host, but generally without causing the host's death), and *mutualism* (where both species gain some advantage from a relationship).

Toxicants can affect community structure and function in several ways. Effects on a particularly sensitive population may not be limited to that population, but may affect other populations with which that species interacts (Figure 14.3). These interactions are not, however, always easily predictable. Elimination of species through effects of toxicants will, in many cases, lead to a decrease in biodiversity within a community. Sometimes, however, this is not the case. For example, elimination of a dominant insect competitor through pesticide use may actually increase biodiversity by allowing a more equitable competition for (and thus sharing of) resources by a greater number of species.

Predator–prey interactions can also be affected by the introduction of toxicants into the environment. Reduction in the number of individuals in a prey population, for example, may also adversely affect one or more predators that depend on that particular prey for a substantial portion of their diet. Likewise, loss of a major predator species can also have a significant impact on community structure. Removal of predation pressure can lead to either an increase in biodiversity, if more species are allowed to flourish, or a decrease in biodiversity, if the absence of predation allows one species to become dominant. Sublethal effects of toxicants on predator or prey may also perturb normal predator–prey dynamics.

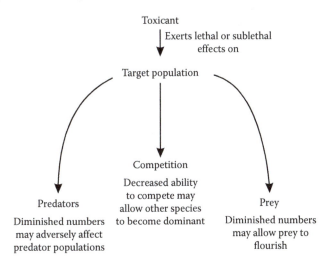

FIGURE 14.3   Effects of toxicants on communities.

EFFECTS OF TOXICANTS AT THE ECOSYSTEM LEVEL

A community together with its physical environment comprises an *ecosystem*. There are many different types of terrestrial and aquatic ecosystems, each with its own unique properties that must be considered when studying the effects of pollutants. Two processes that are commonly studied in ecosystems are *energy flow* and *material cycling*. Toxicants can disrupt either one of these.

ENERGY FLOW IN ECOSYSTEMS

The initial source of energy for ecosystem processes comes from the electromagnetic radiation emitted by the sun. *Autotrophs* are organisms with the capability of trapping this electromagnetic energy and storing it in the chemical bonds of molecules such as glucose. These organisms (plants, protists, bacteria) that build energy-storing molecules function as *producers* in the ecosystem.

The rate at which energy is stored by producers is called the *primary productivity* of the ecosystem. Of this energy, some is used by the producers themselves to maintain physiological processes necessary for their own survival. The excess energy remains stored in the molecules that make up the structure of the organisms, or in other words, in the *biomass* of the organisms.

The second level of organisms in an ecosystem is the organisms that feed directly on the consumers. These organisms are the *primary consumers* or *herbivores*, such as some insects, mammals, and birds. Of the energy these organisms consume, some goes to maintaining their physiological processes and some is stored in their own biomass. *Secondary consumers* or *carnivores* (such as many amphibians and some mammals), in turn, feed on primary consumers. *Omnivores* may feed on both producers and primary consumers. The energy remaining in organic waste products (including dead organisms) is used by *detritivores* such as bacteria and fungi. Graphic representations of the feeding relationships between different organisms in an ecosystem are called *food webs*.

Each of these levels of organisms (producers, primary consumers, secondary consumers, etc.) is called a *trophic level*, and the biomass at each trophic level is less than that at the level below. This is because the organisms at each level use some of the energy they receive from the level below and therefore have less to store as biomass. This concept can be visualized with an *energy pyramid*, showing the relative amount of stored energy available at each level (Figure 14.4).

Because of the tight interrelationships between levels, the impact of a toxicant that affects organisms at one level has the potential to spread to other levels as well. For example, toxicant exposure has the potential to decrease productivity in both terrestrial and aquatic ecosystems. This toxicant-induced biomass reduction in producers may then lead to even longer-lasting and more significant biomass reductions at higher trophic levels. This enhanced effect is partly because producers tend to be r strategists, while secondary consumers tend to be K strategists. Toxicants that affect detritivores may also impact the entire ecosystem by preventing metabolism and release of nutrients for use by producers. For example,

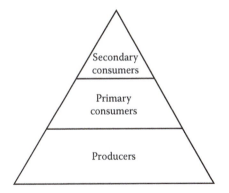

FIGURE 14.4 The trophic pyramid.

studies have shown that metal-contaminated leaf litter is broken down at a slower rate than noncontaminated litter. Changes in soil pH can also impair detritivore function.

MATERIAL CYCLING IN ECOSYSTEMS

The cycling of substances such as water, nitrogen, carbon, and phosphorus through an ecosystem is also critical to ecosystem health. In the *hydrologic cycle*, water molecules cycle between the ocean, ice, surface water, groundwater, and the atmosphere. In the *nitrogen cycle*, nitrogen in the atmosphere is converted by bacteria into ammonia, nitrite, and nitrate. Plants can absorb ammonia and nitrate and incorporate them into proteins. Animals then get their necessary nitrogen by consuming plant products. Other soil bacteria convert nitrogen-containing organic wastes back into ammonia. Some bacteria are also capable of converting nitrogen-containing organic compounds back into gaseous nitrogen. Other important cycles are the *carbon cycle* and the *phosphorus cycle*. These cycles can be disrupted by alterations in the environment. The carbon cycle, for example, has been altered by the increased input of carbon dioxide resulting from combustion of fossil fuels (Figure 14.5). This may ultimately lead to significant changes in global climate.

An understanding of material cycling is important in ecotoxicology because toxicants that are released into ecosystems also cycle. The study of how chemicals move through the environment is called *chemodynamics*. Toxicants may be transported as gases or particulates through the air, dissolved or adsorbed on the surface of particles in the water, or may leach through soils. *Residence times* may be calculated by dividing the total mass of a toxicant by the rate of change (input or output). Residence times for toxicants in the atmosphere are often only a few days, while toxicants in water may have residence times of weeks or months. The residence times for toxicants in soils, however, tend to be much longer, often for hundreds or thousands of years. Sediments on the bottoms of lakes, streams, and oceans may become sinks for pollutants, potentially exposing bottom-dwelling organisms to toxicants for many years.

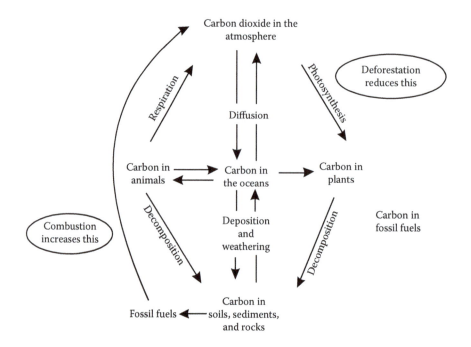

FIGURE 14.5 The carbon cycle, along with two potential environmental impacts that can alter it.

Toxicants in the environment may also be carried by or concentrate in biological tissues through physical processes (such as filter feeding) or chemical processes. The degree to which a chemical is available for uptake by organisms is called its *biological availability*, or *bioavailability*.

Nonpolar, lipophilic compounds in particular tend to undergo *bioconcentration* (or as it is also called, *bioaccumulation*). The bioconcentration factor of a toxicant is the ratio of its concentration in a particular organism to its concentration in the environment. In general, the higher up its position in a food web is, the more susceptible an organism may be to effects due to bioconcentration.

One example where bioconcentration played an important role is in the actions of the pesticide DDT. Although levels of DDT in small aquatic species were lower than 1 ppm, levels in birds of prey (carnivores at the top of the food web) reached as high as 25 ppm. These levels were sufficient to interfere with eggshell formation, dramatically reducing the numbers of viable offspring in affected populations. Increasing bioconcentration through the food web does not always occur, however. For some metals, bioconcentration is highest in the producers and declines at higher trophic levels.

DDT

See also:

Effects of Toxicants on Electrical Conduction

                                            Chapter 10

Pesticides

                                            Chapter 17

Organochlorine Pesticides

                                            Appendix

Toxicants may move through the environment unchanged, or may be altered through chemical or biological interactions. Some toxicants undergo abiotic transformation, reacting with chemicals in the environment, while others may be metabolized by bacteria or other species. For some compounds, these changes may lead to detoxification, but for other relatively nontoxic compounds, the end result may be activation or transformation to a more toxic form. Metals, for example, may be methylated by microorganisms to form more toxic organometals.

Biotransformation

See also:
  Biotransformation
                                   Chapter 3

EXAMPLES OF ECOSYSTEMS AND VULNERABILITY TO IMPACT BY TOXICANTS

MARINE ECOSYSTEMS

Water in the oceans is in the form of salt water, with a salt concentration of 35%. Water temperature can range from very cold for water near the poles and near the ocean bottoms to very warm for surface waters near the equator. Waves and currents move both surface and deep waters. The oceans can be divided into zones (Figure 14.6). The *littoral* and *neritic* zones are the most productive. The greatest biodiversity is found here, where many different species of *phytoplankton* (photosynthetic protists), *zooplankton* (herbivorous protists), and *nekton* (free-swimming organisms) live. The *benthos*, too, supports a high degree of biodiversity, even in very deep regions.

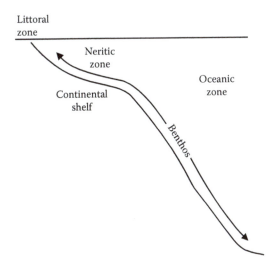

FIGURE 14.6 Zones in the ocean.

The open ocean is vulnerable to several different types of pollutants. Some of the most serious problems are accidental oil spillage and leakage, deliberate offshore dumping of hazardous and radioactive wastes, and disposal of nondegradable plastics such as fishing line and nets or plastic soda can rings.

Oil Spills

See also:

    Petroleum Products as Water
      Pollutants

                    Chapter 17

    Petroleum Products

                    Appendix

Probably the most significant pollution problems in the ocean, though, occur in the areas where aquatic and terrestrial ecosystems meet—the shorelines. *Rocky shores* and *sandy shores*, particularly in popular resort areas, suffer from impacts such as habitat destruction and sewage disposal. The same threats that affect the open ocean (oil, plastic debris, hazardous waste) can also wash up onshore, causing problems. Specialized shoreline ecosystems such as *coral reefs* (tropical offshore structures built from the skeletons of animals called corals) and marine wetlands such as *salt marshes* or *mangrove forests* can also be affected. Estuaries (areas, typically at the mouth of a river, where freshwater meets salt water) are particularly vulnerable to pollution. They may receive heavy pollutant loads from upriver, from the ocean, and finally from municipal, agricultural, and industrial activities concentrated around the estuary itself. Because many important commercial fish- and shellfish-harvesting operations are located in estuaries, pollution of these areas can have significant economic as well as ecological impact.

FRESHWATER ECOSYSTEMS

Lakes and *ponds* are examples of *lentic*, or still water, ecosystems. They are inland depressions filled with water, which may be formed when glaciers gouge out the land or when streams become dammed (by either natural or man-made processes). Lakes, like the oceans, can be divided into zones (Figure 14.7). As in the ocean, the most productive and diverse zone is the *littoral*. Emergent and floating plants, insects, and fish dominate this zone. The *limnetic* zone contains mostly phytoplankton and zooplankton, along with some fish.

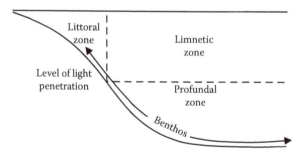

FIGURE 14.7 Zones in lakes.

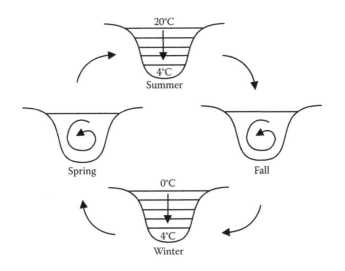

FIGURE 14.8   Lake stratification and overturn in temperate climates.

In *temperate climates* (climates that have changing seasons), lakes go through seasonal changes (Figure 14.8). The density of water varies with temperature, with the highest density at 4°C. Because of this, in the summer lakes become *stratified*, with the warmest (lowest density) water at the surface and progressively cooler, denser water at greater depths (down to 4°C on the floor). Because mixing does not occur, oxygen remains highest at the surface, where it enters the water through diffusion or is produced by phytoplankton. Organic material and nutrients, on the other hand, become concentrated near the bottom. Pollutants may also not disperse evenly through the lake, but may become concentrated in a particular layer, leading to the development of higher concentrations of toxicants than might otherwise be predicted.

In the fall, the water on the surface cools, becomes more dense, and sinks. The next warmest layer is then moved to the surface, where it, in turn, cools and sinks. Eventually, the whole lake approaches the same temperature and waters throughout the lake mix. This is called *overturn*. In the winter, stratification occurs again, except with the coldest waters (0°C) at the surface and the progressively warmer, more dense layers below (again, down to 4°C on the bottom). Then, in the spring, as the surface waters warm, overturn occurs again, mixing oxygen, nutrients, and also toxicants throughout the lake.

As lakes age, they tend to become more shallow. Runoff from surrounding areas and silt from streams bring in sediment and organic matter, which is deposited on the bottom of the lake. This input adds nutrients to the ecosystem. Young lakes are usually *oligotrophic* or nutrient-poor. Older lakes are *eutrophic* or nutrient-rich. This input of

Eutrophication

See also:
   Phosphorus and Nitrogen
                                   Chapter 17

   Cyanobacterial Toxins
                                   Chapter 20

nutrients generally stimulates plant growth, and eutrophic lakes are frequently characterized by a heavy growth of algae. Excessive input of sediments and nutrients as a result of human activities around a lake can lead to accelerated aging of a lake—a phenomenon called *cultural eutrophication*. This phenomenon is accentuated in glacial lakes, which, in pristine condition, are especially oligotrophic; therefore, human inputs, such as lawn fertilizer, can result in dramatic eutrophication.

The qualities of *lotic* (moving water) ecosystems such as rivers and streams are determined by several different variables, including the characteristics of the channel, the nature of the surrounding terrain, and climate conditions such as annual rainfall. Upstream, near the source, rivers and streams tend to be faster moving, eroding sediment from the bottom that will later be deposited downstream where velocity slows. Typically, rivers and streams consist of alternating *riffles*, which are areas of rapid and turbulent flow, and *pools*, which are deeper and have slower flow. Moss and algae can be found attached to rocks in the streambed, and insects and fish live in both fast- and slow-moving areas.

In the processes of material cycling and energy flow, lotic ecosystems receive a good deal of input from outside sources (organic matter falling or being washed into the water, overland runoff, precipitation, or seepage from the ground). Toxicants also enter the ecosystem in this manner. Historically, industrial and municipal waste has been dumped into nearby rivers with little or no treatment. Agricultural runoff containing pesticides and fertilizers has also been a problem. Some of these compounds also act as endocrine disrupters, affecting reproductive function of wildlife. Now, concerns have also arisen over nanoparticles, and their effects on algae and other aquatic life, as well as pharmaceutical products which enter the waterways (see the case study at the end of this chapter).

Transport through the atmosphere also plays a role in the movement of toxicants from one aquatic system to another. Global transport of organochlorine pesticides is likely to have resulted from volatilization into the atmosphere, followed by falling in rain at distant locations. Another example of this pathway was observed in the case of radioactivity from the Chernobyl fire, which was carried in a plume over Stockholm, but contaminated more northern areas of Sweden, where the plume was mixed with rainfall.

Endocrine Disruption

See also:
   Endocrine Disrupters and
   Interference with Hormonal
   Controls
                             Chapter 7

Water Pollution

See also:
   Water Pollution
                             Chapter 17

Freshwater wetlands are also impacted by water pollution. Freshwater wetlands can be defined as land areas that are periodically saturated with water. Examples include *marshes* (characterized by grasses as dominant vegetation), *swamps* (wooded wetlands), and *bogs* (wetlands dominated by sphagnum moss). Wetlands serve as important wildlife habitats, and high concentrations of toxicants can threaten the survival of many species. Wetlands are often destroyed to provide land for agricultural or other commercial uses.

TERRESTRIAL ECOSYSTEMS

The *tundra* occurs both at high latitudes (*arctic tundra*) and at high altitudes (*alpine tundra*), where the weather is so cold that much of the ground remains frozen all year round. In the summer, the upper layers of the ground may thaw, but the lower levels remain frozen. The water released by the thaw then remains on the surface, forming wetlands. Because the ground is so cold, activity of the decomposers in the soil is much slower than in warmer climates. Thus, nutrient turnover is limited. Dominant vegetation consists of moss, lichens, and grasses. In the summer, insects are abundant, as are birds (many of which nest in the wetlands) and mammals such as hare, caribou, and wolf.

The tundra is an example of a fragile ecosystem, meaning that it is slow to recover from perturbations. Several of the world's oil fields occur under the tundra, and as a result, this ecosystem has been exposed to significant pollution problems. Oil spills and leaks are always a threat. Evidence has shown, however, that tundra vegetation can recover, although slowly, from such spills. Another threat is the additional organic waste produced by human colonization of once remote areas (remember that the rate of decomposition is quite slow). The *Arctic National Wildlife Refuge* (ANWR) is an example of a tundra ecosystem that some individuals would like to open up for oil exploration and others would prefer to see remain as wilderness.

The dominant feature of *grasslands* is grass. This ecosystem is also the home to many insects and small, burrowing mammals. Grassland ecosystems are characterized by moderate rainfall that is frequently seasonal in nature. Fire is often an important force in maintenance of grasslands.

Unfortunately, much of the world's original grassland ecosystems have been converted to farming and grazing lands. The *tallgrass* and *shortgrass prairies* of North America, the *pampas* of South America, the *veld* of Africa, and the *steppes* of Eurasia have all been severely impacted. As far as effects of toxicants, probably the greatest threat to remaining grasslands is the contamination by pesticides used to manage adjacent agricultural lands.

Deserts are characterized by low rainfall, typically occurring as infrequent heavy cloudbursts. Temperatures in a desert may fluctuate widely during the day, ranging from very warm during the day to very cool at night. Dominant vegetation types are cacti and shrubs, and animals include insects, reptiles, birds, and mammals. Deserts are fragile ecosystems and in many parts of the world are threatened by the pollution that accompanies human activities such as oil drilling. People also frequently attempt to farm and graze on the marginal lands surrounding deserts. Overgrazing or loss of topsoil there can lead to *desertification* of these lands. Improper irrigation can also create problems. Evaporation of irrigation water can leave a residue of salts that are toxic to many desert plants.

Finally, there are many different types of *forests*. The *taiga* is a forest found at high altitudes and latitudes that is dominated by coniferous trees. *Temperate forests* occur in temperate zones and may be coniferous, deciduous, or mixed. *Tropical rain forests* occur where rainfall is heavy and even and the temperature is warm year-round. These ecosystems are all subjected to destruction through logging and also to air and water pollution caused by industrial activities (including logging, mining, and operation of power plants). They are also vulnerable to erosion.

CLIMATE CHANGE AND ECOTOXICOLOGY

One of the most challenging environmental issues that future generations will need to face is the issue of the changing climate and its impact on both human and ecological health. Estimates based on current models call for significant increases in average temperatures worldwide, along with changes in precipitation and storm patterns. A rise in sea level is predicted, as well, along with acidification of sea water. Obviously, these changes are going to have a significant impact, both on the distribution of environmental pollutants and on their effects on ecosystems. Exactly what those impacts will be is something that is much more difficult to predict.

In terms of atmospheric changes, several studies have estimated increases in ozone in the lower atmosphere in many regions, even while global tropospheric ozone levels decline. Particulate pollutants are likely to decrease in areas with increased rainfall and increase in areas with decreased rainfall. Changes in distribution of organic compounds (including pesticides) in water and soils may be complex, with increased degradation and volatilization due to increased temperatures weighed against increased deposition associated with greater rainfall. Changes in rainfall and runoff patterns, as well, have the potential to alter toxicant distribution. Of course, shifting patterns of agriculture (and geographical shifts in pest populations) will also alter patterns of pesticide use, adding another level of complexity.

Another pattern of complex changes will be likely to be seen in the oceans, where acidification secondary to increased atmospheric carbon dioxide levels has the potential to interact with pollution-related impacts on ocean life (particularly phytoplankton, the major producers in that ecosystem). Particularly in areas where eutrophication is a concern, increases in both respiration and production may lead to both a net decrease in carbon incorporation into plant matter and the development of hypoxic zones. Studies have also implicated the combination of climate change and eutrophication in blooms of harmful cyanobacteria, with potential negative effects on both aquatic and human life.

Effects of global warming on the salinity of both marine and freshwater systems are particularly difficult to determine. While sea level rise may lead to the incursion of higher salinity waters further upstream in some estuarine systems, increased precipitation and snowmelt could actually reduce salinity in others. Either way, the potential for impact on aquatic life is large.

One further issue of concern is the dual impact of toxicant exposure and the physiological stress caused by climate change on wildlife populations. Studies have indicated that increases in ambient temperatures can not only increase uptake of some toxicants from the environment, but can also increase toxicity. This increase in toxicity may be related to alterations in biotransformation or alterations in homeostatic mechanisms.

ECOTOXICOLOGICAL TESTING METHODS

SINGLE-SPECIES TESTING

Classic single-species toxicology testing, of the type discussed throughout this book, plays an important role in ecological toxicology. The difference is that instead of pursuing a goal of better understanding the effects of toxicants on human health (a direction in which most toxicological research is focused), ecotoxicologists are interested in better understanding the effects of toxicants on a variety of species. As such, it is more appropriate to work with a variety of species, including nonmammalian species such as insects, mollusks, amphibians, fish, or birds. Typical test organisms may include algae, daphnids, shrimp, honeybees, quail, trout, and fathead minnows.

Aquatic toxicity tests are often somewhat difficult to design, due to the complex chemistry of water. Water temperature, pH, ion concentrations, suspended solids, and dissolved gases, among other factors, must be closely monitored in order to accurately model real-world conditions. Systems can be either *static*, where water in the system is not changed during the test, or *flow-through*, where water is constantly removed and replenished. Flow-through systems, although more difficult to set up and maintain, are better both for providing acceptable water quality and for maintaining stable toxicant concentrations.

Most ecotoxicological testing to this point has focused less on mechanisms of action of toxicants and more on identifying endpoints with which to quantify toxicity. Although measurement of the relationship between dose and mortality (the classic LD_{50}) is usable

---
LD_{50}

See also:
The LD_{50} Experiment

Chapter 1

---

in many situations, a more straightforward and directly applicable correlation in ecological toxicology is the one between environmental concentration and mortality. The LC_{50} measures the concentration of toxicant in the environment (often an aquatic environment) that is necessary to produce mortality in 50% of the test population. LC_{50} tests are typically conducted for exposures ranging from 24 to 96 h in length. Sublethal effects such as changes in behavioral patterns (activity, feeding, reproductive, etc.) and effects on oxygen utilization and respiration can also be measured.

Many studies focus on identifying species that are particularly sensitive to the effects of a toxicant. These critical organisms (sometimes called *sentinel species*) would be expected to be among the first components of an ecosystem to be affected by a toxicant. Therefore, monitoring the well-being of sentinel species can be used as an early-warning system for detecting toxicant effects on ecosystem health. This concept is called *biomonitoring*.

Different developmental stages may also have different sensitivities to toxicants. In early-life-stage toxicity testing, organisms are exposed from fertilization through early juvenile stage, and growth and survival are quantified.

Finally, to complicate matters further, real-world ecological exposures often include exposures to many different toxic chemicals, sometimes simultaneously. Therefore, sophisticated techniques for analyzing and predicting effects of *mixtures* must often be used.

MICROCOSMS

Single-species tests are, however, insufficient for ecological toxicology testing. Because of their single-species design, they are, by definition, unsuitable for measuring community- and ecosystem-level interactions. These effects may be investigated in the laboratory setting by the use of *microcosms*—artificial ecosystems designed to model real-world processes. Microcosms are generally much less complex than a complete ecosystem, containing only a few selected species in an environment generally limited by size.

Terrestrial microcosms (consisting of soil along with resident microorganisms and invertebrates) are often used to study the fate and transport of pollutants (including microbial metabolism), along with pollutant effects on detritivore function. More complex systems may include plants and even some small vertebrates (typically amphibians such as salamanders or toads). Setting up an aquatic microcosm involves the same complications discussed earlier for single-species aquatic testing. Systems may be static or flow-through and can be used to study fate and transport of toxicants as well as predator–prey interactions and behavior.

Field Studies

Study of an actual toxicant-contaminated ecosystem is probably the best way to study the full set of complex interactions that characterize such a system. Samples of biotic and abiotic components can be taken and analyzed by gas chromatography, HPLC, atomic absorption spectroscopy, etc., for toxicant levels. Population sizes can be estimated and monitored through various ecological sampling methods. However, the actual effects of the toxicant can be difficult to determine without either (1) historical data on the area dating back to before contamination occurred or (2) a similar, uncontaminated ecosystem to use as a basis for comparison.

Mathematical Modeling

Finally, there are a number of ecotoxicological processes that can be modeled mathematically. For example, the fate and transport of a toxicant in an ecosystem may be predicted by using structure, lipid/water partition coefficient, and other physical or chemical properties in conjunction with ecosystem properties, such as soil and water chemistry, population levels, and predator–prey relationships. One type of tool that can be helpful in this sort of analysis is the development of *quantitative structure–activity relationships* (QSARs) (see Case Study in Chapter 6). QSAR methods first require databases relating chemical structure of compounds with their known endpoints (such as fate and transport parameters). Mathematical methods can then be developed to predict endpoints for untested chemicals.

One issue with environmental models is the fact that, due to complexity of ecosystem processes, they can very quickly become tremendously complex. Often, in order to simplify them, assumptions and estimations are made that may or may not be totally valid. Mathematical models, of course, are developed and validated through the use of field studies and thus cannot completely replace these sources of information.

Molecular and Cellular Ecotoxicology: A New Direction

New tools and techniques in the areas of cellular and molecular biology have led to advances in ecotoxicology as well. One such new direction is in the identification of *biomarkers*, which are measurable changes in cellular or biochemical processes or functions in ecosystem components. The identification and measurements of appropriate biomarkers can help to predict potential ecosystem dysfunction—much like taking a person's temperature can help predict the state of his health. Examples of biomarkers include measurement of DNA damage in blood cells, measurement of activity of pesticide-sensitive enzymes such as plasma cholinesterases, or measurement of inducible cytochrome P450 levels in the organisms of an ecosystem.

Gene Expression

See also:
   Sorting Out Genes
                                                                       Chapter 5

Another new direction in ecotoxicology is that of *gene expression profiling.* Using the techniques of genomics, expression of genes can be compared between organisms in various populations, or between organisms in the same population but sampled at different times. Among the genes whose expression may relate to ecosystem stress are genes involved with the stress response (such as stress protein genes) and genes involved in xenobiotic metabolism.

PCR

See also:
   SNPs
                                                                       Chapter 5

Microarrays

See also:
   Toxicogenomics
                                                                       Chapter 5

Techniques that can be used to study gene expression include quantitative RT-PCR as well as cDNA microarrays. The ability to measure gene expression has the potential, in fact, to provide numerous biomarkers for ecosystem stress. For this to become a truly useful technique, however, additional work needs to be done in sequencing genomes and identifying genes in a wider variety of organisms.

CASE STUDY   Plastic Debris in the Marine Environment

Plastics, which are synthetic polymers, have been mass produced for approximately 70 years. Thus, the problem of plastic debris in the ocean is not a new one. However, the rate at which plastics continue to be produced (hundreds of millions of tons annually worldwide), as well as the extended lifespan of many plastics in the environment, ensure that plastic pollution will continue to be a problem for years to come. It has been estimated that up to 10% of the plastics which are produced ultimately make their way to the ocean, either in the form of *macroplastics* (large plastic debris) or *microplastics* (smaller plastic fragments). Plastic waste enters oceans through a variety of sources including wastewater treatment effluents (many microplastics are not trapped by water treatment filtration systems), terrestrial runoff, and marine activities such as boating, fishing, and shipping.

Macroplastics can be problematic in many ways: marine life can become entangled in plastic debris, and ingestion of macroplastics has been observed in many marine species. Plastic debris can also serve as a vehicle for introduction of invasive species into new locales, or as vectors for marine pathogens.

Recently, however, more attention has become focused on microplastics. These are generally defined as pieces of plastics in which at least two out of three dimensions are measured as less than 5 mm in length. However, this encompasses quite a large size range of particles. Some microplastics are fragments of larger plastics which have been broken down through chemical and physical processes such as photodegradation from exposure to sunlight, and erosion by wind and water. These *secondary microplastics* may degrade rapidly in warm, shallow waters, but much more slowly in darker, colder benthic environments. *Primary* microplastics, on the other hand, are plastics which are already small by design and would include plastic particles used in scrubs and cleansers or in industrial processes such as air-blasting.

Assessing levels of both macroplastics and microplastics in the ocean is, of course, difficult. Techniques that have been used include measurement of debris on beaches, sampling of marine sediments, and trawling of marine waters. Plastics can be separated from other materials through filtration or flotation and identified visually or through microscopy. Some plastics can also be dyed to assist in ease of separation. Of course, uptake of plastics by marine life can also be studied. In fact, much of the concern surrounding plastics in the marine environment has been focused on the ingestion of plastics by marine biota.

The distribution of plastics in the oceans is not uniform; it is driven by winds and currents into accumulation zones. These accumulation zones (sometimes termed "garbage patches") tend to be found near the center of ocean *gyres*. Gyres are large, circulating currents driven primarily by the Coriolis effect, which is an apparent deflection in the path of surface wind or water caused by the rotation of the Earth. There are five major ocean gyres, two in the Atlantic Ocean, two in the Pacific Ocean, and one in the Indian Ocean. Plastics can also be a problem in near-shore waters and beaches, at least in part because those are the areas nearest to the sources of the pollution.

As for the effects of plastics on marine life and ecosystems, research has only just begun to identify the potential problems. Multiple studies have documented the discovery of macro- and microplastics in the gastrointestinal tract of seabirds, resulting in physiological problems ranging from physical obstruction to toxicity from chemicals leached from the debris. This potential chemical exposure would include not only chemicals originating in the plastic itself, but also organic pollutants and metals adsorbed onto the plastics from the environment. Evidence also exists for the ingestion of microplastics by a wide array of other organisms including plankton (who may mistake microplastic particles for prey organisms), crustaceans and other invertebrates, and fish. Microplastics can also be passed up the food chain, as larger organisms consume smaller organisms with microplastics in their gut. Some organisms may be able to be excrete the microplastics themselves along with other inedible materials; however, this does not eliminate the potential problems associated with absorption of integral and adhered chemicals. At any rate, there is a good deal of research remaining to be done in order to characterize the impact of both macroplastics and microplastics in marine systems.

BIBLIOGRAPHY

Barthalmus, G.T., Terrestrial organisms, in *Introduction to Environmental Toxicology*, Guthrie, F.E. and Perry, J.J., Eds., Elsevier, New York, 1980.

Cole, M., Lindeque, P., Halsband, C., and Galloway, T.S., Microplastics as contaminants in the marine environment: A review, *Mar. Pollut. Bull.*, 62, 2588, 2011.

Connell, D.W. and Miller, G.J., *Chemistry and Ecotoxicology of Pollution*, John Wiley & Sons, New York, 1984, chaps. 4 and 16.

DiGiulio, R.T. and Newman, M.C., Ecotoxicology, in *Casarett and Doull's Toxicology*, Klaassen, C.D., Ed., McGraw-Hill, New York, 2013, chap. 30.

Eriksen, M., Maximenko, N., Thiel, M., Cummins, A., Lattin, G., Wilson, S., Harner, J., Zellers, A., and Rifman, S., Plastic pollution in the South Pacific subtropical gyre, *Mar. Pollut. Bull.*, 68, 71, 2013.

Escher, B.I.K. and Hermens, J.J.M., Modes of action in ecotoxicology: Their role in body burdens, species sensitivity, QSARs, and mixture effects, *Environ. Sci. Technol.*, 36, 4201, 2002.

Fairbrother, A., Lewis, M.A., and Menzer, R.E., Methods in environmental toxicology, in *Principles and Methods of Toxicology*, 4th ed., Hayes, A.W., Ed., Taylor & Francis, Philadelphia, 2001, chap. 37.

Freedman, B., *Environmental Ecology*, Academic Press, San Diego, CA, 1994.

Gressel, J., Gene flow of transgenic seed-expressed traits: Biosafety considerations, *Plant Sci.*, 179, 630, 2010.

Grimm, V. and Thorbek, P., Population models for ecological risk assessment of chemicals: Short introduction and summary of a special issue, *Ecol. Model.*, 281, 1, 2014.

Heinemann, J.W., Agapito-Tenfen, S.Z., and Carman, J.A., A comparative evaluation of the regulation for GM crops or products containing dsRNA and suggested improvements to risk assessments, *Environ. Int.*, 55, 43, 2013.

Jha, A.N., Genotoxicological studies in aquatic organisms: An overview, *Mutat. Res.*, 552, 1, 2004.

Kendall, R.J., Anderson, T.A., Baker, R.J., Bens, C.M., Carr, J.A., Chiodo, L.A., Cobb, G.P., III, et al. Ecotoxicology, in *Casarett and Doull's Toxicology*, Klaassen, C.D., Ed., McGraw-Hill, New York, 2001, chap. 29.

Kwit, C., Moon, H.S., Warwick, S.I., and Stewart, C.N., Jr., Transgene introgression in crop relatives: Molecular evidence and mitigation strategies, *Trends Biotechnol.*, 29, 284, 2011.

Lavers, J.L., Bond, A.L., and Hutton, I., Plastic ingestion by Flesh-footed Shearwaters (*Puffinus carneipes*): Implications for fledgling body condition and the accumulation of plastic-derived chemicals, *Environ. Pollut.*, 187, 124, 2014.

Mackay, D. and Webster, E., A perspective on environmental models and QSARs, *SAR QSAR Environ. Res.*, 14, 7, 2003.

Mulvaney, D. and Krupnik, T.J., Zero-tolerance for genetic pollution: Rice farming, pharm rice, and the risks of coexistence in California, *Food Policy*, 45, 125, 2014.

Nandi, S., Maurer, J.J., Hofacre, C., and Summers, A.O., Gram-positive bacteria are a major reservoir of class 1 antibiotic resistance integrons in poultry litter, *Proc. Natl. Acad. Sci. USA*, 101, 7118, 2004.

Nikinmaa, M., Climate change and ocean acidification—Interactions with aquatic toxicology, *Aquat. Toxicol.*, 126, 365, 2013.

Noyes, P.D., McElwee, M.K., Miller, H.D., Clark, B.W., Van Tiem, L.A., Walcott, K.C., Erwin, K.N., and Levin E.D., The toxicology of climate change: Environmental contaminants in a warming world, *Environ. Int.*, 35, 971, 2009.

Paerl, H.W. and Paul, V.J., Climate change: Links to global expansion of harmful cyanobacteria, *Water Res.*, 46, 1349, 2012.

Pretti, C. and Cognetti-Varriale, A.M., The use of biomarkers in aquatic biomonitoring: The example of esterases, *Aquat. Conserv.*, 11, 299, 2001.

Rowe-Magnus, D.A., Guerout, A.M., Ploncard, P., Dychinco, B., Davies, J., and Mazel, D., The evolutionary history of chromosomal super-integrons provides an ancestry for multiresistant integrons, *Proc. Natl. Acad. Sci. USA*, 98, 652, 2001.

Snell, T.W., Brogdon, S.E., and Morgan, M.B., Gene expression profiling in ecotoxicology, *Ecotoxicology*, 12, 475, 2003.

Travis, C.C., Bishop, W.E., and Clarke, D.P., The genomic revolution: What does it mean for human and ecological risk assessment? *Ecotoxicology*, 12, 489, 2003.

Wright, S.L., Thompson, R.C., and Galloway, T.X., The physical impacts of microplastics on marine organisms: A review, *Environ. Pollut.*, 178, 483, 2013.

15 Applications
Pharmacology and Toxicology

While toxicology is the science that focuses on the adverse effects of biologically active substances, its sister science, *pharmacology*, focuses instead on the potential therapeutic benefits that can be derived from deliberate administration of those substances. In fact, these two sciences are opposite sides of the same coin, applying the same basic tools and techniques to understanding the impact of xenobiotic chemicals on body functions. Indeed, toxicologists play a major role in the discovery, development, and testing of both over-the-counter and prescription drugs.

BASIC PRINCIPLES OF PHARMACOLOGY

Most of the basic principles of pharmacology are identical to the basic principles of toxicology. Whether you are concerned about adverse effects or therapeutic effects, you still want and need to know, for example, how a substance enters and leaves the body, what modifications are made to it while it is there, and how it interacts on a molecular basis with body tissues.

PHARMACOKINETICS AND DRUG DELIVERY

Pharmacologists and toxicologists working on developing new drugs are vitally interested in the *pharmacokinetics* of those drugs. The principles of pharmacokinetics and toxicokinetics are virtually identical and have been covered in Chapter 2. There are some applications of those principles, however, that are relatively unique to pharmacology, specifically to drug development. These applications primarily deal with designing and implementing effective delivery methods for drugs.

For any drug, there is typically a minimum plasma concentration that must be reached in order for the drug to produce its therapeutic effect. This concentration is sometimes known as the *minimum effective concentration* (MEC). At the same time, there is also a plasma drug concentration at which unacceptably toxic effects begin to occur. This is sometimes known as the *minimum toxic concentration* (MTC). The goal of pharmacokinetics is to develop a delivery method and dosing regimen that produces plasma concentrations that quickly rise above the MEC but do not reach the MTC (Figure 15.1).

There are a number of ways to deliberately modify the pharmacokinetics of a drug. One is by route of administration. Drugs can be administered through the three

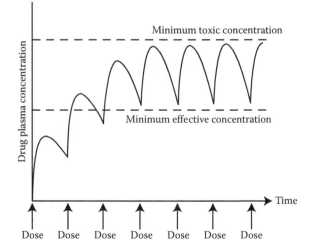

FIGURE 15.1   The changes in plasma concentration of a drug with time on a repeated dos-ing schedule. The goal is to produce concentrations that do not drop below the MEC, but do not rise above the MTC.

major routes already discussed: oral, respiratory, or dermal. Other options include *subcutaneous injection* (injection underneath the skin), *intramuscular injection* (injection into the muscle), or *intravenous injection* (injection directly into a vein). Rates of absorption of a drug through these routes will vary, depending on the route and the physical and chemical characteristics of the drug.

Of course, drug absorption and delivery can also be modified through a number of mechanisms. The use of coatings, for example, can influence the rate of drug release in the gastrointestinal tract, and attachment of other molecules such as anti-bodies can target the drug to a particular tissue. New drug delivery systems are not only an important aspect of the development of many new drugs, but they can also enhance the effectiveness of existing drugs (and at a cost that is typically less than the cost of developing completely new agents).

New directions in drug delivery include the use of biodegradable polymers to form *microparticles*, structures that can be used to encapsulate drugs, protecting them from immune system attack, moving them across the blood–brain barrier, and even regulating the rate of release of the drug into the bloodstream. *Nanoparticles* are similar structures that are small enough to be taken up intact into cells or even into intracellular compartments. *Peptide segments* that can cross cell membranes have also been discovered and have the potential to serve as carriers for other pro-teins and peptides that cannot cross lipophilic barriers. *Dendrimers*, which are highly branched globular polymers, can be synthesized to have specific molecular weights, backbones with the desired level of biodegradability, and the ability to bind to and carry specific therapeutic molecules.

New technologies have also been developed to help deliver drugs *transdermally* (across the skin). Because of the barrier properties of the intact skin, only very lipid-soluble molecules can normally be delivered in this fashion. Timed-release patches, however, have been developed that can deliver these drugs at a constant rate. Patches that generate small electric fields have also been harnessed to force higher-charged molecules across the skin in a process called *iontophoresis*.

Another drug-delivery area in which technology has advanced significantly is the *osmotic pump*. These semi-permeable devices contain the active ingredient in either liquid or solid form and then depend on body fluids to enter the device and provide the vehicle for delivery. If the pump is properly configured, then the process of osmosis will lead to a steady rate of release (zero-order kinetics as long as the solution within the pump remains saturated with respect to the drug; first-order kinetics after). Alternatively, a two-compartment device can be designed, where the entry of fluids into one compartment leads to compression of a second compartment containing the drug, thus forcing its release.

THE MAGIC BULLET: MECHANISMS OF ACTION AND SIDE EFFECTS

Drugs, like toxicants, bind to molecular sites within the cell to exert their effects. Basically, drugs act through the same basic mechanisms of action as described in Chapter 4. The ability of a drug to produce a desired physiological effect is known as its *efficacy*, while the relationship between the dose of the drug and the response describes its *potency* (the lower the dose that is required to produce the desired effect, the higher the potency of the drug).

Some drugs are rather specific in acting on one intracellular target; others are more general. Researchers are always searching, of course, for the proverbial *magic bullet*, a drug that has the desired therapeutic effect and no other. Most drugs, however, far from being magic bullets, have a wide range of less desirable effects that accompany the desired effect. These *side effects* arise either from interaction of the drug with other molecular targets or from secondary effects accompanying interaction with the primary target. One role of toxicologists in the area of pharmacology is to work to understand and minimize potentially hazardous side effects.

The relationship between the dose of a drug necessary to produce therapeutic effects and the dose that produces toxic effects is an important consideration in evaluating the usefulness of a drug. This relationship can be quantified by what is known as the *therapeutic index*, which is the ratio between the LD_{50} of a drug and the ED_{50} (which is the dose that produces the desired therapeutic effect in 50% of a test population). The higher the therapeutic index, the safer the drug. Drugs with a narrow therapeutic index include *lithium* (used to treat manic-depressive disorder), *warfarin* (an anticoagulant), *phenytoin* (an anticonvulsant and antiarrhythmic), and *digoxin* (used to treat congestive heart failure). The therapeutic index of tetrodotoxin, which is currently in foreign trials as an analgesic, is also likely to be very narrow.

LD_{50}

See also:
The LD_{50} Experiment
Chapter 1

Tetrodotoxin

See also:
Effects of Toxicants on Voltage-Activated Ion Channels
Chapter 4

Effects of Toxicants on Electrical Conduction
Chapter 10

TTX, STX
Appendix

Like toxicants, drugs can also interact together in additive, synergistic, or antagonistic fashion. One of the challenges of physicians and pharmacists, in fact, is to coordinate and manage the drugs a patient may be taking and make sure that the interactions between them do not produce unintended consequences. This may particularly be a problem when patients self-medicate (with herbal medicines, for example) without informing their physician. Warnings of known negative interactions are provided on pharmaceutical labels under "contraindications." Some contraindications also arise from genetic or disease-related variations in the biotransformation of the drug.

Induction

See also:
The Role of Cytochrome P450
Chapter 3

Stress, Repair, and Recovery
Chapter 4

Patients can also develop *tolerance* to the effects of many drugs. This means that a higher dose of the drug is required in order to produce the same effects. There are two primary mechanisms of tolerance: pharmacokinetic and pharmacodynamic. *Pharmacokinetic tolerance* occurs when a drug is inactivated at an increased rate due to the induction of cytochrome P450 enzymes or other enzymes involved in xenobiotic metabolism. *Pharmacodynamic tolerance* occurs when cellular-level changes (such as up- or downregulation of the number of receptors) affect the body's response to the drug. Once tolerance develops, there is a risk of *withdrawal* symptoms if the drug is suddenly discontinued, as the body readjusts to its absence.

DRUG DEVELOPMENT AND THE ROLE OF TOXICOLOGY

Prior to the federal Food, Drug and Cosmetics Act in 1938, pharmaceuticals were not registered by the government. Drugs were prescribed at the discretion of the physician, usually a general practitioner, and were prepared and sold at the discretion of the pharmacist, usually the owner of the pharmacy. Pharmacists were educated to identify medicinal plants, plant products, and powders by taste and smell, and to prepare powders, tinctures, and other formulations of the drugs as remedies. Prescriptions were usually written by hand in Latin by the physician, and it was the duty of the pharmacist to verify and, when necessary, correct the prescription.

Most drugs were natural products, usually of botanical origin. Botanicals were complemented by inorganics, such as lithium citrate, and a few simple synthetic organics such as aspirin. Raw materials were often imported directly to the pharmacy, for example, dried opium poppies would sometimes arrive from the Middle East with the small harvesting knife still embedded in the material. Most pharmacies formulated private brands of cough syrups, often with codeine, and provided other common remedies, such as an ethanol tincture of cannabis to be applied to corns on the feet.

This system of pharmacy based on natural products was highly developed, and nearly every small town or neighborhood in the United States was served by a general practitioner and an independent pharmacy. In the 1930s, however, a chemical revolution of sorts occurred. Chemically synthesized organic drugs and pesticides began to be available, and synthetic drugs replaced the uses of many botanical medicines. And even where natural products continued to be used, powerful, pure active ingredients began to be isolated and marketed by pharmaceutical companies, replacing the simpler preparations of the hometown pharmacist. This led to a greater role by the government, both in overseeing the registration process for pharmaceutical manufacturers and in mandating the recording and reporting of drug sales by pharmacies.

Now, before a drug can be marketed, it must go through a rigorous set of testing procedures that are specified by the *Food and Drug Administration* (FDA), the agency that is in charge of drug safety. Typically, a *new drug application* (NDA) may take about 2 or 3 years to be approved. This, of course, is in addition to the 5 or 6 years it may take a pharmaceutical company to develop a drug to the point of being ready to file an NDA. The estimated costs involved in drug development vary significantly depending on the type of the drug and other factors (including how the estimate is calculated), but have been reported to be anywhere from 100 million to over a billion dollars.

Drugs, of course, have a *chemical name* (IUPAC name) that exactly describes the chemical structure of the compound. In addition, once a drug appears destined for the market, a *nonproprietary* name is assigned (which becomes the *official* name once the drug is officially included in the U.S. Pharmacopoeia). Finally, companies that manufacture drugs choose their own trademarked *proprietary name* (or *trade name*) under which to market the drug. It is possible for the same drug to have several different proprietary names if it is marketed by several different manufacturers.

PRECLINICAL STUDIES

The first phases of drug development typically take place in the laboratory and involve *in vitro* testing in animal or human molecular or cellular systems, as well as *in vivo* testing in animal models. Initially, the drug is considered to be in what is known as the *discovery phase*. In this early phase, the potential drug or *new chemical entity* (NCE) is put through a series of screening tests measuring both efficacy and toxicity. Screening tests are typically relatively low-cost procedures that can be carried out fairly rapidly and are designed to quickly determine which NCEs are worth carrying forward to the next phase of development.

A growing area of concern is the testing of *new biological entities* (NBEs), which are typically products of recombinant technology and which may have additional safety issues, such as their potential to invoke immunological responses or the risk of contaminants arising from the recombinant process. Drugs produced by biotechnology are often proteins of two categories: vaccines and antibodies. While the registration of modern conventional drugs has been constrained primarily by safety, with toxicity being the failure of many candidates, biotechnological drugs are more often constrained by lack of efficacy or inconsistency thereof, and candidates are rarely eliminated due to toxicity.

Once an NCE has passed through the initial screening process, requirements for toxicity testing become much more stringent. Laboratories are required to follow a set of standard practices and protocols known as *good laboratory practices* (GLPs). Toxicity testing focuses on the questions of identification of target organs and of development of the dose–response relationship in terms of toxic effects. Dosing protocols that may be used include acute, subchronic, and chronic, and a variety of parameters can be assessed, including body weight, blood chemistry, urinalysis, pathology and histopathology, and behavior. Specialized tests for mutagenicity and carcinogenicity, as well as reproductive and developmental effects, should also be carried out.

CLINICAL STUDIES

When sufficient information from nonclinical testing has been gathered to indicate that an NCE is both reasonably efficacious and reasonably nontoxic, the NCE may enter *clinical trials*. There are many ethical issues involved with clinical trials. One of the main issues centers on the idea of *informed consent*, which means that volunteers for a clinical study must be fully informed as to both the risks and the benefits that may result from their participation in the study. Volunteers also must be free to leave the study at any time. In addition, clinical trials are carefully monitored as they proceed and may be stopped if the treatment being tested proves to be associated with an unforeseen and unacceptable level of risk. Occasionally, trials may also be stopped if the treatment proves so efficacious that it becomes unethical to withhold it from the control participants. Just because a drug enters clinical trials, of course, its success is not guaranteed. It has been estimated, in fact, that only 10%–30% of drugs entering the clinical phase will go on to be successfully developed and marketed.

There are typically three phases to clinical trials. In *phase I clinical trials*, the NCE is tested on a small population of healthy volunteers. The primary goals of phase I testing are to study the pharmacokinetics of the NCE, to identify a safe dose range for the NCE, and to identify potentially problematic side effects.

It is not until *phase II clinical trials* that individuals with the condition the NCE is designed to treat are actually included. In phase II, the focus is on determining efficacy of the NCE as well as confirming the safety. Typically, several hundred patients may be involved, and the trial takes the form of a comparison between two groups. In some cases, one group receives the NCE, while the other group, the *control group*, receives a *placebo* (an inactive substance). In other cases, if receiving a placebo instead of treatment would be hazardous to the health of the control group, that group will receive whatever treatment is the current standard for that condition.

Phase II trials are generally conducted blind, meaning that the patients do not know whether they are receiving the new treatment or a placebo. This *single-blind* design guards against biases on the part of the trial participants as they report on their experiences during the trial. In a *double-blind* design, not only do the participants not know which group they are assigned to, but the researchers also do not know. This avoids intentional or unintentional biases not only on the part of the patients, but also on the part of the researchers.

If the results from phase I and phase II are positive, the NCE can move on to *phase III clinical trials*. These trials are much larger (they may involve thousands of patients) and are designed to confirm the safety and efficacy of the NCE as indicated in phase II. Following this, application may be made to the FDA to market the drug. Of all the NCEs that enter the discovery pipeline, very few (only around 1 of every

10,000) will make it through to FDA approval. Even then, testing may not be complete. The FDA can require further monitoring (*phase IV clinical trials*) to ensure safety of the public.

Recently, rofecoxib and celecoxib (cyclooxygenase inhibitors) were found to have side effects that were not identified in the FDA registration process. This has resulted in public criticism of the process and a request from the FDA for recommendations from the National Research Council. At the same time, the cost and duration of the drug development process is leading companies and investors to seek

---
Rofecoxib and Celecoxib

See also:
   Case Study: Cyclooxygenase
   Inhibitors
                              Chapter 4
---

less expensive alternatives. And since drug manufacturers are increasingly multinational companies with worldwide marketing strategies (in which the United States may not be the priority market), drugs may begin to be initially tested and marketed in countries with a more streamlined registration process, less expensive clinical trials, and with a larger potential market, for example, in China or India.

GENERIC DRUGS

During initial drug development, a pharmaceutical manufacturer will typically apply for and receive a patent on any promising drug. Currently, the lifetime of a patent in the United States is 20 years. Also, following approval of a drug, the FDA will grant the manufacturer *exclusivity*, which gives the manufacturer exclusive rights to market the new drug for a limited period of time (usually around 5 years). Both types of protection run concurrently, and both protect the investment that the manufacturer has made in the development process.

Pharmaceutical manufacturers may be able to extend patent protection on a drug through developing modifications to the drug structure, new formulations, new routes of administration, or new uses. Once patent and exclusivity protections expire, however, other companies are entitled to develop and market their own versions of the drug. To be approved, a *generic* drug must be shown to be bioequivalent to the original patented drug in terms of both safety and efficacy. A higher standard, though, is that they be *switchable*; in other words, a patient should be able to switch from the name brand to the generic with no clinical effects. This is actually the area where most consumer issues with generics arise. Of course, the issue is complicated by the fact that even within a brand name drug there may be some variability from manufacturing lot to manufacturing lot.

TOXICOGENOMICS AND DRUG SAFETY

The relatively new field of toxicogenomics has the potential to greatly enhance the ability to predict human toxicity of an NCE. *Marker genes* can be identified that are variably expressed under different conditions (including conditions associated with toxicity). Then techniques such as RT-PCR

Toxicogenomics

See also:
   Toxicogenomics
                              Chapter 5

(real-time PCR) or microarrays can be designed to measure expression of these genes. In fact, microarrays designed to assay genes expressed during response to toxicants are already available commercially. Genetics of nonresponding or hyper-responding strains of model organisms can be used to identify candidate targets in the genome by map position. Once a mapped marker is found near the trait of drug response, the genome around the marker can be scanned for genes encoding candidate proteins such as receptors, enzymes, and transporters. Completion of both the mouse and rat genome has enhanced this approach. This *positional cloning* is the same approach used to find genes for disease.

THE RETURN OF NATURAL PRODUCTS: REGULATORY ISSUES

There is a recent upward trend in interest in botanical drugs and other alternative medicines; however, these substances are frequently marketed as dietary supplements and thus are not regulated as stringently as pharmaceuticals. The Food and Drug Administration does monitor product safety, product quality, and labeling issues for dietary supplements; however, the products do not need FDA approval before being put on the market. If these products make a health-related claim, they are required to state on the label that the FDA has not evaluated the claim and that the product "is not intended to diagnose, treat, cure, or prevent any disease." Whether these regulations do an adequate job of protecting the interests of consumers is an issue that continues to be debated.

Regulatory responsibility aside, attempting to ensure the safety and efficacy of natural products also poses some difficult technical questions. A primary consideration is that the formulation of natural products can vary widely in composition, as they often consist of multiple active ingredients and those can vary with cultivation and harvesting. Another consideration is shelf life, because many natural products are sensitive to decomposition due to storage conditions. This is in contrast to modern synthetic organic drugs, which are typically composed of a single chemical compound that is usually designed to be stable to oxidation. All of this means that analytic chemistry of natural products for quality control is possible, but sometimes complex and expensive.

BIBLIOGRAPHY

Barry, B.W., Novel mechanisms and devices to enable successful transdermal drug delivery, *Eur. J. Pharm. Sci.*, 14, 101, 2001.

Bell, J.I., The double helix in clinical practice, *Nature*, 421, 414, 2003.

Benet, L.Z., Kroetz, D.L., and Sheiner, L.B., Pharmacokinetics: The dynamics of drug absorption, distribution, and elimination, in *Goodman and Gilman's: The Pharmacological Basis of Therapeutics*, 9th ed., Hardman, J.G., Limbird, L.E., Molinoff, P.B., Ruddon, R.W., and Gilman, A.G., Eds., McGraw-Hill, New York, 1996.

Boehm, G., Yao, L., Han, L., and Zheng, Q., Development of the generic drug industry in the US after the Hatch–Waxman Act of 1984, *Acta Pharmaceut. Sin. B*, 3, 297, 2013.

Castle, A.L., Carver, M.P., and Mendrick, D.L., Toxicogenomics: A new revolution in drug safety, *Drug Discov. Today*, 7, 728, 2002.

Chapman, K.L., Holzgrefe, H., Black, L.E., Brown, M., Chellman, G., Copeman, C., Couch, J., et al. Pharmaceutical toxicology: Designing studies to reduce animal use, while maximizing human translation, *Regul. Toxicol. Pharmacol.*, 66, 88, 2013.

Dorato, M.A. and Vodicnik, M.J., The toxicological assessment of pharmaceutical and biotechnology products, in *Principles and Methods of Toxicology*, 4th ed., Hayes, A.W., Ed., Taylor & Francis, Philadelphia, 2001.

Gillies, E.R. and Frechet, J.M.J., Dendrimers and dendritic polymers in drug delivery, *Drug Discov. Today*, 10, 35, 2005.

Gupta, H., Kumar, S., Roy, S.K., and Gaud, R.S., Patent protection strategies, *J. Pharm. BioAllied Sci.*, 2, 2, 2010.

Hellmold, H., Nilsson, C.B., Schuppe-Koistinen, I., Kenne, K., and Warngard, L., Identification of end points relevant to detection of potentially adverse drug reactions, *Toxicol. Lett.*, 127, 249, 2002.

Herrlich, S., Spieth, S., Messner, S., and Zengerle, R., Osmotic micropumps for drug delivery, *Adv. Drug Deliv. Rev.*, 64, 1617, 2012.

Morgan, S., Grootendorst, P., Lexchin, J., Cunningham, C., and Greyson, D., The cost of drug development: A systematic review, *Health Policy*, 100, 4, 2011.

Nies, A.S. and Spielberg, S.P., Principles of therapeutics, in *Goodman and Gilman's: The Pharmacological Basis of Therapeutics*, 9th ed., Hardman, J.G., Limbird, L.E., Molinoff, P.B., Ruddon, R.W., and Gilman, A.G., Eds., McGraw-Hill, New York, 1996.

Orive, G., Gascon, A.R., Hernandez, R.M., Dominguez-Gil, A., and Pedraz, J.L., Techniques: New approaches to the delivery of biopharmaceuticals, *Trends Pharmacol. Sci.*, 25, 382, 2004.

Rados, C., Inside clinical trials. Testing medical products in people, *FDA Consumer*, September/October 2003.

Redfern, W.S., Wakefield, I.D., Prior, H., Pollard, C.E., Hammond, T.G., and Valintin, J.-P., Safety pharmacology: A progressive approach, *Fundam. Clin. Pharmacol.*, 16, 161, 2002.

Ross, E.M., Pharmacodynamics: The mechanisms of drug action and the relationship between drug concentration and effect, in *Goodman and Gilman's: The Pharmacological Basis of Therapeutics*, 9th ed., Hardman, J.G., Limbird, L.E., Molinoff, P.B., Ruddon, R.W., and Gilman, A.G., Eds., McGraw-Hill, New York, 1996.

Temsamani, J. and Vidal, P., The use of cell-penetrating peptides for drug delivery, *Drug Discov. Today*, 9, 1012, 2004.

Valenta, C. and Auner, B.G., The use of polymers for dermal and transdermal delivery, *Eur. J. Pharm. Biopharm.*, 58, 279, 2004.

16 Applications
Forensic Toxicology

Forensic toxicology involves the application of toxicological principles to problems and questions pertaining to the legal system. Among the issues that forensic toxicologists deal with are testing for alcohol and drugs in individuals accused of criminal behavior (including drunk driving) and detection of the use of poisons in criminal acts. Both of these activities typically involve detection of chemical substances in human tissues and thus are dependent on techniques borrowed from analytical chemistry. In this chapter, we will look at the application of analytical toxicology to the problem of testing biological samples for alcohol, drugs, and other poisons. We will also look at examples of poisons used in the commission of crimes.

ANALYTICAL TOXICOLOGY

The role of an analytical toxicologist employed in the field of forensics is typically the detection, identification, and quantification of poisons in human tissues. This is important because the presence of most drugs and poisons cannot be identified simply by physiological signs or symptoms. In fact, prior to the development of analytical methods in the early 1800s, poisoners were quite likely to get away with their crime, and deliberate poisonings are thought to have been quite common.

The first issue that must be dealt with is the collection and handling of the samples to be tested. In working with living subjects, blood, urine, and even hair samples can be taken and have proved useful. In *postmortem* (after death) cases, these samples would most likely be supplemented by tissue samples from a number of organs, including brain, liver, and kidney. Sometimes bone marrow, muscle, or even vitreous humor may be collected. Of course, the possibility of postmortem redistribution of drugs and toxicants in tissues exists and must be accounted for in any analysis of results. If available, gastrointestinal contents might also be sampled, and there have even been cases where the chemical analysis of larvae that have been feeding on a badly decomposed body has been found to be useful. If the compound of interest is volatile, solid materials may be incubated in a sealed container, so that the gas that collects in the headspace of the container may be sampled.

In forensic matters, the concept of "chain of custody" is also important. This involves the careful documentation of the passage of samples from collection site to laboratory, demonstrating that there have been no opportunities for substitutions or tampering with the evidence. Also, the American Board of Forensic Toxicologists accredits forensic laboratories, ensuring that best practices are used once samples reach the forensic laboratory.

Preparation of the sample depends in part on the analytical technique to be employed, but in general involves using the physical and chemical properties of compounds to achieve separation. Forensic toxicologists classify poisons into several groups according to these characteristics and according to methods necessary to analyze the substances. One grouping that is often used is

- Group I: Gases
- Group II: Volatile substances
- Group III: Corrosive agents
- Group IV: Metals
- Group V: Anions and nonmetals
- Group VI: Nonvolatile organics
- Group VII: Miscellaneous poisons

A variety of analytical tools are used in the forensics laboratory to detect and quantify substances from these categories. The most common include thin-layer chromatography, high-performance liquid chromatography, gas chromatography–mass spectroscopy (GC–MS), and immunoassays.

THIN-LAYER CHROMATOGRAPHY

Thin-layer chromatography (TLC) is a separation technique that offers simplicity and economy as advantages; also, many samples can be analyzed simultaneously. The technique was described and illustrated in extraordinary detail by Egon Stahl (1969). In TLC, a *mobile phase* (in normal-phase TLC, this is a mixture of organic solvents, such as chloroform and methanol) is run across a *stationary phase*, which consists of silica gel spread on a glass plate. The samples to be analyzed are spotted near the bottom portion of the plate and allowed to dry. Then the plate is placed upright into a chamber, with the bottom of the plate (below where the samples have been spotted) in contact with the mobile phase. The mobile phase will then be drawn up across the plate by capillary action.

As the solvent moves past the samples, the components of the samples will migrate, with the speed of migration dependent upon the relative affinity of the components for the mobile phase compared to the stationary phase. When the leading edge of the solvent reaches the top of the plate, it is removed from the solvent and allowed to dry. The location of the sample components can then be visualized. It is convenient to expose the plate to iodine vapors for initial visualization of purple spots; the iodine will sublime after a few hours, allowing another method to be used. Stahl provided methods for 264 stains or dyes that can be applied to react with the component of interest (e.g., the dye *ninhydrin* will react with amphetamines to produce a pink color). Alternatively, a fluorescent dye can be incorporated into the solid phase, so that ultraviolet light will reveal the sample components as dark spots against a bright background due to quenching of the fluorescence. Note that a compound in the sample that also fluoresces, rather than quenches the background fluorescence, will be invisible in the background.

The results of TLC can be quantified. The *retention factor* (Rf) is the ratio of the distance that a sample component moves to the distance that the leading edge of the solvent moves. Sample components can be tentatively identified by comparing their Rf to the Rf of known substances (usually standards that are run at the same time as the samples).

GAS CHROMATOGRAPHY–MASS SPECTROMETRY

In *gas chromatography* (GC), the stationary phase is a liquid and the mobile phase, or *carrier gas*, is generally an inert gas such as helium or nitrogen. There are two main types of columns used in GC. In a *packed column*, the liquid is coated onto particles packed into a stainless steel or glass column; in a *capillary column*, the liquid is coated onto the walls of the column itself, which is very narrow and typically made of glass. Samples are injected into a heated port, where they are vaporized and carried into the column along with the carrier gas. A detector then produces a signal as sample components exit the column.

One common type of detector used with GCs is a *flame ionization detector*, which uses a hydrogen/air flame to combust organic materials. The ions generated in this process then produce an electronic signal that can be measured. When the detector is hooked up to a recorder, a recording called a *gas chromatogram* can be produced. This is a plot of the electronic signal vs. time, and it typically shows a series of peaks that correspond to the components of the sample. The time it takes for a substance to pass through the column (*retention time*, Rt) can be compared to standards in order to tentatively identify that substance. Also, since the area under each peak is proportional to the concentration of that substance, comparison with a standard of known concentration allows estimation of concentrations. *Flame photometric* detectors offer an increase in sensitivity over flame ionization. Ever greater sensitivity is possible using specialized detectors for halogenated compounds of interest or those containing nitrogen or phosphorus. The *electron capture detector*, which uses a radioactive source, can detect picograms of DDT due to the presence of five chlorine atoms in the molecule.

Although the various forms of chromatography allow tentative identification of substances based on comparison of Rf or Rt with standards, definitive identification requires additional analysis. One of the most effective techniques is *mass spectrometry*, which is often used in combination with GC in forensics laboratories. As the sample components exit the GC column, they are routed into a vacuum chamber in the mass spectrometer, where they are hit with a beam of electrons. This knocks electrons off of the sample molecules, creating positive ions and breaking them into fragments. These ionized fragments are then passed through an electromagnetic field, which separates them by their mass/charge ratio. The resulting spectrum plotting the abundance and mass/charge ratio of each fragment is quite specific for a given substance. Figure 16.1 shows a GC–MS analysis of an herbal infusion containing the alkaloids atropine, harmine, and scopolamine. (This infusion was given to around 30 individuals participating in a meditation session; all recovered without incident.)

FIGURE 16.1 GC–MS analysis of an herbal infusion including (1) atropine, (2) harmine, and (3) scopolamine. The top panel is the gas chromatographic separation of the components. Individual mass spectra are the panels below. (Reprinted from *Forensic Sci. Int.*, 128, 50, Balikova, M., Collective poisoning with hallucinogenous herbal tea, Copyright 2002, with permission from Elsevier.)

HIGH-PERFORMANCE LIQUID CHROMATOGRAPHY

High-performance liquid chromatography (HPLC) is similar to TLC and GC in that there is both a mobile phase and a stationary phase, and sample components are separated based on their relative affinity for the two phases. In HPLC, however, the stationary phase is a column packed with solid particles and the mobile phase is a liquid solvent. As the mobile phase is pumped through the column, the sample is injected. A detector then identifies the presence of components as they exit the column. Components are identified by their Rt, the length of time it takes them to pass through the column. As with GC, the Rt of an unknown can be compared with the Rt of a known standard for tentative identification. Figure 16.2 shows an HPLC chromatogram of steroids isolated from the skin and parotid glands of toads.

The choice of column material, solvent, and detector in an HPLC setup is made based on the components to be analyzed. *Normal-phase* columns use polar silica particles as the stationary phase, and a nonpolar organic solvent is pumped through

FIGURE 16.2 HPLC chromatogram of steroids isolated from the skin and parotid gland of toads. (From Wang, Z. et al., Simultaneous determination of four bufadienolides in human liver by high-performance liquid chromatography, *Biomed. Chromatogr.*, 2004. 18, 318. Copyright Wiley-VCH Verlag GmbH & Co. KGaA. Reproduced with permission.)

as the mobile phase. Dissolved chemicals of interest passing through the column are retained differentially by adsorption on the silica, thus affecting a separation. *Reverse-phase columns* contain nonpolar, derivatized silanes bonded to the silica particles, and a polar solvent, such as water with methanol, is pumped through as the mobile phase. Chemicals of interest are partitioned differentially between the bonded reverse-phase and the aqueous mobile phase, again effecting a separation. Columns are also available that can separate materials based on size (*size exclusion*) or on interaction with negative or positive charged groups bonded to the stationary phase (*ion exchange columns*). Enantiomers of chiral (right- and left-handed) chemicals can also be separated using specialized chiral stationary phases.

HPLC columns may be coupled with any one of a number of different detectors that identify the presence of compounds eluting from the column. *Refractive index detectors* compare refraction of light between the solvent alone and the solvent coming off the column, while other detectors measure light absorption or fluorescence. *Electrochemical detectors* can detect compounds that can undergo redox reactions. HPLC, like GC, can also be coupled with mass spectroscopy to create a liquid chromatography–mass spectroscopy (LC–MS) combination.

Immunoassays

Antibodies

See also:
Humoral Immunity
Chapter 13

Immunoassays utilize the ability of specific antibodies to bind with high affinity to a substance of interest, such as a drug or toxicant. These antibodies are produced by combining the drug or toxicant with a protein (to ensure an immune reaction), injecting it into an animal (such as a rabbit or goat), and then isolating the resulting antibodies from the animal's serum.

One commonly used immunoassay is the enzyme-multiplied immunoassay technique (EMIT). This is frequently used to detect the presence of drugs in urine. Two other methods, the fluorescent polarization immunoassay (FPIA) and the radioimmunoassay (RIA), are also used by forensic scientists. In competitive binding immunoassays such as EMIT, the first step is to prepare a labeled version of the drug or toxicant of interest. In the case of EMIT, the label is an enzyme that is attached to the drug or toxicant. The labeled drug or toxicant is then combined with an antibody prepared against it, and binding is allowed to occur (which generally inactivates the enzyme). Then the sample being tested is added, and if it contains the drug or toxicant of interest, it will displace some of the labeled drug or toxicant from binding to the antibody. This displacement will then be detectable by an increase in activity of the bound enzyme. Other competitive binding assays use radioactive or fluorescent tags and measure the decreases in radioactivity or fluorescence with displacement.

FORENSIC TOXICOLOGY AND ALCOHOL USE

Ethyl alcohol, or *ethanol*, is the primary legal recreational drug in this country. Ethanol is an addictive central nervous system (CNS) depressant, hypnotic, and sedative, with dose-related neurological effects. Therefore, measurement of ethanol concentration in the body is one important method for quantifying the level of impairment of individuals who are under the influence of the drug.

Ethanol is relatively quickly absorbed through the gastrointestinal tract, although the rate of absorption does depend on stomach contents (alcohol is more slowly absorbed in the presence of other substances). Once absorbed, ethanol is distributed throughout most body tissues, with the exception of adipose tissue, bone, hair, and other tissues that are low in water content. Elimination of ethanol occurs through excretion and metabolism and follows *zero-order kinetics*: it is eliminated at a constant rate. As measured by blood alcohol levels, this corresponds to a decrease of approximately 0.015%–0.020% (w/v) (weight per volume, or g/100 mL) per hour.

About one-half of alcohol detoxication to acetaldehyde is via *peroxisomal oxidation*. Large peroxisomes are found in the liver and in the kidney, and inside these peroxisomes, the enzyme *catalase* is found. Catalase uses the hydrogen peroxide (H_2O_2) generated by other peroxisomal enzymes to eliminate both alcohol and hydrogen peroxide in the following reaction:

$$H_2O_2 + R'H_2 \rightarrow R' + 2H_2O,$$

where $R'H_2$ would include phenols, formic acid, formaldehyde, and alcohol.

Another route by which ethanol is eliminated from the body is through the respiratory system. This is the basis for the *breathalyzer test*, one of the commonly used tests for ethanol intoxication. This instrument collects a sample of air from the alveoli (individuals taking a breathalyzer test are always instructed to breathe deeply) and exposes it to a potassium dichromate, silver nitrate, and sulfuric acid mix. Ethanol reacts with potassium dichromate and sulfuric acid (silver nitrate acts as a catalyst), resulting in the breakdown of potassium dichromate. Since potassium dichromate absorbs light at 420 nm, the disappearance can be monitored and the concentration of ethanol in the air sample estimated. And since the relationship between ethanol concentration in exhaled air and blood alcohol levels is known, blood alcohol levels can then be calculated. The legal standard for intoxication in most states is a blood alcohol level of 0.10% (100 mg/100 mL).

FORENSIC TOXICOLOGY AND ILLEGAL DRUG USE

THE CONTROLLED SUBSTANCES ACT

There is a network of laws regulating drug use, and it frequently falls to forensic toxicologists to help identify violations of those laws. The *Controlled Substances Act* defines five categories (*schedules*) of drugs, based on their potential for abuse as well as their potential for legitimate medical use. Schedule I drugs have a high potential for abuse and no current medical use in the United States. Schedule II drugs also have a high potential for abuse, but do have accepted medical uses. Schedule III, IV, and V drugs have progressively lower potential for abuse, as well as progressively lower tendencies for physical or psychological dependence to develop. Examples of drugs in each of these categories are shown in Table 16.1.

The Controlled Substances Act is enforced by the *U.S. Drug Enforcement Administration* (DEA). The DEA can prosecute for unauthorized manufacture, sale, or possession of regulated drugs, their precursors, or *designer drugs*, which are chemically related compounds that share similar physiological effects with regulated drugs.

DRUG IDENTIFICATION

The identification of drugs, whether it is of the drug itself or its presence in body tissues, is typically performed using the analytical techniques described earlier. These would include TLC, HPLC, and GC–MS. Other techniques can include *UV–VIS spectroscopy* (some drugs absorb light at specific wavelengths) or *IR spectroscopy* (which produces a characteristic IR fingerprint for each drug). For some drugs, there are also *color tests* that can be used for identification. For example, the *Marquis reagent* (2% formaldehyde in sulfuric acid) turns purple in the presence of opiates.

TABLE 16.1

Examples of Drugs within the Various Schedules of the Controlled Substances Act

Schedule	Examples
I	Heroin
	Lysergic acid diethylamide (LSD)
	Marijuana
II	Amobarbital
	Methamphetamine
	Cocaine
	Morphine
III	Anabolic steroids
IV	Phenobarbital
	Diazepam
V	Preparations with low concentration of codeine

MAJOR CATEGORIES OF ILLEGAL DRUGS: NEUROACTIVE DRUGS

Neuroactive Drugs

See also:

Effects of Toxicants on the Nervous System: General Principles

Chapter 10

Most illegal drugs are *neuroactive*, exerting their effects on the nervous system. *Central nervous system depressants* produce feelings of relaxation, remove inhibitions, and at high enough doses act as *sedatives* (induce sleep). Examples of CNS depressants include ethanol, *barbiturates* such as *amobarbital* (blue heavens), *pentobarbital* (yellow jackets), and *secobarbital* (Christmas trees), and *benzodiazepines* such as *diazepam* (trade name *Valium®*).

Barbiturates stimulate $GABA_A$ receptors, one subclass of receptors for the inhibitory neurotransmitter GABA, and also seem to block AMPA receptors (a type of glutamate receptor). Benzodiazepines also act on $GABA_A$ receptors, but through a different mechanism, potentiating the binding of GABA itself. Overdoses of barbiturates can lead to coma and death through depression of respiration; benzodiazepines have a much higher margin of safety than the barbiturates. Interestingly, the GABA antagonist Ro 15-4513, an analog of flumazenil, was found to antagonize behavioral effects of ethanol, suggesting that ethanol acts, at least in part, on GABA receptor–chloride ion channels. *Methaqualone* (ludes) is another CNS depressant.

On the other side of the coin are the *central nervous system stimulants*. This category includes *amphetamines* such as *methamphetamine* (crank) and *cocaine*. Amphetamines can be taken orally, intravenously, or smoked, and act by stimulating the release of dopamine from presynaptic neurons. Cocaine, one of the most addictive drugs known, is generally snorted (taken intranasally). Crack cocaine is cocaine that has been heated with baking soda and water, dried and broken into pieces, and then smoked. Cocaine acts by blocking reuptake of dopamine, norepinephrine, and serotonin. Side effects of cocaine use include increased risk for cardiac arrhythmias.

The *opioids*, or *narcotics*, are drugs that either are found in *opium* (a mixture of around 20 alkaloids found in the juice of the opium poppy, *Papaver somniferum*) or are synthetic derivatives of those alkaloids. The primary alkaloid in opium is *morphine*; others include *codeine* and *papaverine*. Synthetic opioids include *heroin*, formed by reacting morphine with acetic anhydride or acetyl chloride.

Opioids interact with a set of receptors known as the *opioid receptors*. These receptors mediate pain and other pathways and produce feelings of euphoria. Three groups of peptide neurotransmitters (the *enkephalins*, the *endorphins*, and the *dynorphins*) serve as endogenous ligands for these receptors. Physiological tolerance develops to the opioids, leading to physical dependence on the drug and making it difficult to discontinue use. *Methadone* is a synthetic opiate that is sometimes administered to heroin users to help suppress the withdrawal symptoms associated with cessation of heroin use. An overdose of opioids causes pupillary constriction and depresses respiration. In fact, death from overdose is generally due to respiratory arrest.

Two final categories of neuroactive drugs are the *psychedelic drugs* and *marijuana*. Psychedelic drugs are sometimes also called *hallucinogens* and produce distortions of perception and thinking. *Lysergic acid diethylamide* (LSD) is the most potent of the psychedelic drugs, producing perceptual distortions such as blurring of the self–nonself boundary, or *synesthesias*, the sensation that one hears colors or sees sounds. Although LSD often produces feelings of elation and arousal, it can also engender panic and depression (a "bad trip").

LSD interacts with a subclass of serotonin (5-HT) receptors, and as little as 25 micrograms of the drug is sufficient to produce effects. Other psychedelic drugs include *phencyclidine* (PCP), which binds to and blocks NMDA receptors. PCP was developed as an anesthetic but proved to be somewhat less than ideal due to the often violent and bizarre state of intoxication it produces. *MDMA* (Ecstasy) is another psychedelic.

Marijuana, the most widely used illegal drug in the United States, is derived from the hemp plant *Cannabis sativa*. The major drug found in marijuana is Δ-9-*tetrahydrocannabinol* (THC). The effects of THC include a relaxed "high" sensation accompanied by impairment of coordination, as well as cognitive functions. These effects are unique from the effects of other drugs, and cross-tolerance to other drugs is not generally seen. Given this, it was not surprising to researchers to discover that THC binds to a unique set of receptors in the brain that are now referred to as *cannabinoid receptors*. The endogenous ligand for the receptors is a derivative of arachidonic acid called *anandamide*, which may play a role in learning and memory as well as motor functions.

ANABOLIC STEROIDS

The *anabolic steroids* form a very different category of illegal drug. These compounds were originally developed in an effort to separate the *anabolic*, or muscle-building, effects of steroids from the *androgenic*, or sex-related, effects. Complete separation of these aspects, however, appears impossible to achieve. Thus, although anabolic steroids do, in fact, facilitate increase in muscle mass, they also produce side effects, including masculinization in women, feminization and infertility in men, liver toxicity, and premature cessation of bone growth in adolescents.

Pharmacogenetics

See also:
Personalized Susceptibility
and Tailored Therapeutics
Chapter 5

One new consideration in drug analysis that should be mentioned is the emerging field of pharmacogenetics. This is the study of the genetic factors which result in individual differences in drug absorption, distribution, and metabolism. As knowledge in this area progresses, traditional forensic drug analyses may well be supplemented with analysis of the potential genetic variations found in enzymes relevant to drug handling in the body. These would include, of course, the enzymes included in the cytochrome P450 system, as well as drug transporters such as the P-glycoprotein transporter, among others.

CRIMINAL POISONINGS

Of course, toxicants have also been used in criminal poisonings. One historically notorious poisoner was the Roman emperor Nero, who probably used cyanide to dispatch several problematic family members. In Italy, in the Middle Ages, the Borgia family used arsenic and phosphorus, while Catherine de Medici experimented with a variety of toxicants. In seventeenth-century France, Antonio Exili and colleagues plied a thriving trade in poisonings, and Catherine Deshayes, also known as "La Voisine" (the Neighbor), provided a mixture of arsenic, aconite, belladonna, and opium to her clients.

Arsenic

See also:
Airborne Toxicants
Categories of Waste
Chapter 17

Arsenic
Appendix

Even in modern times, the poisons used by these historical figures continue to be used in criminal poisonings. The most commonly used poison is *arsenic*. Available as a pesticide, arsenic binds to sulfhydryl-containing proteins and disrupts cellular respiration. Acute exposure to arsenic can result in gastrointestinal distress, cardiomyopathy and arrhythmias, anemia, and peripheral neuropathies. Chronic exposure results in peripheral neuropathy and hepatotoxicity.

The presence of arsenic in blood or urine can be detected by *atomic absorption spectroscopy*. In this technique, the ability of metals to absorb ultraviolet light at

Cyanide

See also:
Effects of Toxicants on
Enzymes
Chapter 4

Cyanide
Appendix

characteristic wavelengths is used to measure the concentration of metal in a sample. A system consisting of a *hollow cathode lamp* (with a filament made of the metal being measured) and a *monochronometer* (basically a filter that only allows a narrow band of light through) is used to generate the proper wavelength of light. The light is passed through the sample, which has been vaporized either by a flame or in a *graphite furnace*, and light absorption is measured. Comparison of sample results with a standard curve generated from known concentrations of the metal allows calculation of the concentration of the metal in the sample.

One poisoner who used arsenic in commission of her crimes was Nannie "Arsenic Annie" Doss, who poisoned 11 people (including five husbands).

The second most common poison used in criminal poisonings is *cyanide*. A naturally occurring substance (found in the pits of a variety of fruits), it is also available for use as a pesticide. Cyanide blocks the respiratory chain in mitochondria, preventing cells from carrying out oxidative phosphorylation. Following exposure, victims quickly develop neurological and cardiovascular symptoms (these, of course, are two of the organ systems that are most dependent on oxidative phosphorylation), and death can occur within a few minutes. Cyanide can be detected by colorimetric reaction.

Cyanide was the poison used in the mass murder/suicide of 913 individuals that was presided over by Rev. Jim Jones at his People's Temple in Guyana in 1978. Cyanide was also the poison found in Tylenol® capsules in the tampering incident of 1982 that led to significant changes in drug packaging. Unfortunately, although seven people died as a result of the tampering, no arrests have been made in connection with that case.

The third most common substance used in criminal poisoning is *strychnine*, a naturally occurring alkaloid found in a South Asian plant and now commonly used as a rodenticide. Strychnine blocks the inhibitory neurotransmitter glycine, leading to overstimulation of neurons controlling muscle con-

Strychnine

See also:
    Strychnine
                 Appendix

traction. This produces severe muscle contraction and convulsions within minutes of exposure, with death generally due to paralysis of the diaphragm. Strychnine can be detected by colorimetric reaction or by GC.

Ricin, a toxic lectin found in the castor bean seed, was apparently used in a particularly notorious case in 1978 in which the exiled Bulgarian journalist Georgi Markov was struck in the leg by an umbrella and died a few days later. A platinum capsule was found in his leg that contained traces of what may have been ricin. Another case was discovered when a fellow Bulgarian who had been injured and fallen ill in a similar manner (but survived) was subsequently found to have a similar capsule embedded in his back. Other substances involved in criminal poisonings have included antimony, chloroform, insulin, morphine, paraquat, mercury, thallium, and most recently TCDD, which was apparently used to poison the Ukranian presidential candidate Victor Yushchenko in 2004.

Organophosphorus and *carbamate* antiacetylcholinesterase insecticides and chemical warfare agents have also been used in criminal poisonings, including in the execution of civilian prisoners in German concentration camps of World War II and the attacks of Um Shin Rikio in Tokyo. Due to high rates of chemical reactivity, these classes of agents can be difficult to detect. Some aliphatic antiacetylcholinesterase agents, for example, are particularly difficult to detect because they present neither chromophore for ultraviolet detection nor halogen for electron capture GC; however, some can be derivatized following separation to introduce a fluorescent tag.

Biological assays using birds, mice, mosquitoes, or other sentinels can be used to indicate the presence of these extremely toxic poisons. Birds and insects, in particular, are highly susceptible, in part, due to a lack of arylester hydrolase (paraoxonase).

TCDD

See also:

The Role of Cytochrome P450

Chapter 3

Stimulation of Cell Growth and Division

Chapter 6

Toxicant-Induced Immunosuppression

Chapter 13

Pesticides

Chapter 17

TCDD

Appendix

Another approach is to measure inhibition of acetylcholinesterase *in vitro*. Blood of the poisoned subject can be inspected for reduced levels of acetylcholinesterase activity of erythrocytes or reduced serum carboxylesterase hydrolase activity.

It must be considered that some organophosphorus pesticides and chemical agents are chiral, and one enantiomer is much more toxic than its opposite enantiomer. This phenomenon is due to the right- or left-handedness of the active site of acetylcholinesterase and of competing enzymes, for example, acetylcholinesterase and chymotrypsin are opposites in reactivity with chiral organophosphinates. Separation and detection of specific enantiomers is achieved using derivatived supports for HPLC.

BIBLIOGRAPHY

Balikova, M., Collective poisoning with hallucinogenous herbal tea, *Forensic Sci. Int.*, 128, 50, 2002.

Brown, T.M. and Grothusen, J.R., High-performance liquid chromatography of 4-nitrophenyl organophosphinates and chiral-phase separation of enantiomers, *J. Chromatogr.*, 294, 390, 1984.

Bryson, P.K. and Brown, T.M., Reactivation of carboxylesterhydrolase following inhibition by 4-nitrophenyl organophosphinates, *Biochem. Pharmacol.*, 34, 1789, 1985.

Drummer, O.H., Postmortem toxicology of drugs of abuse, *Forensic Sci. Int.*, 142, 101, 2004.

Fisher, B.A.J., *Techniques of Crime Scene Investigation*, 6th ed., CRC Press, Boca Raton, FL, 2000.

Goldberger, B.A. and Wilins, D.G., Analytical and forensic toxicology, in *Casarett and Doull's Toxicology*, Klaassen, C.D., Ed., McGraw-Hill, New York, 2013, chap. 32.

Grothusen, J.R. and Brown, T.M., Stereoselectivity of acetylcholinesterase, arylester hydrolase and chymotrypsin toward 4-nitrophenyl alkyl(phenyl)phosphinates, *Pest. Biochem. Physiol.*, 26, 100, 1986.

Hobbs, W.R., Rall, T.W., and Verdoorn, T.A., Hypnotics and sedatives: Ethanol, in *Goodman and Gilman's: The Pharmacological Basis of Therapeutics*, 9th ed., Hardman, J.G., Limbird, L.E., Molinoff, P.B., Ruddon, R.W., and Gilman, A.G., Eds., McGraw-Hill, New York, 1996.

Introna, F., Campobasso, C.P., and Goff, M.L., Entomotoxicology, *Forensic Sci. Int.*, 120, 42, 2001.

Musshoff, F., Stamer, U.M., and Madea, B., Pharmacogenetics and forensic toxicology, *Forensic Sci. Int.*, 203, 53, 2010.

O'Brien, C.P., Drug addiction and drug abuse, in *Goodman and Gilman's: The Pharmacological Basis of Therapeutics*, 9th ed., Hardman, J.G., Limbird, L.E., Molinoff, P.B., Ruddon, R.W., and Gilman, A.G., Eds., McGraw-Hill, New York, 1996.

Olsnes, S., The history of ricin, abrin, and related toxins, *Toxicon*, 22, 361, 2004.

Poklis, A., Forensic toxicology, in *Introduction to Forensic Sciences*, 2nd ed., Eckert, W.G., Ed., CRC Press, Boca Raton, FL, 1997, chap. 8.

Poklis, A., Analytic/forensic toxicology, in *Casarett and Doull's Toxicology*, Klaassen, C.D., Ed., McGraw-Hill, New York, 2001, chap. 31.

Reisine, T. and Pasternak, G., Opioid analgesics and antagonists, in *Goodman and Gilman's: The Pharmacological Basis of Therapeutics*, 9th ed., Hardman, J.G., Limbird, L.E., Molinoff, P.B., Ruddon, R.W., and Gilman, A.G., Eds., McGraw-Hill, New York, 1996.

Saferstein, R., *Criminalistics*, 7th ed., Prentice-Hall, Upper Saddle River, NJ, 2001.

Skopp, G., Preanalytic aspects in postmortem toxicology, *Forensic Sci. Int.*, 142, 75, 2004.

Stahl, E., Ed., *Thin Layer Chromatography, A Laboratory Handbook,* 2nd ed., Springer, Berlin, 1969.

Trestrail, J.H., III, *Criminal Poisoning*, Humana Press, Totowa, NJ, 2001.

Wang, Z., Wen, J., Zhang, J., Ye, M., and Guo, D., Simultaneous determination of four bufadienolides in human liver by high-performance liquid chromatography, *Biomed. Chromatogr.*, 18, 318, 2004.

17 Applications
Environmental Toxicology and Pollution

Many human activities in our industrialized society produce the unpleasant by-product of pollution. This chapter covers the types and sources of air pollution, water pollution, and hazardous waste, and the role of toxicology in investigating and identifying the potential effects of pollution on human health and ecosystem function.

AIR POLLUTION

TYPES AND SOURCES OF AIR POLLUTANTS

Since the route of exposure for most air pollutants is respiratory, toxicologists tend to categorize air pollutants in the same manner as other inhaled toxicants: as either *gases* or *particles*. *Primary pollutants* enter the atmosphere directly as a result of some natural or man-made activity or process; *secondary pollutants* are formed when primary pollutants and other atmospheric constituents undergo chemical reactions in the atmosphere.

The major gaseous air pollutants include *carbon oxides* (carbon monoxide and carbon dioxide), *sulfur oxides*, *nitrogen oxides*, *ozone*, and volatile *hydrocarbons* (benzene, methane, and a special class of halogenated molecules called CFCs). Major particulates include *dusts*, *pollen*, and *heavy metals*. Most of these are produced during the process of *combustion*, the burning of organic matter. In *more developed countries*, fossil fuels such as oil, gas, and coal are burned for energy and heat, while in *less developed countries* wood and other crop matter is burned. The clearing of forests and grasslands by burning also releases pollutants into the atmosphere.

Countries making the transition from less developed to more developed frequently suffer problems of their own, as industrial and economic expansion often outstrips pollution control. China, for example, has some of the most severe air pollution problems of anywhere in the world. Increases in use of fossil fuels in transportation against a background of primarily coal-fired power plants have led to extremely poor air quality in cities such as Beijing and Shanghai. Unfortunately, pressures to sustain economic growth in these developing countries, however, are often seen as more immediate than the need to balance long-term health and ecological risks.

General Effects of Air Pollutants

Exposure to air pollution has been related to increased risk for a number of different adverse effects. Analyzing human health or ecosystem-level effects of pollutants is difficult, though, due to the large number of different pollutants and the potential for interactions between them. Effects of two irritant pollutants, for example, are frequently additive. Also, some pollutants may affect the mucociliary escalator, macrophage activity, or one of the other defense mechanisms of the respiratory tract, and thus exacerbate the effects of others.

Respiratory Effects

See also:
*Effects of Toxicants on
the Respiratory System:
General Principles*
Chapter 8

Levels of pollutants high enough to produce immediate adverse effects on human health frequently occur during a weather condition called a *thermal inversion.* Thermal inversions occur when a layer of warmer air traps a layer of colder air near the surface. This also traps airborne emissions, leading to the development of very high concentrations of pollutants (particularly if the inversion lasts for several days). The most serious inversion-related air pollution episode in this country occurred in Donora, PA, in 1948 and led to the death of over 20 individuals. Most of these deaths were due to exacerbation of pre-existing respiratory diseases (bronchitis, emphysema, or asthma, for example). A similar event occurred in London in December 1952, where burning of low-grade coal and an inversion produced pollutant levels that were increased by a factor of at least 10 from normal average values (which were already quite high by modern standards). It has been estimated that as many as 4000 individuals may have died of pollution-related causes during this 4-day episode.

Long-term exposure to lower levels of pollutants has also been implicated in the development of chronic respiratory diseases. Epidemiological evidence points to higher rates of chronic respiratory problems in areas with high levels of pollution. Some cases of lung cancer, too, may be attributable to exposure to air pollutants. Again, firm conclusions are hard to draw, due to confounding factors such as smoking.

Recently, questions have also been raised as to the effects of air pollution on neurological health. Epidemiological studies have linked exposure to air pollutants to decreases in cognitive function, developmental delays, and increased risk for autism. The mechanisms behind these effects are not well understood to this point, but studies indicate that it is possible for particulate matter to enter the brain via olfactory neurons, where it might contribute to inflammation, perhaps through interaction with microglial cells, astrocytes, and/or oligodendroglial cells. Some studies also hint at a connection between air pollution exposure, accumulation of abnormal neuronal proteins (amyloid beta protein, tau protein), and development of neurodegenerative diseases. More research into this area is certainly needed.

In terms of ecosystems, many studies of the effect of air pollutants on ecosystems have focused on effects of pollutants on plant growth and survival. On a molecular level, pollutants interact with plants in much the same way as with humans, producing cellular dysfunction and eventual physiological impairment. In plants, injury to tissues typically occurs in leaves and needles. Injury to plants, which, of course, as producers, function as the base of the trophic pyramid, can have repercussions throughout an ecosystem.

CARBON OXIDES

Carbon monoxide and carbon dioxide make up the carbon oxides. The most significant health effects of *carbon monoxide* (CO) are produced as a result of the high-affinity binding of carbon monoxide to the oxygen-carrying molecule hemoglobin. When 2% of circulating hemoglobin is converted into carboxyhemoglobin (the form that is unable to carry oxygen), neurological impairment can be measured. At 5% conversion, cardiac output increases and other cardiovascular changes are noted. CO also binds to another heme-containing molecule: cytochrome P450.

Levels of CO in the atmosphere vary by location, time of day, and time of year, and range from only a few parts per million in non-industrialized areas to 40 or 50 ppm or higher in urban areas. Levels are particularly high inside automobiles and in tunnels and garages. Exposure to environmental levels of 30 ppm CO over 8 hours results in the conversion of 5% of circulating hemoglobin.

Increasing levels of *carbon dioxide* (CO_2) in the atmosphere lead to a different set of problems. Carbon dioxide is one of the gases known as greenhouse gases, which are capable of absorbing the infrared radiation reflected by the earth. Thus, instead of escaping into space, this radiation heats the lower levels of the atmosphere, in what has been called the *greenhouse effect* (Figure 17.1). To some extent, the greenhouse effect is necessary to support life, because without it the temperatures near the earth's surface would very likely be too cold for life to exist. The burning of fossil fuels, however, is releasing CO_2 to the atmosphere at a much higher rate than ever before, and at the same time deforestation is reducing the number of plants able to remove CO_2 from the atmosphere through photosynthesis. The result has been an increase in atmospheric CO_2 over the last 100 years.

Most scientists predict that the increase in levels of CO_2 and other greenhouse gases will lead to an increase in surface temperatures (this predicted increase has been termed *global warming*). One unanswered question is: How high will temperatures go? Some mathematical climatological models have indicated that average temperatures could rise as much as 5°C overall, with greatest increases at the poles. Other models predict less warming due to offsetting factors such as increased reflection of incoming

---

Carbon Oxides

See also:
  *Effects of Toxicants on
    Transport Proteins*
                    Chapter 4

  *Vascular Spasms and Blood
    Pressure*
                    Chapter 9

  Carbon Monoxide
                    Appendix

Climate Change

See also:
  *Climate Change and
    Ecotoxicology*
                    Chapter 14

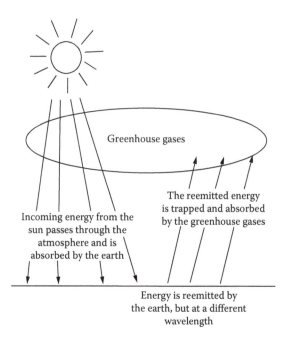

Greenhouse gases

The reemitted energy
is trapped and absorbed
by the greenhouse gases

Incoming energy from the
sun passes through the
atmosphere and is
absorbed by the earth

Energy is reemitted by
the earth, but at a different
wavelength

FIGURE 17.1   The greenhouse effect.

solar energy by pollutants in the atmosphere. Recent analysis of the global fluc-
tuating temperature over the last 600 years, from ice cores, tree rings, and direct
surface temperature measurements suggests that temperature has been rising more
rapidly in the present industrial age in what has been called the hockey stick graph
(Mann et al., 1998).

If this warming trend is caused by human activities and extends beyond natural
fluctuation (and strong evidence indicates that it is), significant changes may result.
Melting of polar ice would lead to rising of the oceans (perhaps by as much as
several feet), potentially putting many coastal areas below sea level. Changes in
rainfall patterns are also likely, which could adversely affect agricultural areas.
Shifts in forest composition would be seen, as cold-adapted species die out in
southern regions and spread into northern regions. Populations of animals, too,
may migrate, moving either to more northern regions or to higher altitudes. The
rapidity of the potential change is a cause for concern for many biologists, however, who wonder if species will have the time to make these adaptations in range
and habitat. Many fear that biological diversity will be significantly diminished
worldwide.

In 1997, an international agreement to reduce greenhouse gases was negotiated
in Kyoto, Japan. Over 140 nations have agreed to the *Kyoto Protocol*. The US government, however, has declined to ratify the agreement, citing economic hardships
imposed by the restrictions as well as objections to differential treatment of less- and
more-developed countries.

Sulfur Oxides and Nitrogen Oxides

Both sulfur oxides and nitrogen oxides are pulmonary irritants that are released during the burning of coal and petroleum products. Increases in airway resistance (due to irritation of upper airways) in humans can be seen at exposure levels as low as 5 ppm of sulfur dioxide (SO_2). Individuals with asthma or other chronic respiratory conditions may, however, be much more sensitive to the irritant, displaying increases in airway resistance at exposure levels of only 0.5–1.0 ppm SO_2.

Nitrogen dioxide (NO_2), on the other hand, is an irritant that affects the lower airways primarily. There is also some evidence that exposure to NO_2 may cause increased susceptibility to respiratory infection. However, concentrations necessary to produce pulmonary effects are higher than are generally encountered even in polluted atmospheres.

Both sulfur and nitrogen oxides can react with hydroxyl radicals in the atmosphere to produce sulfuric and nitric acids, which then dissolve into water droplets. These acids are not only pulmonary irritants, but are also the major components of the phenomenon called *acid rain*. Although rain is by nature somewhat acidic, high levels of sulfuric and nitric acids in the atmosphere can produce rain with a pH of 5.5 or lower (rain with a pH as low as 2.6 has been measured). Acid rain typically does not fall near the source of the sulfur and nitrogen oxides that produce it, but instead the pollutants may travel hundreds of miles (during which time the acids are produced) to fall on areas far downwind.

The degree to which a body of water is affected by acid rain depends on its surroundings. Lakes in areas rich in limestone and other water-soluble alkaline rocks generally contain bicarbonate and other ions that can neutralize the acid precipitation. Also, if the lake's substrate contains metals such as calcium or magnesium, cation exchange can occur. This process decreases the water's hydrogen ion concentration (and thus raises the pH), but at the same time increases the concentration of heavy metals.

Lakes that lack these natural buffering systems, however, may undergo changes in pH. This acidification in turn affects the many non-acid-tolerant species and causes a shift in community composition, potentially reducing biodiversity of fish, zooplankton, and phytoplankton. Lakes in the Adirondack region of New York, for example, have been particularly hard hit by a combination of high levels of acid precipitation (produced by electric/utilities and industry in the upper Midwest) and a lack of natural buffers. Fortunately, studies have indicated that reducing acid input will allow lakes to recover, although return to the initial pH may take many years.

Acid rain can affect terrestrial ecosystems, too. In recent years, forests in Europe and the United States have been undergoing a decline that has been

Sulfur Dioxide

See also:
Immediate Responses:
Upper Airway Effects
Chapter 8

Sulfur Dioxide
Appendix

Nitrogen Dioxide

See also:
Immediate Responses:
Lower Airway Effects
Chapter 8

Nitrogen Dioxide
Appendix

attributed at least in part to effects of acid rain. A wide variety of species of trees are affected, in many different locations. Particularly hard hit, however, are the coniferous forests found at high elevations. The mechanism of this effect is not clear, but several hypotheses have been advanced. Through the process of cation exchange, acid rain may leach calcium and magnesium (which are necessary for tree growth) from the soil or from needles and leaves. Also, soil pH may become lowered to the point that soil bacteria and other decomposers are unable to break down decaying organic matter and release the nutrients for uptake by trees and other vegetation. These stresses may then combine with other environmental stresses (including other types of air pollutants) to make the trees more susceptible to disease or injury. Lichens and moss are also susceptible to damage from acid rain.

On a hopeful note, studies have indicated that global sulfur emissions have been declining since 1990, primarily due to technological successes in pollution control. Since the late 1990s, this trend has been seen not only in more-developed countries, but also in less-developed countries, indicating that effective pollution control is possible, even in less economically advantaged areas of the world. Also, as a result of reduced emissions, evidence indicates that some affected lakes in North America have stabilized, and, in fact, a few have begun to show increased pH as well as some biological recovery.

HYDROCARBONS AND THE FORMATION OF SECONDARY POLLUTANTS (INCLUDING OZONE)

Ozone

See also:
  Immediate Responses:
  Lower Airway Effects
                Chapter 8

  Ozone
                Appendix

Incomplete combustion of fossil fuels and other organic materials leads to the release of hydrocarbons into the atmosphere. There are also some natural sources of atmospheric hydrocarbons, including release of terpenes by plants and the release of methane by decomposers. Hydrocarbons react in the presence of sunlight with oxygen or nitrogen oxides to produce a number of secondary pollutants in the atmosphere. These secondary pollutants include ozone, *aldehydes*, and *peroxyacetylnitrate* (PAN), a mixture commonly referred to as *photochemical smog*.

Although ozone is a necessary component of the stratosphere or upper atmosphere, it is a pollutant in the troposphere or lower atmosphere. Ozone is a respiratory irritant that affects the lower airways, producing inflammation directly (perhaps through the process of lipid peroxidation) and also causing an increase in reactivity to other irritants (such as other air pollutants and some allergens). Short-term exposure (a few hours) to ozone concentrations on the order of 0.10 ppm has been shown to produce temporary decreases in measured lung volumes in humans. Ozone also affects plants, probably reacting with unsaturated lipids in cell membranes to damage leaves and needles and ultimately to reduce growth. Effects of ozone and PAN on membranes are similar.

CHLOROFLUOROCARBONS

Chlorofluorocarbons (CFCs) are a group of stable compounds with several different uses in commercial and industrial processes, including use as propellants in aerosol cans, as refrigerants, and in the making of Styrofoam and other polystyrene products. Due to their chemical stability, CFCs do not react with other molecules in the lower atmosphere

---

Chlorofluorocarbons

See also:
   Chlorofluorocarbons
          Appendix

(the troposphere); instead, they travel to the upper atmosphere (the stratosphere), where they catalyze the breakdown of ozone (Figure 17.2). As a catalyst, the chlorine in CFCs participates in the reaction but yet remains unchanged. Thus, one CFC molecule may ultimately catalyze the breakdown of tens of thousands of ozone molecules.

The ozone layer in the stratosphere absorbs ultraviolet radiation, protecting life on earth from its adverse effects. Exposure to ultraviolet radiation can increase the risk of skin cancer and cataracts and can slow plant growth. Scientists have measured large holes in the ozone layer over Antarctica, and reductions in thickness of the layer in other parts of the world (including over North America).

The *Montreal Protocol* is a landmark pollution control agreement developed in 1987 and since signed by 188 countries (including the United States) in the attempt to control ozone degradation. Signatories have agreed to reduce production and consumption and eventually to phase out usage of chlorofluorocarbons and related ozone-depleting compounds. Although ozone recovery is difficult to measure, the concentrations of ozone-depleting gases are now declining, and scientists are cautiously optimistic about the effectiveness of the measures taken by the parties to the protocol.

PARTICULATES

Particles found in the atmosphere can be divided into two classes on the basis of size and chemical composition. Small particles (<1 mm in diameter) are produced primarily by combustion processes. These particles contain high levels of sulfate and carbon and tend to be acidic. Larger particles come from mechanical processes, such as weathering of rock and soil, and are not acidic. Particulates act as irritants, and many of the smaller particulates may contain carcinogenic components.

AIRBORNE TOXICANTS

Other air pollutants that do not fall into the above categories are considered airborne toxicants (sometimes referred to as *air toxics*). One main group of airborne toxicants is the heavy metals, including arsenic, mercury, lead, and cadmium, which are emitted from smelting and other industrial operations. In Anaconda, MT, and the surrounding areas, a very large hazardous waste site was the result of deposition onto the soil of airborne toxicants, including arsenicals, from a smelter used to process ore for gold, copper, manganese, and other metals mined in Butte, MT. Other airborne toxicants include various pesticides and solvents. Global transport of

Oxygen molecules are dissociated
by ultraviolet light

$$O_2 \xrightarrow{\text{hv}} O + O$$

An oxygen atom combines with an
oxygen molecule to make ozone

$$O + O_2 \longrightarrow O_3$$

A chlorine atom is released from
a CFC by ultraviolet light

$$CFC \xrightarrow{\text{hv}} CFC + Cl$$

A chlorine atom combines with ozone to form
chlorine monoxide and an oxygen molecule

$$Cl + O_3 \longrightarrow ClO + O_2$$

Chlorine monoxide reacts with an oxygen
atom to regenerate a chlorine atom

$$ClO + O \longrightarrow Cl + O_2$$

FIGURE 17.2   The chemical reactions involved in the breakdown of ozone in the atmosphere.

organochlorine pesticides, for example, is likely to have resulted from volatilization into the atmosphere, followed by falling in rain at distant locations.

There have been several major releases of highly toxic compounds into the atmosphere in recent decades. The most notorious examples include the emission of 2,3,7,8-tetrachlorodibenzo-*p*-dioxin (*TCDD*) from a hexachlorophene manufacturing plant in Seveso, Italy, on July 10, 1976. Remediation of the area involved removal of topsoil, and there has been a long-term epidemiological study of the population. In another event, the respiratory irritant *methyl isocyanate* was released from an underground storage tank in a facility manufacturing carbamate pesticides from this precursor. This accident in Bhopal, India, on December 3, 1984, resulted in many fatalities. And in yet another catastrophe, a large explosion in a nuclear power plant in Chernobyl, Ukraine, on April 26, 1986, led to severe contamination of the local area with radionuclides as well as radioactive fallout from the airborne plume in parts of Eastern Europe and Scandinavia. Rainfall through the atmospheric plume increased the fallout of cesium, and the plume was detected around the world in about 2 weeks.

INDOOR AIR POLLUTION

Another growing problem is that of indoor air pollution. As energy conservation measures have led to the construction of "tighter" homes and offices with lower air exchange rates, pollutants that once would have been vented regularly to the outside are now trapped inside for longer periods. Cigarette smoke, formaldehyde, and solvents such as trichloroethylene and benzene have all been postulated to be health threats.

In recent years, attention has also focused on *radon*, a radioactive gas released by the decay of radium and uranium. Radon gas escapes to the surface from the underground rocks and soils in which it is formed—an event that only poses problems if it is then confined by a structure such as a house. In areas where radon release is high, concentrations of radon in homes may reach unacceptable levels (particularly in basement areas). Currently, the EPA defines acceptable levels as below 4 picocuries of radiation per liter of air. There is, however, some discussion among scientists regarding whether the significance of health risks from radon exposure may have been somewhat overestimated.

CONTROL OF AIR POLLUTION

The main piece of legislation that regulates levels of air pollutants is the *Clean Air Act*, most recently reauthorized in 1990. The Clean Air Act divided the United States into air quality control regions and then set emission standards to regulate the amount of pollutants that an industry can release, and ambient air standards to specify maximum allowable levels of pollutants in each region. The latest reauthorization set new timetables for meeting these air quality goals and set up, in addition, an emissions trading policy. This policy allows companies to buy and sell permits that allow them to emit sulfur dioxide. Thus, the biggest polluters would have to buy extra permits at market value, while companies that pollute less could sell their additional permits for a profit. Limits on this trading are necessary, though. Additional emissions could not be purchased for use within an air quality region if it meant that ambient air standards would be exceeded. Industries also use many other methods to control emissions of air pollutants. Some of these methods are illustrated in Figure 17.3.

WATER POLLUTION

Almost everywhere water is found, it is vulnerable to pollution. Water pollution is a complex topic—there are many sources and types of water pollutants. Also, water pollutants may react together or with water itself, resulting in altered chemical forms. Because of this, the toxicity of pollutants in the aquatic environment varies with a number of factors (e.g., water temperature, hardness, and pH) and may in fact be very difficult to predict. Protecting our water resources is, however, critical to both ecological and human health.

Water pollutants are generally classified as belonging to one of several broad groups. Organic substances include dead and decaying plant and animal matter and wastes as well as organic compounds such as petroleum products, solvents, pesticides, and polymers. Inorganic substances include metals, nitrates, and phosphates. Biological agents include viruses, bacteria, protozoans, and other parasites that can cause disease. Suspended matter includes large, insoluble particles of soil and rock. Finally, radioactive materials and heat each constitute their own group.

Water pollutants can come from industrial, municipal, or agricultural sources. Industrial sources tend to be of the type called *point sources*, meaning that the emission of pollutants occurs at one or more very specific locations (a discharge pipe,

Prevention
- Conserve energy
- Burn cleaner fuels (low-sulfur coal,
    methanol), use cleaner solvents
- Improve fuel efficiency

Improve combustion process
- Improve fuel efficiency
- Use new techniques to remove/reduce pollutants
    during combustion process (better fuel mix,
    lower temperatures, etc.)

Remove pollutants from emissions

| Use catalytic converters on automobiles (these convert CO and hydrocarbons to CO_2 and water) | Use flue gas scrubbers for gases (these use chemical reactions to remove sulfur and nitrogen oxides from emissions) | Use electrostatic precipitation, filters, or cyclones for particulates (these methods use charged plates which attract oppositely charged particles, filters, or rapid spinning which sediments out particles) |

FIGURE 17.3 Methods for reducing emissions of air pollutants.

for example). Some municipal sources, such as discharges from sewage treatment plants, are also point sources. Most agricultural sources, on the other hand, are *nonpoint sources*. This means that the emission of pollutants occurs over a wide area, not just at a single point. One example of nonpoint source pollution is the runoff of fertilizer and pesticide-contaminated water from croplands. Urban storm drains are another example. Storm runoff can carry rain-borne pollutants that are less likely to be bound to soil particles than are agricultural chemicals, and thus may pose an even greater threat. Most water pollution is of the nonpoint source variety. Unfortunately, this is the most difficult type to control.

WATER IN THE ECOSYSTEM

Most (over 97%) of the world's water is in the oceans. The remaining water is found in ice and glaciers, in the ground, in the atmosphere, in lakes and rivers, and in living organisms. The structure and function of the aquatic ecosystems found in oceans, lakes, and rivers have been discussed in Chapter 14. All of these surface waters are vulnerable to pollution from chemicals deliberately discharged into them or from pollutant-contaminated runoff from surrounding terrestrial areas.

Water pollution affects not only surface water systems, but also water found beneath the surface. The pollution of *groundwater* is a special case for concern.

Groundwater is found below the surface of the earth, contained in a porous rock structure called an *aquifer*. Groundwater is a major source of irrigation and drinking water. Aquifers are recharged by surface water that percolates down through the soil, and if the water passes through contaminated areas (hazardous waste disposal sites, near leaking underground tanks, etc.), toxicants may leach out and be carried into the aquifer. Water in the aquifer flows, as does surface water, from higher to lower elevations, but at a rate as slow as a few inches a day. With little mixing and diluting, and few bacteria to decompose waste, toxicant levels may become quite high.

Polluted groundwater may contaminate the lakes and streams that it feeds into and can lead to human health problems if used for drinking water. The EPA has estimated that almost half of the public water supply systems that rely on groundwater as a source are contaminated with one or more potentially hazardous toxicants.

ORGANIC WASTES AS WATER POLLUTANTS

As discussed in the chapter on ecological toxicology, energy flows and materials cycle through aquatic ecosystems. Aquatic ecosystems contain producers, consumers, and of course decomposers. Decomposers are the bacteria and other organisms that break down dead and decaying organic matter. In the process, they obtain energy and release nutrients for reuse by other organisms. Some of these bacteria (particularly those that dwell in the sediments) are anaerobic, which means that they perform their metabolic processes without needing oxygen. Typically, the energy repackaging pathways for these bacteria involve either fermentation (the conversion of sugar to either lactic acid or ethanol) or a variation of oxidative phosphorylation in which sulfates or nitrates substitute for oxygen as an electron acceptor. Other bacteria, however, are aerobic, which means that they require oxygen for their metabolic activities.

Normally, this system is balanced, with the bacterial population size limited by the supply of waste from which they obtain energy. If, however, additional organic waste material is added to the ecosystem (through influx of sewage, or organic waste-containing sediments), the bacterial population may undergo rapid growth. The respiratory activities of all these bacteria can then lead to oxygen depletion, particularly in lakes (the greater surface area of streams combined with the motion of the water help keep stream oxygen levels high). If oxygen levels drop to lower than around 5 mg/L, the survival of species with high oxygen needs may be threatened.

Respiratory activities of the decomposers can be assessed through a measurement called *biological oxygen demand* (BOD). A high BOD indicates that there is a high level of decomposer activity as a result of high levels of organic waste (animal wastes, fertilizer, etc.) in the water.

PETROLEUM PRODUCTS AS WATER POLLUTANTS

Several million tons of petroleum are spilled or leaked into the oceans every year. Part of this petroleum comes from land-based sources such as municipal and industrial wastes that are dumped into rivers and streams and ultimately find their

Petroleum Products

See also:
   Petroleum Products
                    Appendix

way into the ocean. The remainder comes from tankers (both accidental spills and routine releases) and leakage from offshore drilling sites. Among the most significant tanker accidents are the wreck of the *Torrey Canyon* off the southern coast of England in 1967, the grounding of the *Amoco Cadiz*, which dumped 230,000 metric tons (MT) of oil into the English Channel in 1978, and the *Exxon Valdez* accident in Alaska in 1989. In 1996, the tanker *Sea Empress* lost 72,000 tons of crude oil off of the southwest coast of Wales, badly damaging shorelines and impacting the physical and psychological health of individuals living in the spillage area. In 1999, the tanker *Erica* spilled 5.8 million gallons of fuel oil off the coast of France; in 2002, the *Prestige* sank off the coast of Spain. One of the biggest single accidents was the IXTOC I drilling rig accident in 1979, which was responsible for the release of 400,000 MT into the Gulf of Mexico. But the biggest oil disaster was the spill that occurred in 1991 during the Persian Gulf War, when an estimated 250 million gallons escaped (twice the volume that was released in the IXTOC accident and 20 times the volume released by the *Exxon Valdez*). Challenging, but not quite surpassing, that record was the more recent Deepwater Horizon oil spill in the Gulf of Mexico in 2010.

Petroleum is a complex substance consisting of hundreds of different compounds. The bulk of crude petroleum (also called *crude oil*) is made up of *aliphatic* (straight-chain) *hydrocarbons* with backbones of anywhere from 1 to 20 or more carbons. Mixed aliphatic hydrocarbons are refined to products including natural gas (1 or 2 carbons), bottled gas (3 or 4 carbons), gasoline (5–10 carbons), kerosene (12–15 carbons), fuel and diesel oil (15 or more carbons), and lubricating oils (19 or more carbons). Crude oil also contains *cyclic hydrocarbons*, *aromatic hydrocarbons* (with structures based on benzene), sulfur, nitrogen, and a variety of trace metals.

When oil is spilled or leaks into a waterway, it initially spreads across the surface (forming an *oil slick*). Some of the more volatile components may then evaporate, and because the rate of evaporation depends directly on temperature and surface area, the process is more rapid in warm areas and in rougher seas (which promote formation of droplets of spray). Other components (particularly the aromatics) may dissolve into the water. The remaining heavier material forms an emulsion sometimes called *mousse*, or may eventually form lumps called *tar balls*. Some of the oil is eventually broken down by microorganisms or by photochemical processes. Oil released at depth, however, forms what is often termed a *plume* that spreads in a complex manner with only some actually reaching the surface.

As with other compounds, toxicity of crude oil components in general correlates well with lipid solubility, because lipid-soluble compounds are (1) better able to cross membrane barriers and enter organisms, (2) more likely to be widely distributed in the body (including into the brain), and (3) more likely to be retained in depot fat tissues over time and to bioconcentrate through the food web. Studies of oil spills along rocky shores (the *Torrey Canyon*, *Amoco Cadiz*, and *Exxon Valdez*) have indicated that following release of oil, immediate effects are seen on community structure, with species of green algae replacing the more sensitive red and brown algae. Many populations of invertebrates (including crustaceans, mollusks, and starfish) may be completely destroyed. Some fish populations may also be impacted. A year and a half after the Deepwater Horizon spill, studies have indicated reductions in biomass of salt marsh vegetation in affected areas of the Gulf Coast, although other studies have indicated relatively little incorporation of carbon from the oil into estuarine food webs. For the Deepwater Horizon event (which was not a surface discharge, but occurred at a depth of approximately 1500 m), impact on benthic communities is also an issue, but is difficult to assess given the lack of baseline, pre-spill data. However, some studies have indicated negative impacts on benthic areas predicted to have been in the path of the heaviest plume.

Probably the most dramatic and immediate effects of oil spills, though, are on the birds and mammals. The feathers of seabirds become coated with oil, causing them to lose their capacity to insulate the animal, resulting in death from hypothermia. In addition, oiling of feathers causes loss of buoyancy, potentially resulting in drowning. Ingestion and systemic toxicity may also occur during attempts by the bird to clean the feathers. Marine mammals such as otters and seals may suffer the same fates, due to similar effects of the oil on their fur. Although estimates are difficult to make, the *Exxon Valdez* spill killed probably close to half a million birds and several thousand marine mammals. The results for the impact of the Deepwater Horizon spill on birds and mammals are not yet in.

Ironically, analysis of the aftermath of the *Exxon Valdez*, Deepwater Horizon, and other spills has indicated that attempts to clean fouled shores may not, in fact, be ecologically beneficial. Some studies have indicated that recovery is quicker on beaches that have not been cleaned than on beaches that have. For example, scientists have pointed out that cleaning with detergents may do more harm than good, as the detergent–oil mix may produce greater mortality than the oil alone. Also, using hot water to blast beaches may not only drive oil deeper into the sediments, but may also kill invertebrates. Thus, there is considerable debate about whether the $2.5 billion spent by Exxon on cleanup was particularly effective. The use of dispersants in both the *Exxon Valdez* and Deepwater Horizon spills was controversial, also. Dispersants may dilute the spill, but may also result in spreading the damage to areas that otherwise might have remained unaffected. Finally, there has been some discussion as to potential toxicity of the dispersants themselves to marine algae, zooplankton, and other marine organisms.

One promising group of experimental techniques for cleaning up spills is grouped together under the term *bioremediation*. One form of bioremediation consists of adding nutrients to the water, thereby ensuring that the natural oil-metabolizing microorganisms have sufficient quantities of these nutrients. This allows the population to grow more rapidly and hopefully to metabolize more of the oil. Initial studies of the effectiveness of this technique in Prince William Sound, however, were inconclusive. Another bioremediation option is the addition of nonnative oil-metabolizing (perhaps even genetically engineered) microorganisms. Probably the best solution, however, to the oil spill problem is to prevent the spill through use of double-hulled ships (which could contain their oil cargo even if the outer hull is damaged) and better training of crews. The Deepwater Horizon incident, of course, has also focused attention on the issue of safety (in terms of both equipment design and drilling procedures) on offshore drilling platforms.

Although many of the most visible effects of an oil spill may disappear after a few months or years, the impact lingers on, particularly in relatively isolated, protected shoreline ecosystems. Because degradation is slow and incomplete, crude oil components find their way into the food chain and may bioaccumulate to toxic levels. Shellfish, for example, from polluted areas may be unsuitable for consumption. Evidence has shown that traces of oil can be found in sediments and biological tissues for as long as 20 years after the initial spill.

Also, most oil pollution occurs in regions that suffer from chronic exposure to low levels of crude oil (areas near refineries, for example). Again, effects from this type of exposure may not be as dramatic as that resulting from a single incident, but over the long-term bioaccumulation and resulting systemic toxicity can affect community and ecosystem structure as well as pose potential health hazards for humans who harvest fish and shellfish from the area.

PESTICIDES

Pesticides are widely used in our society. They play a role in blocking transmission of vector-borne diseases such as malaria, in controlling insects and weeds in farming, and in maintaining relatively pest-free homes and yards. Because of their widespread use, however, they frequently find their way into aquatic ecosystems. In many cases, pesticides enter waterways as a component of nonpoint source runoff from agricultural lands. Aerial application of pesticides can also result in water pollution, as pesticides may drift downwind and be deposited in lakes or rivers. Pesticides may even be deliberately introduced, in order to limit growth of algae or control insects.

Both routine industrial effluent emissions and accidental spills from pesticide manufacturing facilities can also contaminate aquatic ecosystems, as can urban runoff. In 1986, for example, a fire at the Sandoz warehouse near Basel, Switzerland, led to the release of more than 1000 tons of pesticides into the already polluted Rhine River. Pesticides may also leach out of hazardous waste disposal facilities and enter groundwater.

The behavior of pesticides in an aquatic ecosystem depends mainly on the chemistry of the pesticide itself. Pesticides vary widely, for example, in lipid solubility as well as persistence in the environment (a measure of the amount of time the pesticide remains in the environment before being broken down). Pesticides may dissolve in the water, may adsorb to the surface of particles in the water, or may be absorbed by aquatic organisms. Of course, the more lipid-soluble the pesticide, the more likely it is to be absorbed by an organism, partition into fat tissues, and be passed along through the food chain in the process of bioconcentration. Many pesticides undergo metabolism by bacteria and other organisms as well as other nonenzymatic photochemical alterations.

Organochlorines

See also:
Effects of Toxicants on Electrical Conduction
Chapter 10

Material Cycling in Ecosystems
Chapter 14

Organochlorines
Appendix

One major category of pesticides is the *organochlorine* or *chlorinated hydrocarbon insecticides*. Pesticides in this group include the dichlorodiphenylethanes, the cyclodienes, and the hexachlorocyclohexanes (Table 17.1). *DDT*, a dichlorodihenylethane, is a persistent insecticide. This biologically active

---

TABLE 17.1

Organochlorine Insecticides

Dichlorodiphenylethanes

  DDT

  Methoxychlor

Cyclodienes

   Chlordane

   Heptachlor

   Aldrin

   Dieldrin

Hexachlorocyclohexanes

   Lindane

---

compound is initially metabolized to breakdown products that are also biologically active and that are only slowly converted into inactive forms. The half-life of DDT in the environment is generally estimated at between 2 and 3 years. Because of its low water solubility and high lipid solubility, DDT does not dilute out in aquatic environments, but instead tends to adsorb onto organic particles and to bioconcentrate in organisms. As is typical with other lipophilic environmental toxicants, highest concentrations of DDT are observed in organisms at the top of the food web. In some studies, levels as high as 10–20 ppm were measured in some seabirds, and high levels have also been found in larger fish and marine mammals. While DDT is no longer used in the United States, its use continues in other parts of the world, and residues have been found even in Antarctic birds.

Although mammalian toxicity is relatively low (oral LD_{50} of around 200 mg/kg), DDT is, of course, toxic to many insect species, including flies, beetles, and mosquitoes (the vector for the malaria-causing organism *Plasmodium*). For insects, though, pesticides are just another selection pressure. Because insects are rapidly reproducing r strategists, insect populations can adapt very quickly to the presence of pesticides through the evolution of resistant populations. Fish are also affected, with the young of many species (such as salmon, for example) being particularly sensitive. While there is evidence that fish may suffer from behavioral and reproductive changes following exposure to DDT, the effects of DDT on fish populations do seem to be reversible. DDT is also toxic to several species of algae.

The most significant effects of DDT on wildlife are on birds of prey, especially those species that prey on fish. As K strategists, large birds are slower to reproduce and thus slower to develop adaptive mutations. The combination of these slow reproductive rates with the tendency of DDT to bioaccumulate has led to severe impacts on bird populations. Populations of grebes, bald eagles, peregrine falcons, and osprey have all reportedly been affected by DDT. DDT causes changes in reproduction characterized by alterations in mating and parental behavior, decreases in number of eggs laid, decreases in number of eggs successfully hatched (due at least in part to decreased eggshell thickness), and increased mortality rates among chicks. Although the survival of many populations in the United States was threatened, some are now recovering following the ban on DDT use, first by Wisconsin and Michigan and then by the EPA.

Chronic exposure to organochlorine pesticides has been associated in epidemiological studies with increased risk for development of a variety of cancers, as well as reproductive and developmental disorders. Due to their tendency to bioaccumulate, as well as their toxicity, many organochlorine pesticides (including chlordane, heptachlor, aldrin, and dieldrin) are either banned or have severe restrictions on their use in the United States.

Organophosphates

See also:
  Serine Hydrolases
                Chapter 3

  Effects of Toxicants on
    Enzymes
                Chapter 4

  Acetylcholine
  Axonopathies
                Chapter 10

  Organophosphates
                Appendix

Carbamates

See also:
  Carbamates
                Appendix

Two other major classes of insecticides (which share a basic mechanism of action) are the *organophosphates* and *carbamates* (Table 17.2). These are the current insecticides of choice for many uses, because they are relatively nonpersistent (with half-lives measured in weeks rather than years) and are considerably more water-soluble than the organochlorines. Organophosphates tend to be more toxic to invertebrates than to fish. For example, the 96-h LC_{50} for methyl parathion ranges around 10 mg/L for invertebrates, compared to over 5000 mg/L for fish. Discussion of mammalian neurotoxicity of organophosphates and carbamates can be found elsewhere in this text.

The *pyrethroids* are another class of insecticides (Table 17.2). These compounds have low persistence, partially due to rapid metabolism and detoxification by cytochrome P450 systems. They also bind so strongly to soil particles that availability to nontarget organisms is relatively low, although they are toxic to fish in the low parts per billion range.

Herbicides are often used to control aquatic plants and algal growth. Some common herbicides that are potential pollutants are listed in Table 17.3. Among the most widely used herbicides are the chlorphenoxy

TABLE 17.2
Other Insecticides

Organophosphates
  Parathion
  Fenitrothion
  Malathion
Carbamates
  Carbaryl
  Aldicarb
  Carbofuran
Pyrethroids
  Cypermethrin

TABLE 17.3

Herbicides

Chlorphenoxy acids

    2,4-D

    2,4,5-T

Bipyridals

    Paraquat

    Diquat

acids *2,4-dichlorophenoxyacetic acid* (2,4-D) and *2,4,5-trichlorophenoxyacetic acid* (2,4,5-T). These compounds typically provide fewer water pollution problems than DDT because: (1) their much higher water solubility encourages dilution rather than bioconcentration and (2) their residence times in the environment are much shorter, in the order of weeks rather than years. Also, LD_{50} values for most species are high (in the order of several hundred mg/kg). Many of the reported adverse effects (such as reproductive and teratogenic effects and immunosuppression) related to exposure to these compounds may instead be due to a contaminant, *2,3,7,8-tetrachlorodibenzo-p-dioxin* (TCDD). The range of LD_{50} values for TCDD is much lower, with values for some species of less than 1 μg/kg. Because of its low solubility and its tendency to bind tightly to soils, TCDD is rarely detected in water.

A different class of herbicide includes the bipyridyl herbicides, *paraquat*, and *diquat*. Diquat is commonly applied to aquatic environments and has an LD_{50} of around 100–200 mg/kg in mammalian species and causes free radical-induced damage to liver and kidney (in contrast to paraquat, which accumulates in and preferentially affects the lung). Both these compounds bind tightly to soils and have very long half-lives (more than 5 years).

Finally, two organometallic compounds frequently used for pest control in aquatic environments are: (1) *copper sulfate*, a commonly used algicide, and (2) *tributyltin*, an antifoulant commonly used to control growth of barnacles and other organisms on the hulls of ships. The toxicity of these compounds will be discussed under the section on metals.

2,4-D, 2,4,5-T

See also:

    2,4-D, 2,4,5-T

        Appendix

2,3,7,8-TCDD

See also:

    The Role of Cytochrome P450

        Chapter 3

    Stimulation of Cell Growth
    and Division

        Chapter 6

    Endocrine Disrupters
    and Interference with
    Hormonal Controls

        Chapter 7

    Toxicant-Induced
    Immunosuppression

        Chapter 13

    TCDD

        Appendix

Paraquat

See also:

    Immediate Responses:
    Lower Airway Effects

        Chapter 8

    Paraquat

        Appendix

OTHER ORGANIC COMPOUNDS

Plastics

See also:

   Case Study: Plastic Debris in
   the Marine Environment
              Chapter 14

   Other Additives, Including
   Indirect Additives
              Chapter 19

Pollution by *plastics* has become a significant problem, primarily because of the chemical stability of most plastics. Every year, tons of plastic trash is dumped into bodies of water where it floats free or washes up onto beaches and shores. Problems arise particularly in the ocean, where buoyant plastic trash such as plastic netting, fishing line, and six-pack rings entangle marine birds, reptiles, fish, and mammals. Some studies estimate, for example, that tens of thousands of seals are killed each year by plastic debris, and even whales may be fatally tied up in discarded netting. Plastic may also be mistaken for food and consumed by these animals, blocking and causing irritation of the gastrointestinal tract.

Another group of organic compounds of particular concern are *trihalomethanes* (chloroform, bromodichloromethane, and dibromochloromethane). Some of these compounds appear naturally in water, whereas others are thought to be produced by reaction of chlorine (added as a disinfectant) with organic matter in the water. Formation of trihalomethanes is limited if chlorine is added as the final step in water treatment (after most organic compounds have already been removed). The presence of these compounds in drinking water has been associated with an increased risk of several types of cancer.

PCBs

See also:

   Toxicant-Induced
   Immunosuppression
              Chapter 13

   PCBs
              Appendix

A group of synthetic organic chemicals, *polychlorinated biphenyls* (PCBs) have been linked to several ecological and health-related effects. These highly lipophilic, persistent compounds are primarily used as insulating material in electrical components and enter the water supply through discharge of industrial effluents and leaching from landfills and hazardous waste dumps. Also, because very high temperatures are required to destroy these compounds, incomplete incineration may also lead to their release. One of many areas that has been polluted with PCBs is the upper Hudson River in New York, where two factories that manufactured capacitors dumped PCBs into the river for more than 30 years.

Due to their lipophilicity and persistence, PCBs bioaccumulate and can be found in aquatic organisms in virtually all parts of the world. Levels of PCBs in some species of fish collected from the upper Hudson in 1977 averaged as high as 6000 µg/g. The levels of PCBs in this ecosystem are now declining (slowly) due to slowing of rate of discharge, dredging and removal of some contaminated sediments, and the gradual degradation of the compounds.

PCBs are more toxic to fish than to birds or mammals and have been found to interfere with reproductive function in a number of different species. PCBs lead to induction of cytochrome P450 in mammals and have been reported to cause liver cancer in rats. In addition, they are immunosuppressants.

PHOSPHORUS AND NITROGEN

Phosphorus and sometimes nitrogen are what are known as *limiting nutrients*, meaning that they are among the first necessary nutrients to become depleted and thus to limit the rate of growth of a plant population. These two essential nutrients occur in many chemical forms. Most of the available phosphorus in aquatic environments is in the form of phosphate (PO_4^{3-}). Primary forms of nitrogen include nitrate (NO_3) or ammonium ion (NH_4^+). When human activities around a body of water result in the addition of phosphorus and nitrogen, rapid bursts of algal and plant growth (sometimes called blooms) can occur. Lakes are particularly susceptible. Algae levels increase, activity of decomposers must increase (to break down dead algae), and as a result, oxygen levels become depleted, affecting many species of fish. This input of additional nutrients, and the resulting ecological changes, is called *cultural eutrophication*.

Eutrophication

See also:
   Freshwater Ecosystems
              Chapter 14

   Cyanobacterial Toxins
              Chapter 20

There are many activities that can contribute to cultural eutrophication. Municipal wastewaters account for most of the input, carrying nitrogen- and phosphorus-rich sewage as well as phosphates from detergents (phosphates are used in detergents to bind ions that may interfere in the cleaning process). Agricultural runoff may also contain phosphates and nitrates leached from the soil following fertilizer application.

One of the areas in the United States that has suffered most from eutrophication is the Great Lakes. Lake Erie, in particular, has been affected as a result of the many urban and agricultural activities on its shores. At one point, one region of Lake Erie had phosphorus concentrations that were eight times those of the much less affected Lake Superior. Algae levels increased, mayfly larvae and other benthic insects disappeared, and populations of trout and whitefish declined.

Cultural eutrophication is fortunately reversible with time. Reduction of phosphate and nitrate input will eventually restore lakes to their original, more oligotrophic condition. Recently, due to the Great Lakes Water Quality Agreement signed by the United States and Canada, phosphorus input into Lake Erie has decreased and algal growth has dramatically declined. Oxygen levels near the bottom of the lake also appear to have improved.

Nitrates and Nitrites

See also:
   Effects of Toxicants on
    Transport Proteins
              Chapter 4

   Vascular Spasms and Blood
    Pressure
   Effects of Toxicants on
    Hemoglobin
              Chapter 9

   Preservatives and
    Antioxidants
             Chapter 19

   Nitrates, Nitrites
              Appendix

The presence of nitrates in drinking water has also been associated with a human health problem called methemoglobinemia. In this condition, nitrates are converted in the body into nitrites, which oxidize hemoglobin and prevent the molecule from carrying oxygen. Infants are particularly susceptible.

METALS

Although metals are normally present in the environment, human activities may increase metal concentrations in aquatic environments to levels that may be hazardous to ecological systems or human health. Metals interact with water and other compounds in a complicated manner, with each metal forming many different *species*, or chemical forms, depending on conditions. Metals can be found in water as the *free metal ion*, or combined with chloride or a variety of other anions to form an *ion pair*. Metals also bind to both small and large organic molecules. Small organometallic molecules are frequently soluble in water, while the larger complexes formed by association of metals with large organic components of soils and sediments are usually not. Factors such as pH and hardness (the concentration of calcium, magnesium, and other cations in the solution) can affect the relative concentrations of different species of a metal.

Speciation of metals is often key in determining effects of metals in an aquatic environment, as different species of the same metal may vary widely in toxicity. For example, some studies have found that copper carbonate ($CuCO_3$) is much less toxic to trout than copper hydroxide ($Cu(OH)_2$) or copper ion (Cu^{2+}). Also, complexation with organic matter generally lowers toxicity by making the metal less available for absorption by organisms. Bioaccumulation of metals does occur, however, particularly in plants. Organisms also have the potential to metabolize some metals, thus altering their chemical species. In fact, genes for copper resistance have evolved in plant pathogenic bacteria (selected for by agricultural use of copper bactericides), suggesting novel means of bioremediation.

One metal with toxic potential is *lead*. Lead enters lakes and streams when it leaches from soils as a component of runoff. It then tends to accumulate in sediments, with freshwater levels generally only reaching a few micrograms per liter. Lead is a much more significant problem in drinking water, where it may leach from lead pipes or solder and reach concentrations of 50 µg/L or higher. Lead produces both neurological and hematological effects.

Many of the various species of *cadmium* are quite soluble in water and also tend to bioaccumulate (*particularly in shellfish, but also in fish and plants*). *Cadmium's potential as a toxic water pollutant* was demonstrated in Japan, in areas where water from the Jinzu River in Toyama Prefecture, Honshu, was used to irrigate rice fields. Many people in these areas,

Lead

See also:
Interference with Cell Division
                                    Chapter 7

Anemias, Hemolysis, and Related Disorders
                                    Chapter 9

Myelinopathies
Other Neurotoxicants
                                    Chapter 10

Toxicant-Induced Immunosuppression
                                    Chapter 13

Lead
                                    Appendix

Cadmium

See also:
Interference with Cell Division
                                    Chapter 7

Vascular Spasms and Blood Pressure
                                    Chapter 9

Damage to the Proximal Tubule
                                    Chapter 12

Cadmium
                                    Appendix

particularly older women, suffered from a disease they called *itai-itai, or "ouch ouch,"* *because of the severe bone pain that accompanied it. Itai-itai* was characterized by severe osteoporosis and kidney dysfunction. The river in the affected area was highly polluted by a mining operation upstream that was discharging tailings that were contaminated with cadmium (among other metals). Cadmium bioaccumulated to high levels in rice from the irrigated fields, and consequently in the people who consumed the rice. Levels of cadmium and other metals in the tissues of afflicted persons were found to be as high as several parts per thousand, confirming metal pollution as the cause.

A second incident in Japan involved pollution by another heavy metal, *mercury.* Mercury, that was released into Minamata Bay in Kumamoto Prefecture, Kyushu, by a plastics manufacturer, was methylated by microorganisms in the bay to form *methylmercury,* a neurotoxicant. Bioconcentration occurred and effects were seen in organisms near the top of the food chain. Large fish were found dead in the bay, seabirds fell into the water, and the cats living in villages around the bay began to stagger around (this led to the nickname of "dancing cat disease" for the condition). Eventually because the inhabitants of bayside villages derived most of their food from the bay, they too were affected.

Mercury

See also:
   Vascular Spasms and Blood
     Pressure
                                  Chapter 9

   Other Neurotoxicants
                                 Chapter 10

   Damage to the Proximal
     Tubule
                                 Chapter 12

   Mercury
                                 Appendix

As mentioned before, another organometal that is a water pollutant is the compound *tributyltin* (TBT). This compound is used to kill barnacles, mussels, and other organisms that attach themselves to the hulls of ships. Usually, TBT is incorporated into paint, which releases the chemical gradually. Unfortunately, many marine species, such as oysters, are extremely sensitive to TBT, with some species showing effects (shell malformations, reproductive effects) at levels as low as a few parts per billion. This has led to more stringent regulation, and in some cases banning of TBT use.

Zinc, although an essential trace element, can produce toxicity through interference with absorption of copper and iron, leading to anemia. Zinc can also cause necrosis of gill tissues in fish. Increased levels of zinc may, however, protect against cadmium poisoning, perhaps through induction of the cadmium- and zinc-binding enzyme *metallothio-*

Metallothionein

See also:
   Distribution
                                  Chapter 2

nein. High concentrations of copper (which is often used as an algicide) can lead to accumulation of the metal in the liver (an effect seen in both mammals and fish).

OTHER WATER POLLUTANTS

Biological agents frequently enter water systems when sewage is inadequately treated. This is particularly a problem in less-developed countries, where contact with contaminated water through swimming, drinking, or eating contaminated fish or shellfish can lead to infections. Diseases that can spread in this manner include bacterial infections such as *cholera* and *typhoid*, as well as diseases such as *amoebic dysentery*, which are caused by protozoans.

When sediment is carried off from land into water, it becomes *suspended matter*. Suspended matter can block sunlight and thus prevent photosynthesis. And because many sediment particles contain organic matter, suspended matter can contribute to the problem of eutrophication.

Thermal pollution is a problem associated most frequently with power plants. These plants bring in cool water from a nearby ocean, lake, or stream, and then discharge it at a higher temperature. The higher the temperature and the greater the volume of the discharged water, the more significant the problem is likely to be. Warm water holds less dissolved oxygen than cooler water, and at the same time the increased temperature causes an increase in body temperature, and thus rates of cellular respiration for many aquatic organisms. The increased oxygen requirements associated with the changes in respiration combine with the lower availability of oxygen to produce oxygen deprivation.

REGULATION AND CONTROL OF WATER POLLUTION

Water pollution in the United States is regulated by two major laws: the *Clean Water Act* (originally written in 1972 and reauthorized as the Water Quality Act of 1987) and the *Safe Drinking Water Act*, written in 1974. The Clean Water Act and its reauthorizations require industries to pretreat wastes before discharge into municipal water systems and require permits for discharges into navigable waters. In addition, federal financial support is provided for construction of better water treatment facilities. The Safe Drinking Water Act requires the establishment of maximum acceptable levels for pollutants (*maximum contaminant levels*, MCLs) in drinking water. The process is slow, however, and standards have not been set for all chemicals known or suspected to be a problem. A general wastewater treatment scheme is shown in Figure 17.4.

TOXIC WASTES

SOURCES OF TOXIC WASTES

One serious consequence of the industrial revolution and modern development has been the generation of *toxic waste* materials. Toxic wastes are produced as a by-product of many activities in modern society, including manufacturing and other industrial processes, and are also produced in the consumption or utilization of manufactured goods. While *toxicity* of waste substances can be estimated by a median lethal dose study or similar experiment, estimating the actual *hazard* posed by a waste involves estimating not only intrinsic toxicity of the substance, but also factors such as the likelihood that exposure to the substance will occur. Toxic waste, when handled or disposed in such a way as to threaten public health or the environment, can be considered *hazardous waste*.

Generators of hazardous waste can be found in every segment of society. Activities that generate hazardous wastes include manufacturing, mining, defense (weapons manufacture), agriculture, utility company operations (power plants), small business

Waste water treatment

Primary treatment

Screens and filters remove solid materials
(which are then sent to sanitary landfill)

Secondary treatment

Small particles settle out in settling tanks
(the sludge which results from this
step must then be disposed of by
incineration, digestion, or burial)
In an aeration tank, aerobic bacteria are then
allowed to digest the remaining organic materials
(the bacteria are then allowed to settle
out and are added to the sludge)

Tertiary treatment

The removal of phosphorus, nitrogen, metals, and
other pollutants requires additional special treatments

FIGURE 17.4   Wastewater treatment.

operations (dry cleaners, paint shops, automobile service, etc.), hospital operations, laboratory research, and even household and municipal activities.

CATEGORIES OF WASTE

The great majority of hazardous waste is produced by large generators in manufacturing, mining, and the military. These sources discard wastes containing toxicants such as organic solvents, polycyclic aromatic hydrocarbons, pesticides, and heavy metals. In a survey of 1189 sites, the most commonly found solvents included 1,1,2-trichloroethylene (found in 401 sites), toluene (found in 281 sites), and benzene (found in 249 sites). Polycyclic aromatic hydrocarbons found included naphthalene (60 sites), phenanthrene (41 sites), and benzo[a]pyrene (37 sites). Among the many toxic metals found were lead (395 sites), chromium (395 sites), and arsenic (187 sites). Polychlorinated biphenyls were found in 185 sites, and the chlorinated organic insecticides DDT and chlordane were found in 50 and 33 sites, respectively. Some Department of Energy sites hold radioactive plutonium, cesium, and uranium in an extremely toxic sludge with no known means of remediation.

Toxic Chemicals

See also:
Halogenated HC Solvents
Appendix

Benzene
Appendix

Arsenic
Appendix

Lead
Appendix

Organochlorines
Appendix

Polycyclic Aromatic HCs
Appendix

PCBs
Appendix

With the dramatic increase over the last few decades in usage of electronic devices (including many with very short useful lifespan), the problem of disposal of *electronic waste* (or *"e-waste"*) has been growing. E-waste contains a variety of heavy metals (lead, cadmium, and mercury, for example) as well as other toxic wastes. These materials from discarded electronic products can wind up in landfills or worse. There is some evidence that scrap or reclaimed materials from the manufacture or disposal of these products may also be used in some parts of the world to produce low-cost goods such as jewelry and toys, resulting in consumer products with dangerously high levels of heavy metals.

Household hazardous waste disposal is also a growing problem. Organic solvents are used in every household; there is methanol or phenol in many bathroom-cleaning agents, chloroform in some toothpastes, and ethyl acetate in fingernail polish. Inorganic cleaning agents include sodium hypochlorite in bleach and ammonium hydroxide in household ammonia. An increasing number of household and personal appliances are powered by small alkaline nickel and cadmium batteries that are often disposed of with garbage. Household paints and home and garden pesticides are used in large volumes, and unused portions often accumulate until they become outdated and thus considered waste. Considered cumulatively, a community of many consuming households may generate a considerable volume of toxic waste and should attempt to manage that waste just as a manufacturing plant must, by law, manage its hazardous waste.

LOVE CANAL AND HAZARDOUS WASTE LEGISLATION

Throughout much of history, waste (even toxic or hazardous waste) was simply disposed of in an expedient manner, often by simply digging a hole and burying it. Such dumping has resulted in landfills that are now slowly leaking undefined mixtures into groundwater. A buildup of waste materials is also contaminating the oceans and, in some cases, washing back onshore. The problems inherent in these practices came to light in the late 1970s at a site called *Love Canal* in New York. Love Canal was an abandoned canal dug in the 1890s and used as a swimming hole until it was purchased by Hooker Electro-Chemical Company in 1942. Hooker used the canal as a dump site, filling the canal with more than 20,000 tons of waste, which included more than 200 different chemicals.

In the 1950s, the dump was covered with a clay cap and sold to the city of Niagara Falls for $1. The city promptly built an elementary school on the site, and a neighborhood soon sprung up around it. Additional construction disrupted the cap, allowing rainfall and groundwater to penetrate into the site, which soon began to overflow. This carried toxic chemicals into basements and onto surfaces and into contact with the Love Canal residents. Following a series of protests by the affected individuals, an emergency was declared and a number of homes were eventually purchased and residents evacuated.

Prior to Love Canal, there was little or no public awareness of the problems posed by haphazard hazardous waste disposal. This and other incidents, however, made it clear that from an environmental perspective, disposal is only temporary because, in many situations, the air, surface water, or groundwater can easily become contaminated from the disposed hazardous waste in the future. As a result of these concerns, legislation was passed that defined hazardous waste and required the use of approved disposal methods.

In the United States, hazardous waste is regulated by the Environmental Protection Agency. The *Resource Conservation and Recovery Act of 1976* (RCRA) with its Hazardous and Solid Waste Amendments of 1984 established requirements for management of hazardous waste generated through many activities (although some sources were exempted).

One major contribution of RCRA was the definition of hazardous waste as ignitable, corrosive, reactive, or toxic substances. All solid (defined in the act as solid, liquid, semisolid, or gas in a container) hazardous waste was to be treated prior to being placed in specially constructed, secure landfills. The act exempted industrial effluent and irrigation return flows (which are regulated under the Clean Water Act), nuclear waste (regulated under the Atomic Energy Act), household waste, coal combustion waste, agricultural fertilizers, petrochemical drilling muds, some mining wastes, and cement kiln dust. However, these categories may be added by future amendments. Remedial action to prevent environmental damage from existing hazardous waste dump sites was directed under the *Comprehensive Environmental Response and Recovery Act of 1980* (CERCLA), also known as *Superfund*. Superfund addressed issues such as how abandoned waste sites should be handled and who is liable for the cost. It also established a fund that could be used to pay for cleanup if no liable parties could be identified. The trust fund is generated by a tax on manufacturers, cost recovery through litigation, federal appropriation, and interest. CERCLA was amended by the *Superfund Amendments and Reauthorization Act of 1986* (SARA).

In 1991, candidate sites for a *National Priorities List* were rated by a Hazard Ranking Assessment (HRA) System for the relative threat to human health or the environment, and the 1189 considered most hazardous were listed. By 2005, the National Priorities List included 1244 uncontrolled hazardous waste sites eligible for remedial action by the federal government. These included federal sites (mostly nuclear [Department of Energy] or defense facilities), municipal landfill sites, commercial and industrial landfills and surface impoundments (evaporation ponds), and chemical manufacturing and processing areas filled with containers or drums. Many sites were associated with electroplating (chromium, copper, mercury), military testing and ordnance (chemical warfare agents, explosives), wood preserving (pentachlorophenol), waste oil processing (lead, benzene, etc.), ore processing (fluoride, arsenic, etc.), battery recycling (lead, cadmium), and incineration (metals, PAHs, PCBs).

Sites on the National Priorities List are found throughout the Unites States, but tend to cluster around major cities. In 2014, out of a total of 1318 sites there were 199 sites in New York and New Jersey (EPA region 2), and 35% of the sites were in the combined EPA regions 1, 2, and 3, which encompass the 13 northeastern states from Maine to Virginia. There were also 244 sites listed in the six states of EPA region 5, which border the Great Lakes. Therefore, approximately 54% of the total sites were in 19 states of the northeast and upper Midwest.

Among the highest ranked nonfederal sites in terms of the HRA system in 2014 were the Big River Mine Tailings/St. Joe Minerals Corp. site in Desloge, MO; two sites in the Washington County Lead District (also in Missouri); the Lipari Landfill, Pitman, NJ; the McCormack and Baxter Creosoting Company site, Portland, OR; and the Helen Kramer Landfill, Mantua Township, NJ. The highest ranking federal site was the Pearl Harbor Naval Complex in Hawaii.

Many problems have complicated the task of cleaning up hazardous waste sites. At hundreds of sites, leaking containers have deteriorated so badly that the site has become difficult to stabilize without causing more leakage. Also, these conditions tend to produce mixtures of diverse organic and inorganic toxic compounds, which present complex technical difficulties in designing detoxication strategies. Because of the need for many preliminary investigations and plans to be made for each site in order to deal with these factors, the cleanup rates have been very slow. Political pressures and diversity of opinion over treatment strategies have also led to delays.

Also, the most hazardous (and therefore the most complex) sites have required the most planning and thus have not necessarily been the first to be cleaned. This slowness of action has generated a great deal of criticism for the EPA and the program as a whole. Remediation has, however, been completed to the point of formal deletion of several hundred sites from the original list.

Waste Management: Reduce, Recycle, Treat, Store

A plan for the management of hazardous waste may employ various techniques that may be classified into the following categories: reducing the volume generated, recycling, treatment, and storage. Storage of an accumulating volume of hazardous waste, however, must be viewed as the least desirable way to manage the problem and should be considered as a temporary approach, while other ways are found to eliminate the hazardous waste.

Reducing the volume of hazardous waste is becoming a major emphasis in industry as a result of increasing treatment and disposal costs. Waste minimization programs involve quantification of the waste produced, identification of the source of waste in the process, and engineering to improve the efficiency of the manufacturing process, or to substitute less toxic compounds for the end product and for the intermediates in the process stream. Finding alternatives to those products generating toxic waste is the ounce of prevention that, in this case, is worth much more than a pound of cure. Alternatives have been found for PCBs, DDT, ozone-depleting chlorofluorocarbons, and other environmental pollutants.

A large industry has also developed for the recycling of waste, including toxic substances. For example, industrial process and cleaning solvents are now routinely recycled so that approximately two-thirds of solvent used is recovered. Solvent recycling is accomplished by sequential settling, filtering, and distillation, so it is a relatively simple process. Rather than disposal of the contaminated solvent, contaminants collected as filter cake and still bottoms are treated for detoxication or decomposed, and recycled solvent is obtained at high yield.

Detoxication is the reduction of toxicity of waste through physical, chemical, or biological treatment. Ultraviolet irradiation in methanol is used to detoxify TCDD and pesticides by dechlorination. In an aqueous stream containing hazardous waste, chemical oxidation, reduction, or hydrolysis may be used to detoxify the waste. In some cases, separation of the hazardous waste from the liquid process or waste stream may be accomplished by precipitation, centrifugation, or filtration. This results in a concentrated cake or solid that can be collected for detoxication. Separation depends on the chemical characteristic of the waste constituents. Useful filtration matrices include silica for organic solvent streams, ion exchange resins, and activated carbon.

Biodegradation has the advantage of being inexpensive because microorganisms are used to degrade the waste. Enzymes within or secreted from the microorganism catalyze detoxication reactions such as hydrolysis, oxidation, or reduction. Recent advances in biotechnology have succeeded in engineering microorganisms for more efficient waste degradation. The treatment of existing toxic waste sites by this technique falls under the category of bioremediation (discussed previously).

Incineration is the burning of waste composed of organic compounds in the presence of oxygen at temperatures of up to about 1500°C. High-temperature combustion will detoxify many toxic organic compounds by oxidation to carbon dioxide and water. As the waste burns, the heat evolved may be used for the manufacturing process or for some other purpose. Incineration does not produce detoxication of all types of hazardous wastes, however, and may even produce added toxic oxides as the waste is oxidized. Also, incineration at reduced temperature may simply volatilize the waste. For these reasons, incineration smoke may be toxic and, if so, must be scrubbed before entering the atmosphere. In addition, the Supreme Court has ruled that ash from municipal incinerators must be regulated as potentially hazardous waste.

Pyrolysis is a more universally detoxicative process that employs extremely high heat, like in an iron smelter, to decompose organic and inorganic toxic compounds to elemental form. This process is expensive, energy-consuming, and not commonly available; however, new companies specializing in this process are likely to grow rapidly, thus lowering the costs somewhat. This process is most elegant because it is the simplification of toxic wastes to the elements from which they originated.

Storage of hazardous waste in secure landfills is considered by EPA as a temporary approach, while other ways are found to eliminate the hazardous waste. Under the Resource Conservation and Recovery Act, secure landfills must meet construction standards, and disposal must be preceded by characterization and treatment of the waste.

EPA regulations include land disposal restrictions that state that all hazardous waste must be treated to meet best-demonstrated achievable technology prior to land disposal. These regulations require measuring the concentrations of waste constituents by approved test methods before or after a toxicity characteristic leaching procedure that simulates leaching of the constituents from the solid matrix.

Secure landfills must be constructed with liners to prevent leaching down to the groundwater. They must also have a drainage system above the liner to collect leaking waste into a holding system. Furthermore, secure landfills must have adjacent monitoring wells to test for contamination of the surrounding area.

CASE STUDY   Pharmaceuticals in the Water Supply

One growing environmental concern is the problem of increasing levels of pharmaceutical substances (and also what are often termed "personal care products") in both surface waters and groundwater. Some of the most common substances which have been found through environmental sampling include NSAIDS (ibuprofen, acetaminophen, diclofenac), antibiotics (sulfamethoxazole, trimethoprim), psychiatric medications (lorazepam, carbamazepine), beta blockers, calcium channel blockers, and hormones. Measured concentrations are typically in the nanogram/liter range, but in some cases, for some pharmaceuticals, can be much higher. One study, for example, has measured concentrations of acetaminophen in groundwater in the microgram/liter range.

There are multiple potential sources from which these compounds can enter the water supply. One, of course, is at the point of manufacture, where they may enter the water as part of an industrial effluent. Another potential source is landfills, where unused medications and products have been discarded and may leach into nearby ground or surface waters as contaminants. However, evidence indicates that the primary source for these types of compounds may be institutional (e.g., hospitals) and domestic sewage and septic systems, where urine containing parent drugs and/or metabolites is excreted into the wastewater stream.

Unfortunately, standard wastewater treatment methodologies are often not effective at removing these generally hydrophilic and often chemically stable compounds from wastewater. This is complicated by the fact that there are literally thousands of drugs and products (with widely varying structures and chemical properties) which may need to be dealt with.

Although concentrations of pharmaceuticals as currently measured tend to be low, risk assessment is difficult. Little is known, for example, about environmental fate and transport of many of these compounds. Some drugs and/or metabolites (typically, these possess a relatively high octanol–water partition coefficient and low rate of biodegradability) may have the potential to bioaccumulate and/or bioconcentrate, while others may not. Nor are the effects of low-level chronic exposure to these compounds on human health well understood. In particular, there are concerns with the potential for effects on sensitive populations (such as children or the elderly). Likewise, few studies have been carried out on effects on wildlife, although there is evidence for concern over impact on both individual species and ecosystem structure. Some studies have indicated, for example, that environmentally relevant concentrations of antidepressants have the potential to affect reproduction and behavior in a variety of aquatic invertebrates. Antidepressants have also been shown to affect foraging behavior in fish, and alterations in microbial community structure have been recorded following exposure to NSAIDs.

Given the significant lack of knowledge concerning the behavior and effects of most pharmaceuticals in the environment, another problem to be solved is the determination of where research priorities should lie (i.e., with so little known, which compounds should be studied first?). Thus, another area of research deals with the development of models that use known data to predict the relative potential of different pharmaceuticals to pose risks to human health and the environment. One system, for example, proposes setting priorities based on factors including number of prescriptions written and filled, estimated effectiveness of existing water treatment plants in removing the drug, and relative toxicity in terms of known human and ecological effects. Of course, the very scarcity of data which is one motivating factor for developing such a model also hampers effective model development; likewise, as more data is collected through bench and field research, better and more accurate models will be able to be developed and tested.

BIBLIOGRAPHY

Almeda, R., Hyatt, C., and Buskey, E.J., Toxicity of dispersant Corexit 9500A and crude oil to marine microzooplankton, *Ecotoxicol. Environ. Safety*, 106, 76, 2014.

Block, M.L., Elder, A., Auten, R.L., Bilbo, S.D., Chen, H., Chen, J.-C., Cory-Slechta, D.A. et al., The outdoor air pollution and brain health workshop, *NeuroToxicology*, 33, 972, 2012.

Borgmann, U., Metal speciation and toxicity of free metal ions to aquatic biota, in *Aquatic Toxicology*, Nriagu, J.O., Ed., John Wiley & Sons, New York, 1983, chap. 2.

Bronmark, C. and Hansson, L.-A., Environmental issues in lakes and ponds: Current state and perspectives, *Environ. Conserv.*, 29, 290, 2002.

Brown, M.P., Werner, M.B., Sloan, R.J., and Simpson, K.W., Polychlorinated biphenyls in the Hudson River, *Environ. Sci. Technol.*, 19, 656, 1985.

Brown, P. and Clapp, R., Looking back on Love Canal, *Public Health Rep.*, 117, 95, 2002.

Cooter, W.S., Clean water act assessment processes in relation to changing U.S. Environmental Protection Agency management strategies, *Environ. Sci. Technol.*, 38, 5265, 2004.

Corcoll, N., Acuna, V., Barcelo, D., Casellas, M., Guasch, H., Huerta, B., Petrovic, M., Ponsati, L., Rodriguez-Mozaz, S., and Sabater, S., Pollution-induced community tolerance to non-steroidal anti-inflammatory drugs (NSAIDs) in fluvial biofilm communities affected by WWTP effluents, *Chemosphere*, 112, 185, 2014.

Costa, D.L., Air pollution, in *Casarett and Doull's Toxicology*, Klaassen, C.D., Ed., McGraw-Hill, New York, 2001, chap. 28.

DePinto, J.V., Young, T.C., and McIlroy, L.M., Great Lakes water quality improvement, *Environ. Sci. Technol.*, 20, 752, 1986.

Dong, Z., Senn, D.B., Moran, R.E., and Shine, J.P., Prioritizing environmental risk of prescription pharmaceuticals, *Regul. Toxicol. Pharm.*, 65, 60, 2013.

Echobichon, D.J., Toxic effects of pesticides, in *Casarett and Doull's Toxicology*, Klaassen, C.D., Ed., McGraw-Hill, New York, 2001, chap. 22.

Englande, A.J.J. and Eckenfelder, W.W.J., Toxic waste management in the chemical and petrochemical industries, *Water Sci. Technol.*, 25, 286, 1991.

Fong, P.P. and Ford, A.T., The biological effects of antidepressants on the molluscs and crustaceans: A review, *Aquat. Toxicol.*, 151, 4, 2014.

Freedman, B., *Environmental Ecology*, Academic Press, San Diego, 1989, chap. 6.

Fry, B. and Anderson, L.C., Minimal incorporation of Deepwater Horizon oil by estuarine filter feeders, *Mar. Pollut. Bull.*, 80, 282, 2014.

Heckman, C.W., Reactions of aquatic ecosystems to pesticides, in *Aquatic Toxicology*, Nriagu, J.O., Ed., John Wiley & Sons, New York, 1983, chap. 12.

Hedgespeth, M.L., Nillson, P.A., and Berglund, O., Ecological implications of altered fish foraging after exposure to an antidepressant pharmaceutical, *Aquat. Toxicol.*, 151, 84, 2014.

Hillyer, M.M., Finch, L.E., Cerel, A.S. Dattelbaum, J.D., and Leopold, M.C., Multi-technique quantitative analysis and socioeconomic considerations of lead, cadmium, and arsenic in children's toys and toy jewelry, *Chemosphere*, 108, 205, 2014.

Hoffman, A.J., An uneasy rebirth at Love Canal, *Environment*, 37, 4, 1995.

Holloway, M., Soiled shores, *Sci. Am.*, 265, 102, 1991.

Houghton, R.A. and Woodwell, G.M., Global climatic change, *Sci. Am.*, 260, 36, 1989.

Huggett, R.J., Unger, M.A., Seligman, P.F., and Valkirs, A.O., The marine biocide tributyltin, *Environ. Sci. Technol.*, 26, 232, 1992.

Jeffries, D.S., Brydges, T.G., Dillon, P.J., and Keller, W., Monitoring the results of Canada/ U.S.A. acid rain control programs: Some lake responses, *Environ. Monitor. Assess.*, 88, 3, 2003.

Jones, P.D. and Wigley, T.M.L., Global warming trends, *Sci. Am.*, 263, 84, 1990.

Kan, H., Chen, R., and Tong, S., Ambient air pollution, climate change, and population health in China, *Environ. Int.*, 42, 10, 2012.

Kolias, K., Hahladakis, J.N., and Gidarakos, E., Assessment of toxic metals in waste personal computers, *Waste Manage.*, 34, 1480, 2014.

Laxen, D.P.H., The chemistry of metal pollutants in water, in *Pollution: Causes, Effects, and Control*, Harrison, R.M., Ed., Royal Society of Chemistry, London, 1982, chap. 6.

Lewis, M. and Pryor, R., Toxicities of oils, dispersants and dispersed oils to algae and aquatic plants: Review and database value to resource sustainability, *Environ. Pollut.*, 180, 345, 2013.

Lin, G.-H., Sauer, N.E., and Cutright, T.J., Environmental regulations: A brief overview of their applications to bioremediation, *Int. Biodeterior. Biodegrad.*, 38, 1, 1996.

Lyons, R.A., Temple, J.M.F., Evans, D., Fone, D.L., and Palmer, S.R., Acute health effects of the *Sea Empress* spill, *J. Epidemiol. Commun. Health*, 53, 306, 1999.

Makarewicz, J.C. and Bertram, P., Evidence for the restoration of the Lake Erie ecosystem, *Bioscience*, 41, 216, 1991.

Mann, M.E., Bradley, R.S., and Hughes, M.K. Global-scale temperature patterns and climate forcing over the past six centuries, *Nature*, 392, 779, 1998.

Maynard, R., Key airborne pollutants: The impact on health, *Sci. Total Environ.*, 334/335, 9, 2004.

Menzer, R.E., Water and soil pollutants, in *Casarett and Doull's Toxicology*, Amdur, M.O., Doull, J., and Klaassen, C.D., Eds., Pergamon Press, New York, 1991, chap. 26.

Mishra, D.R., Cho, H.J., Ghosh, S., Fox, A., Downs, C., Merani, P.B.T., Kirui, P., Jackson, N., and Mishra, S., Post-spill state of the marsh: Remote estimation of the ecological impact of the Gulf of Mexico oil spill on Louisiana Salt Marshes, *Remote Sens. Environ.*, 118, 176, 2012.

Mostafalou, S. and Abdollahi, M., Pesticides and human chronic diseases: Evidences, mechanisms, and perspectives, *Toxicol. Appl. Pharmacol.*, 268, 157, 2013.

Mrema, E.J., Rubino, F.M., Brambilla, G., Moretto, A., Tsatsakis, A.M., and Colosio, C., Persistent organochlorinated pesticides and mechanisms of their toxicity, *Toxicology*, 307, 74, 2013.

Norby, R.J. and Luo, Y., Evaluating ecosystem responses to rising atmospheric CO_2 and global warming in a multi-factor world, *New Phytol.*, 162, 281, 2004.

Proia, L., Osorio, V., Soley, S., Kock-Schulmeyer, M., Perez, S., Barcelo, D., Romani, A.M., and Sabater, S., Effects of pesticides and pharmaceuticals on biofilms in a highly impacted river, *Environ. Pollut.*, 178, 220, 2013.

Sammarco, P.W., Kolian, S.R., Warby, R.A.F., Bouldin, J.L., Subra, W.A., and Porter, S.A., Distribution and concentrations of petroleum hydrocarbons associated with the BP/ Deepwater Horizon oil spill, Gulf of Mexico, *Mar. Pollut. Bull.*, 73, 129, 2013.

Schindler, D.W., Effects of acid rain on freshwater ecosystems, *Science*, 239, 149, 1988.

Short, J.W., Lindeberg, M.R., Hrris, P.M., Maselko, J.M., Pella, J.J., and Rice, S.D., Estimate of oil persisting on the beaches of Prince William Sound 12 years after the *Exxon Valdez* oil spill, *Environ. Sci. Technol.*, 38, 19, 2004.

Stern, D.I., Global sulfur emissions from 1850 to 2000, *Chemosphere*, 58, 163, 2005.

United Nations Environment Programme, Environmental Effects Assessment Panel, Environmental effects of ozone depletion and its interactions with climate change: Progress report, 2004, *Photochem. Photobiol. Sci.*, 22, 177, 2005.

U.S. Environmental Protection Agency, National Priorities List under the Original Hazard Ranking System 1981–1991, Fact Book 9320.7-08, EPA 540-R-93-079, EPA Office of Solid Waste and Emergency Response, 1993.

U.S. Environmental Protection Agency, National priorities list for uncontrolled hazardous waste sites, *Fed. Reg.*, 59, 27989, 1994.

U.S. Environmental Protection Agency, Superfund website, http://www.epa.gov/superfund/ index.htm.

Valentine, M.M. and Benfield, M.C., Characterization of epibenthic and demersal megafauna at Mississippi Canyon 252 shortly after the Deepwater Horizon oil spill, *Mar. Pollut. Bull.*, 77, 196, 2013.

Weidenhamer, J.D. and Clement, M.L., Leaded electronic waste is a possible source material for lead-contaminated jewelry, *Chemosphere*, 69, 1111, 2007.

White, R.M., The great climate debate, *Sci. Am.*, 263, 36, 1990.

Whittaker, A., BeruBe, K., Jones, T., Maynard, R., and Richards, R., Killer smog of London, 50 years on: Particle properties and oxidative capacity, *Sci. Total Environ.*, 334/335, 435, 2004.

Zenker, A., Cicero, M.R., Prestinaci, F., Bottoni, P., and Carerd, M., Bioaccumulation and biomagnification potential of pharmaceuticals with a focus to the aquatic environment, *J. Environ. Manage.*, 133, 378, 2014.

18 Applications
Radiation

While toxicology is most generally concerned with chemical hazards, there is one type of physical hazard that also falls within its scope: radiation. This chapter will examine the nature of the different types of radiation and the effects of both short- and long-term radiation exposures on biological tissues. We will also examine some of the circumstances under which radiation exposure might typically occur, as well as a case study of accidents leading to radiation exposures.

BASIC TYPES OF RADIATION

When people talk about "radiation," they are typically including two very different things within the same general category. Some types of radiation are electromagnetic waves; others are actual physical particles. We will look at both types, but will tackle *electromagnetic radiation* first.

ELECTROMAGNETIC RADIATION

Electromagnetic radiation is energy, plain, and simple. You'd best, however, not inquire too deeply and philosophically as to exactly what "energy" is, as not even physicists can really answer that one. An operational definition, however, is that energy is a property of objects, which is able to be transferred between them, as well as to be converted from one form to another. There appear to be laws that govern energy's behavior (these are the *laws of thermodynamics*), primary among which is that energy appears to neither be created nor be destroyed.

Electromagnetic radiation can be described as a wave propagating through space. All electromagnetic radiation moves through a vacuum at the same velocity, c, which is around 186,000 miles per second (or almost 300 million meters/second in metric terms). Its fundamental wave nature allows it to be described by characteristics such as frequency (ν) and wavelength (λ) (Figure 18.1). And, since frequency × wavelength = velocity for a wave, and since the velocity of all electromagnetic waves is c, if you know the frequency of electromagnetic radiation you can find the wavelength and vice versa. It is also true, however, that under some circumstances, electromagnetic radiation can also behave as if it were a stream of massless particles, or *photons*. So which is it really, a wave or a particle? According to the quantum mechanical concept of *wave particle duality*, it is both.

The energy carried by electromagnetic radiation is proportional to its frequency: the higher the frequency, the greater the amount of energy carried. The range of frequencies of observed electromagnetic radiation is described by the *electromagnetic*

Wavelength (λ)

Frequency (ν) = the number of wave crests to
travel past a given point in a set time period

FIGURE 18.1 The characteristics of a wave.

spectrum; bands within the spectrum are defined based on frequency (or wavelength); the major bands and their approximate frequency ranges are given in Table 18.1.

Electromagnetic radiation can interact with biological tissues, primarily through imparting some of its energy to the tissue. Radiation with enough energy to knock electrons loose from their orbits (and thus to ionize atoms) is known as *ionizing radiation*. Not all electromagnetic radiation is ionizing; it is primarily the far ultraviolet, x-rays, and gamma rays that have that capability. This is not to imply that other types of radiation cannot interact with biological tissues, also—clearly, they do. In fact, the process of photosynthesis depends on interactions of nonionizing radiation with pigment molecules in plant cells. However, the interactions of nonionizing radiation with tissue tend to be less damaging (although we will also discuss some possible negative effects of nonionizing radiation later in this chapter.)

The sources of electromagnetic radiation are everywhere. Whenever atoms undergo a change in shift in energy from a higher to a lower state, the excess energy is released in the form of electromagnetic radiation. X-rays are typically produced when high-energy electrons are slowed (and energy released) through interactions

TABLE 18.1

Bands of the Electromagnetic Spectrum, Listed from Lowest to Highest Energies

Band Name	Frequency Range (Hz)	
	Low	High
Extremely low frequency (ELF)	3	30
Super low frequency (SLF)	30	3×10^2
Ultra low frequency (ULF)	3×10^2	3×10^3 (3 kHz)
Radio waves	3×10^3	3×10^{11} (300 GHz)
(including microwaves)	3×10^8	3×10^{11}
Infrared	3×10^{11}	4.3×10^{14}
Visible	4.3×10^{14}	7.9×10^{14}
UV	7.9×10^{14}	3×10^{16}
X-rays	3×10^{16}	3×10^{19}
Gamma rays	3×10^{19}	

with a magnetic field or with other particles. Gamma rays are typically produced through a process known as *radioactive decay*. In this process, the nucleus of an unstable atom breaks apart (a process known as *fission*), releasing some of the excess energy in the form of gamma rays.

The stability of a nucleus is determined by the interactions between its constituent protons and neutrons. Many atoms have multiple *isotopes* (atoms of the same element which differ in number of neutrons), some of which are more stable than others. The more unstable the isotope, the shorter its *half-life*, which is the period of time in which the quantity of the isotope drops by 1/2 due to the decay process. Half-lives of unstable isotopes can range from extremely short (in the order of 10^{-24} seconds) to unimaginably long (in the order of 10^{24} years). Common examples of unstable isotopes are uranium-238 and americium-241 (the isotope commonly used in smoke detectors). The numerical designation in these names refers to the atomic mass of that isotope.

PARTICULATE RADIATION

A second basic form of radiation is also produced through radioactive decay, but this type of radiation consists of actual massive particles which carry kinetic energy away from a nucleus as it breaks apart. Among the particles that can be typically be released through nuclear fission are protons, neutrons, *alpha particles* (two protons and two neutrons bound together), and *beta particles,* which can be either electrons or *positrons*, particles with the same mass but opposite charge to the electron. (You may note that there are typically no electrons in a nucleus; beta particles are formed during the decay process itself.) The energy carried by these massive particles varies, but all have the capacity to carry sufficient energy to be considered as ionizing radiation.

Although it may seem obvious, one aspect of exposure to radiation is frequently misunderstood. Exposure to either the electromagnetic radiation or to particulate radiation given off by a radioactive substance does not make an individual radioactive. The only way for a person to become radioactive is to actually ingest the source materials (i.e., unstable isotopes) which are emitting the radiation.

MEASUREMENT OF RADIOACTIVITY

Probably the most straightforward measurement of the level of radioactivity in any material is the measurement of the number of nuclei which decay in one second. One *Becquerel (Bq)* is defined as an amount of material with an activity level of one decay per second. An older unit that may still be seen in some literature is the *Curie (Ci)* which is defined as the amount of material with an activity level of 3.7×10^{10} decays per second (it was defined based on the decay rate of 1 g of radium-226).

There are a number of ways to measure radioactivity in the environment (which of course cannot be directly detected by our senses), including the well-known *Geiger counter*. In this instrument, ionizing radiation induces an electrical current (which can be measured) in a tube of inert gas. The limitations of Geiger counters are that they can measure the activity level (how much radiation is emitted), but not the type of radiation or energy carried by it. They are also less effective at high radiation levels. *Ionization chamber*-type monitors work on a similar principle (production of electrical current in a gas through ionization) but are better able to perform at high radiation levels.

INTERACTION OF IONIZING RADIATION WITH BIOLOGICAL TISSUES

CELLULAR MECHANISMS

In terms of the cellular mechanisms of action, the energy carried by radiation interacts with cellular macromolecules, with DNA as one of the major targets. Evidence indicates that radiation-induced strand breaks in DNA (particularly double breaks, with one break in each strand) may lead to repair errors and deletions. The first cells to be affected by this are most typically cells which are in the process of cell division, as those that are quiescent may be able to repair the damage before there are negative consequences. Thus, the populations of cells most likely to be affected by radiation are those which divide frequently, including cells in the bone marrow, intestinal lining, reproductive tract, and skin.

Bystander effects are those effects which impact cells which are not directly exposed to radiation, but which may be indirectly affected by changes in nearby cells which are exposed. These effects may include various types of damage to DNA (an increase in point mutations is often seen), as well as other damage up to and including cell death. The mechanism behind bystander effects is not understood; however, developing a better understanding of this phenomenon is of particular importance in designing safe and effective radiation therapy.

There is also some evidence that cells exposed to low doses of radiation may, in fact, experience an adaptive response which would lower risk for any subsequent exposures at higher dosage levels. However, the mechanisms behind this are not clear, however, although they probably involve stimulation of DNA repair and other protective cellular mechanisms.

CALCULATION AND MEASUREMENT OF RADIATION DOSAGES

The measurement of radioactive activity itself does not really describe the magnitude of the interaction of the radiation with biological tissues. A unit which does measure total transfer of radioactive energy to biological tissue is the *Gray* (*Gy*), in which 1 Gy corresponds to transfer of 1 J of energy to 1 kg of tissue. An older unit measuring energy absorption is the *radiation absorbed dose* or *rad* (100 rad = 1 Gy).

Although all forms of ionizing radiation have sufficient energy to ionize atoms, their individual physical characteristics determine the manner in which their energies are transferred to biological tissues. One factor that must be considered is the *linear energy transfer* (*LET*). This is a measure of the energy lost per unit path length as the radiation travels through tissue. The LET depends both on the energy carried by the radiation and the path length over which it is dispersed (a shorter path length corresponds to more focused transfer; with a longer path length, the energy is more dispersed). Typically, alpha and beta particles have a high LET, due to the fact that they typically travel a very short path in tissues (only 1–3 cm for alpha particles). X-rays and gamma rays, on the other hand, travel a long path and thus have a low LET.

Several concepts have also been developed to allow comparison of different forms of radiation in terms of their biological effects. One is the *quality factor*, or as it is now most often called, the *weighting factor*, w_R. w_R takes into account a number of factors (including LET) and is defined as the number by which a radiation dose must be multiplied in order for the exposure to be comparable to x-rays and gamma rays. Thus, w_R can range from 1 (for gamma rays) to 20 (for alpha particles). The unit produced when multiplying energy absorption in Gy times the weighting factor is a *Sievert* (*Sv*). Likewise, multiplying the older unit rad times the weighting factor will produce *rems*. Both measures, Sv and rem, are still commonly seen.

There is instrumentation available which can help estimate and quantify the dosage of radiation experienced by exposed individuals. For routine, predicted exposures, *pocket dosimeters* can be worn, with different types of specialized dosimeters available for measuring different types of radiation. These can be monitored regularly and used to quantify exposure and to ensure that exposure limits are not exceeded. Dosimeters can range from simple film badge dosimeters, where unexposed film records individual radiation exposure events, to more complex devices that provide much more information. Some *electronic personal dosimeters*, for example, can engage in real-time monitoring and calculating of dosages, and can emit warning alarms if safe exposure limits are exceeded. There are even dosimeters available that are integrated into wristwatches and that function as an attachment to an iPhone©. In emergency, high-dosage situations, however, doses are typically estimated by combining environmental monitoring with estimated times of exposure.

SOURCES OF IONIZING RADIATION

Cosmic radiation consists of particles such as alpha particles, electrons, and protons which originate both from the sun and from outside of our solar system. The atmosphere shields the surface from cosmic radiation, so at higher elevations (where radiation has a shorter atmospheric path to travel through) exposure to cosmic radiation is higher. Interaction of cosmic radiation with atoms in the atmosphere and on the ground can also produce *cosmogenic* radioactive isotopes. One example of a cosmogenic isotope is *carbon-14*. There are also naturally occurring isotopes such as uranium-238, thorium-232, and potassium-40 that are sources of *terrestrial* radiation. The radioactive isotope of potassium, potassium-40, can be found not only in rocks and soils, but also in living tissues, including the human body. Finally, there are *anthropogenic* sources of radiation, including fallout from nuclear weapons and testing, emissions from nuclear power facilities (including both routine and accidental releases), and, of course, radiation associated with medical testing and treatment.

Routine background levels of radiation exposures vary greatly from individual to individual, but the United States Nuclear Regulatory Commission supports an online calculator at http://www.nrc.gov/about-nrc/radiation/around-us/calculator. html which allows individuals to estimate their annual exposure. The average annual dose according to the USNRC is 350 mrems.

PHYSIOLOGICAL EFFECTS OF EXPOSURE TO IONIZING RADIATION

Obviously, we are all being exposed to background radiation continuously. Toxicology, however, is most concerned with exposures which carry a significant increased risk for adverse health effects. These can be either higher dosage acute exposures, or, as is more common, lower dosage acute or chronic exposures.

ACUTE EXPOSURES: ACUTE RADIATION SYNDROME

Acute whole-body exposures to radiation greater than approximately 1 Gy in a short time period produces what is referred to as *acute radiation syndrome*. In the first few days following exposure, the *prodromal phase* occurs, characterized by anorexia, nausea and vomiting, fever, and headache. In general, the higher the exposure, the more rapid the onset of the prodromal phase is (an indicator that can help medical personnel who lack of accurate data on a patient's exposure level).

At 2–3 Gy or higher, the hematopoietic syndrome presents. This is characterized by reductions in numbers of lymphocytes and other white blood cells, as well as development of anemia. This progresses over the ensuing weeks to bone marrow hypoplasia or aplasia. A dose of 5–12 Gy also produces *gastrointestinal syndrome*, a worsening of GI effects, accompanied by malnutrition and infection, as cells lining the gastrointestinal tract are destroyed. The *cerebrovascular syndrome* is seen at 10–12 Gy and consists of a variety of neurological problems (including headache, severe nausea, disorientation, and seizures) secondary to brain edema and hemorrhaging. *Cutaneous syndrome* may occur at any point and involves development of erythema, edema, and ulcerations.

Treatment of acute radiation syndrome may include replacement of lost body fluids, administration of antibiotics and medications for nausea, blood transfusions, and treatment with cytokines to stimulate white cell proliferation (or bone marrow transplant in the case that a population of stem cells did not survive). The LD_{50} for acute radiation exposure (measured at 60 days post-exposure) ranges from 3.5 to 9 Gy depending on the type of medical care available. Exposure of greater than 10–12 Gy is virtually uniformly lethal.

CHRONIC EXPOSURES AND DELAYED EFFECTS OF ACUTE EXPOSURES

Carcinogenicity Testing

See also:
   Testing Compounds for
   Carcinogenicity
                    Chapter 6

For most individuals, however, the exposure to low to moderate levels of radiation either in an acute exposure or over longer periods of time is a more likely scenario. In these cases, carcinogenesis is the major concern. But because of the difficulties of testing effects of low-dosage exposures in the laboratory, much of the information on development of cancer as a result of chronic radiation exposure and delayed effects from acute exposures come from epidemiological studies.

The largest radiation-exposed population available for long-term study has been the atomic bomb survivors. The Life-Span Study has followed a cohort of over 90,000 atomic bomb survivors for almost 70 years, with radiation doses calculated based on location at the time of the blast and estimated amount of shielding. Attempts have also been made to account for confounding factors such as smoking and other cancer risk factors. Conclusions from that study indicate that the risk of developing cancer, in general, increases with increased radiation exposure. Risk for developing solid tumors (stomach cancer, lung, liver, and others) were seen to increase in a linear fashion with exposure; however, the increase in risk for leukemia (the earliest delayed cancer effect to be seen) appears to be nonlinear, with low exposures producing fewer excess deaths than predicted. Other (non-cancer-related) effects noted by the study have included increases in risk for thyroid, liver, cardiovascular, and kidney disease.

Other cohorts that have been studied for delayed or chronic effects include patients who were irradiated as treatment for a variety of medical ailments. One such condition is anyklosing spondylitis, an inflammatory disease affecting the spine which was often treated with radiation in the 1930s, 1940s, and 1950s. Another disease sometimes treated with radiation during the same time period was scalp ringworm. Occupationally exposed groups have also been a source for epidemiological data, including miners (exposed to high levels of radon), medical workers (radiologists, for example), and nuclear power workers (particularly those working at plants during incidents involving significant releases of radiation, such as occurred at Chernobyl). Also, populations in areas with high natural background radiation have been studied. These areas may have exposures of seven to eight times that of areas with low natural background radiation, although to this point most studies have not indicated significantly increased mortality risks for individuals living in these areas.

Both medical and occupational cases tend to involve exposure to external radiation sources; however, a few have involved actual ingestion of radioactive materials themselves. One such case is that of the radium dial painters, women (usually) who were employed to apply a luminous radium-containing paint to watch dials in order to make them readable at night. Because the painters frequently "tipped" their brushes by tapping them on their tongues to wet them (a common practice when painting fine dots and lines), many of them ended up ingesting a significant amount of radium. Because radium in the body tends to deposit in bone, the results were a significant increase in bone cancers, as well as sinus cancers (likely due to the radon gas released during decay of radon in the body) in this population.

NONIONIZING RADIATION

ULTRAVIOLET RADIATION

There are many examples of exposure to radiation which lacks sufficient energy to produce ionization. For some of these, the potential for negative health effects is clear; for others, less so. The most common source of exposure to nonionizing radiation is, of course, the sun. The component of solar radiation most likely to produce tissue damage is the ultraviolet range, typically divided into UV-A, UV-B, and UV-C bands. With UV-C totally absorbed by the atmosphere (including the ozone layer), it is UV-A (which constitutes 95% of the UV light reaching the surface) and UV-B (which constitutes 5%) which are significant in terms of human health.

The cells which are most typically exposed to UV radiation are the cells of the epidermis and dermis. Within these cells, the radiation can be absorbed by cellular constituents, including DNA, leading to alterations in molecular structures. Specifically, UV radiation can produce mutations through producing damage to DNA bases (often cytosine or thymine) which then often triggers substitutions for individual bases or base pairs. The absorbed energy from UV radiation can also lead to the creation of reactive oxygen species, which are also capable of producing DNA damage (here, guanine is a typical target). The most significant potential consequence of this is, of course, development of skin cancers.

Based on laboratory evidence, the specific type of DNA damage which is most frequently seen in skin cancers is the type that is most typical of UV-B, rather than UV-A-induced damage. UV-A is thought to contribute more to the production of reactive oxygen species, and thus, to indirect damage. There is also evidence that the type of damage produced by UV-B is more resistant to repair than that produced by UV-A.

Another more immediate form of UV-induced epithelial damage is *sunburn*. As anyone who has spent a little too much time at the beach is aware, this condition is characterized by redness (erythema), swelling (edema), and pain. On the cellular and biochemical levels, sunburn is basically a UV-induced inflammatory response, characterized by production of cytokines, prostaglandins, and a variety of other bioactive compounds. Production of melanin is also stimulated, as is hyperplasia, leading to darkening and thickening of the skin. Some UV-damaged cells may be able to effect repairs; badly damaged cells may undergo apoptosis (a protective mechanism which precludes development of a neoplastic phenotype). Severe cases of sunburn (sometimes termed "sun poisoning") can also include systemic effects, including dehydration and electrolyte imbalance. Over the long-term, chronic sun exposure can lead not only to cancer (as previously discussed), but also to premature aging, or *photoaging*, which is characterized by changes in the dermis including disruption of collagen structure (leading to development of leathery skin and wrinkles), irregular pigmentation, and *telangiectasia* (development of small, dilated blood vessels commonly called "spider veins").

RADIOFREQUENCY ELECTROMAGNETIC FIELDS

The debate about potential health effects of radiofrequency (RF) electromagnetic fields continues. One aspect of this topic which has garnered significant attention is the question of potential health risks associated with the use of cellular phones. In 2011, the International Agency for Research on Cancer classified emissions from cellular phones as a "possible" human carcinogen. Some epidemiological studies have, in fact, indicated a potential association between cell phone usage and development

of some neurological cancers, including glio-
mas; however, most studies have found no statistically significant link. The molecular mechanisms
which would be involved in RF effects are unclear,
although there is evidence that cell phone use can
result in an increase in glucose metabolism in brain
tissues near to the antenna, as well as inducing a
stress response.

Stress Proteins

See also:
   Stress Repair and Recovery
                      Chapter 4

EXTREMELY LOW-FREQUENCY ELECTROMAGNETIC FIELDS

Extremely low-frequency (ELF) electromagnetic fields are generated by power lines,
household circuitry, and electrical devices. Some epidemiological studies have indicated a possible link between ELF and the development of childhood leukemia, but
the evidence is far from conclusive due to the difficulties in quantifying exposures
as well as the multitude of confounding factors in the studies. Once again, beyond
transfer of thermal energy, the molecular mechanisms involved in any adverse cellular effects are unclear. There is evidence, however, that ELF exposures may also
induce a stress response, as well as interacting with other signaling pathways within
the cell.

RADIATION SAFETY

The basic concepts of safety in dealing with radiation are relatively simple and come
down to three variables which can be manipulated. These are *time, distance, and
shielding.* First, obviously, the shorter the period of time during which an individual
is exposed, the lower the dose he or she will receive. Second, the greater the distance
of the individual from the radiation source, the lower the dose that he or she will
receive. In fact, exposure decreases by the square of the distance, so that if distance
from the source is doubled, exposure decreases by a factor of 4. Shielding refers to
the practice of inserting materials between an individual and the source in order to
block radiation. Effective shielding depends on the knowledge of the type of radiation being emitted, as different types and thicknesses of materials are required to
block different types of radiation. For example, alpha particles can be blocked by
a sheet of paper. Beta particles are commonly shielded with sheets of aluminum,
plastic, wood, or glass, typically a few millimeters thick. For x-rays and gamma rays,
lead and/or thick layers of concrete are commonly used. Neutron radiation is particularly difficult to shield, although glass with added boron (which absorbs neutrons
well) is sometimes used, as are tanks of water, gravel, or concrete.

   A number of agencies share responsibility for regulating use of radioactive materials and setting limits on exposures to ionizing radiation. These would include
the Nuclear Regulatory Commission, the Environmental Protection Agency, the
Department of Energy, the Occupational Safety and Health Administration, and even
the Department of Transportation. An organization without regulatory authority, but
which issues recommendations on these issues, is the International Commission on
Radiological Protection.

CASE STUDY   Three Mile Island, Chernobyl, and Fukushima

Accidents at nuclear power plants are among the most dramatic incidents resulting in radiation exposure, and the potential threat of these is of significant concern to much of the public. However, the risk of such occurrences is not trivial to estimate, partly because so few of them have occurred.

Nuclear power plants, like coal-fired power plants, generate electricity by heating water, which then turns a turbine, which then generates electricity. In *boiling water nuclear reactors*, the heated water is turned directly into steam which then turns a turbine. *Pressurized water reactors* have two circulating loops which do not mix, but do allow heat transfer. In the first, water heated by the reactor is kept under pressure so that it does not boil; the heat from that water is then transferred to water in a second, separate system, which is not under pressure, and which creates the steam that then turns a turbine. The Three Mile Island plant was a pressurized water reactor; Chernobyl was a boiling water reactor (although of a different design) and Fukushima had three boiling water reactors operating.

Heat in a nuclear power plant is generated in the reactor *core*. The core contains the *fuel*, which is usually in the form of rods, and which consists of *"enriched"* uranium, in which the concentration of the uranium-235 isotope is increased from its naturally occurring levels (generally less than 1%) to somewhere around 3%–5%. What makes uranium-235 special is that it is *fissionable*, which means that when it is struck by a neutron of the correct energy level, it will undergo fission, break into fragments and emit gamma rays as well as additional neutrons. Then, if there are other uranium-235 atoms nearby to be struck by those neutrons, they will also undergo fission, producing yet more neutrons. This, then, is the basis of the nuclear *chain reaction*. Eventually fuel becomes *spent*, and no longer contains sufficient concentrations of uranium-235 to sustain a reaction. The problem of what to do with this spent fuel from nuclear plants is one that has still not been solved.

The goal within the core of a nuclear power plant, then, is to both sustain and manage the reaction in order to produce the desired amount of heat. First of all, the neutrons released by a fissioning uranium-235 atom are actually traveling too fast to interact with other uranium atoms. Thus, to sustain a reaction, the neutrons need to be slowed down. This is accomplished through the presence of a *moderator* in the reactor core. This is a material that actually slows down neutrons so as to make them more likely to be able to initiate fission in another uranium atom. For most reactors, water serves as the moderator. In some reactors, *graphite* is used. Chernobyl, for example, was a reactor type known as an RBMK reactor, in which graphite (not water) serves as the moderator. There can be safety disadvantages to this, as graphite, unlike water, can burn.

It is also important, of course, to be able to slow or shut a nuclear reaction down. To accomplish this, most reactors have *control rods* which are made of a neutron-absorbing substance and which can be raised or lowered into the core depending on whether the plant's operators wish to speed up or slow down the reaction. Keeping the core cool is also important, because if the generated heat is not removed promptly, the vessel in which the core is contained may fracture or melt. This can lead to the release of radioactive materials into the concrete or steel *containment building* that houses the vessel, or worse, into the environment. When the heat in the core rises enough to melt the fuel, it is referred to as a *meltdown*.

So what happens in these systems when something goes wrong? In 1979, a glitch in a routine repair made several hours earlier at the TMI-2 reactor at Three Mile Island Nuclear Power Plant (Middletown, PA) led to a shutdown of the pumps in the secondary water system (that second loop in the pressurized water reactor system). Because heat was no longer being removed from the core, temperatures began to rise, and an automatic emergency shutdown (also known as a "*scram*") occurred, with control rods dropping into the core to terminate the reaction. Temperatures in the core and pressures in the pressurized water loop continued to rise until a relief valve in the pressurized water loop opened, and then stuck in the open position, allowing serious loss of fluid from the pressurized water loop. Badly designed control panels prevented operators from truly understanding the situation, and a number of erroneous decisions were made, leading to a partial meltdown of the core and the escape of contaminated water into the containment building, as well as into an overflow area. Finally, 16 h after the accident began, backup pumps were restarted and coolant began to flow again.

Ultimately, the consequences of this, the worst nuclear power facility accident on US soil, were minor. The only release of radioactivity to the environment was the release of a small amount of radioactive isotopes including inert gases and iodine-131. Estimated increased exposure levels for area residents were minimal, averaging approximately 0.09 mSv, with a maximum of 0.25 mSv. (For comparison, the average exposure for a resident of the United States is around 3.5 mSv/year.) Epidemiological studies have followed the residents living near Three Mile Island closely over the years, and results have indicated little or no increased cancer risks for that population. The Three Mile Island incident did, however, lead to significant changes in nuclear plant design, as well as in operator training.

This event can be compared and contrasted with what happened at Chernobyl, in the former Soviet Union, in 1986. As mentioned before, this was an RBMK reactor, using graphite instead of water as a moderator. The accident actually occurred during a test, when automatic shutdown mechanisms had been disabled and an attempt to rapidly shut down what had become a very unstable reactor situation resulted in a power surge, disruption of the coolant system, and an explosion of steam that exposed the core. A second explosion then dispersed radioactive fuel, graphite, and other debris. The hot graphite and fuel continued to ignite fires and burn, releasing more radiation over a 10-day period and complicating the emergency response. Overall, 13×10^{18} Bq of radiation was released, with released isotopes including inert gases (xenon-33, for example), iodine-131, cesium-134 and cesium-137, and strontium-90. Unfortunately, this radioactivity spreads across a sizeable portion of what are now Belarus, Russia, and the Ukraine as well as into a number of other European countries.

In terms of effects on human life, there were seven deaths the night of the Chernobyl accident; 47 additional deaths from acute radiation sickness followed. Approximately 116,000 individuals had to be evacuated from the surrounding areas, some of which are still not habitable. Estimated exposures for individuals in those areas averaged around 17 mSv (almost 200 times greater than the average exposure from Three Mile Island). All in all, the total population of all the countries that experienced some radioactive contamination is over 5 million people. The International Chernobyl Project has been studying a segment of the affected population (825,000 individuals) and has found the most significant concern to be thyroid cancer (due to the fact that iodine, including iodine-131, tends to accumulate in the thyroid). The likely route of human exposure for iodine-131 was through milk from cows which had grazed in contaminated pastures.

In the wake of Chernobyl, many reactors of the RBMK design have been shut down, and major modifications have been made to the remaining operating RBMK reactors (of which there are several, all in Russia). The importance of international collaboration on nuclear safety was also highlighted.

A third major nuclear accident occurred at the Fukushima Daiichi Nuclear Power Plant (Japan) in 2011. In this case, the precipitating factor was a massive (magnitude 9.0) earthquake and tsunami that struck Japan. The 10-m sea wall which protected the Fukushima reactors, unfortunately, was no match for the 14-m tsunami, even though the seismic waves from the earthquake (a little less than an hour earlier) had already triggered an automatic shutdown of the three (out of six) reactors which were in operation at the time. The problems came when the tsunami flooded the already-operating emergency diesel generators, leaving the cooling systems without power and unable to run. This affected not only the three reactors which still needed to be cooled down, but also the spent fuel pool, which needs to be continuously cooled. Heroic efforts were made to cool the reactors (using fire pumps and sea water), relieve rising pressures, and minimize radioactivity release; however, meltdown of the three reactor cores did occur, along with several hydrogen explosions.

The release of radiation at Fukushima, in spite of the fact that three hot reactors were damaged, has been estimated to be far less than at Chernobyl (probably in the order of 10%–15% of the Chernobyl release). Isotopes of concern included the usual iodine-131, cesium-134 and cesium-137, and strontium-90. Because of geographic location, no countries other than Japan were significantly impacted by released radioactivity, although trace amounts of radioactivity could still be detected worldwide as a result of the accident. Best estimate for human exposures following the Fukushima accident is an excess of approximately 3.5 mSv, which is higher than Three Mile Island, but much lower than Chernobyl. It is too soon to measure excess cancer-related deaths in the region as a result of Fukushima, but estimates range from around 100 to 1000 cases. Amongst the workers, there were no cases of radiation sickness reported. For both Three Mile Island and Fukushima, the reduced spread of radiation following the meltdowns is most likely due, at least in part, to more stringent design of containment structures.

Lessons learned from Fukushima have included a realization that plans and procedures need to be in place to support plant operations during extended power outages, as well as the need to better plan for earthquakes, flooding, and other natural disasters. The importance of better backup plans for safe maintenance of spent fuels was also highlighted.

Because their immediate environmental impact is arguably less damaging than coal-fired plants, there are many reasons to consider nuclear power as a viable, long-term energy source for the world. However, concern over the possibility of accidents, as well as the issue of how to safely store spent fuel will also have to be factored in to any decisions on whether to aggressively pursue nuclear power as an energy source. In either case, toxicologists will undoubtedly be involved in the ongoing debate.

BIBLIOGRAPHY

Besaratinia, A., Kim, S., and Pfeifer, G.P., Rapid repair of UVA-induced oxidized purines and persistence of UVB-induced dipyrimidine lesions determine the mutagenicity of sunlight in mouse cells, *FASEB J.*, 22, 2370, 2008.

Blank, M. and Goodman, R., Electromagnetic fields stress living cells, *Pathophysiology*, 16, 71, 2009.

Calvente, I., Fernandez, M.F., Villaba, J., Olea, N., and Nunez, M.I., Exposure to electromagnetic fields (non-ionizing radiation) and its relationship with childhood leukemia: A systematic review, *Sci. Total Environ.*, 408, 3062, 2010.

Cassidy, D., Holton, G., and Rutherford, J., *Understanding Physics*, Springer, New York, 2002.

Cember, H. and Johnson, T.E., *Introduction to Health Physics*, 4th edition, McGraw-Hill Medical, New York, 2009.

Han, Y.-Y., Youk, A.O., Sasser, H., and Talbott, E.O., Cancer incidence among residents of the Three Mile Island accident area: 1982–1995, *Environ. Res.*, 111, 1230, 2011.

Hardell, L., Carlberg, M., and Mild, K.H., Use of mobile phones and cordless phones is associated with increased risk for glioma and acoustin neuroma, *Pathophysiology*, 20, 85, 2013.

Hoel, D.G., Toxic effects of radiation and radioactive materials, in *Casarett and Doull's Toxicology*, Klaassen, C.D., Ed., McGraw-Hill, New York, 2013, chap. 25.

ICRP, *Relative Biological Effectiveness, Radiation Weighting and Quality Factor*, ICRP Publication 92, *Ann. ICRP*, 33, 2003.

Kadhim, M., Salomaa, S., Wright, E., Hildebrandt, G., Belyakov, O.V., Prise, K.M., and Little, M.P., Non-targeted effects of ionizing radiation-implications for low dose risk, *Mutat. Res.*, 752, 84, 2013.

Levy, B.S., Wegman, D.H., Baron, S.L., and Sokas, R.K., *Occupational and Environmental Health: Recognizing and Preventing Disease and Injury*, Oxford University Press, New York, 2011.

Lopez, M. and Martin, M., Medical management of the acute radiation syndrome, *Rep. Practical Oncol. Radiother.*, 16, 138, 2011.

Matsumaura, Y. and Ananthaswamy, H.N., Toxic effects of ultraviolet radiation on the skin, *Toxicol. Appl. Pharmacol.*, 195, 298, 2004.

Nair, M.K., Nambi, K.S.V., Amma, N.S., Gangadharan, P., Jayalekshmi, P., Jayadevan, S., Cherian, V., and Reghuram, K.N., Population study in the high natural background radiation area in Kerala, India, *Radiat. Res.*, 152, S145, 1999.

Nicolaou, A., Pilkington, S.M., and Rhodes, L.E., Ultraviolet-radiation induced skin inflammation: Dissecting the role of bioactive lipids, *Chem. Phys. Lipids*, 164, 535, 2011.

Ono, T. and Ikehata, H., The mechanisms of UV mutagenesis, *J. Radiat. Res.*, 52, 115, 2011.

Podgorsak, E.B., *Radiation Physics for Medical Physicists*, Springer, Berlin/Heidelberg, 2006.

Rabe, J.H., Mamelak, A.J., McElgunn, P.J.S., Morison, W.L., and Sauder, D.N., Photoaging: Mechanisms and repair, *J. Am. Acad. Dermatol.*, 55, 1, 2006.

Sakata, R., Grant, E.J., and Ozasa, K., Long-term follow-up of atomic bomb survivors, *Maturitas*, 72, 99, 2012.

Steinhauser, G., Brandl, A., and Johnson, T.E., Comparison of the Chernobyl and Fukushima nuclear accidents: A review of the environmental impacts, *Sci. Total Environ.*, 470, 800, 2014.

U.S. Nuclear Regulatory Commission, Office of Public Affairs, Three Mile Island accident, accessed at http://www.nrc.gov/reading-rm/doc-collections/fact-sheets/3mile-isle.pdf on 8/22/2014.

U.S. Nuclear Regulatory Commission, Office of Public Affairs, Chernobyl Nuclear Power Plant accident, accessed at http://www.nrc.gov/reading-rm/doc-collections/fact-sheets/chernobyl-bg.pdf on 8/22/2014.

Volkow, N.D., Tomasi, D., Wang, G.J., Vaska, P., Fowler, J.S., Telang, F., Alexoff, D., Logan, J., and Wong, C., Effects of cell phone radiofrequency signal exposure on brain glucose metabolism, *JAMA*, 305, 808, 2011.

Wakeford, R., The cancer epidemiology of radiation, *Oncogene*, 24, 6404, 2004.

19 Applications
Food Safety

It can be easily argued that one of the most central activities of life is eating. While accurate assessment of the safety of what we propose to consume has always been important, it perhaps has never posed such a complex problem as it does now. In this chapter, we will leave aside the very difficult question of what exactly comprises the optimal healthy diet, a topic which is well beyond the scope of this book. Additionally, we will not cover foodstuffs which are in and of themselves toxic, as many of these will be considered in Chapter 20. Finally, we will not directly address food allergies or idiosyncratic reactions such as lactose intolerance. We will instead focus on two major categories of potential toxicants: food *additives*, substances which are deliberately added to foods, and *contaminants*, substances which inadvertently become incorporated into foods. We will also look at the agencies and regulations governing food safety, particularly in the United States.

FOOD ADDITIVES

Today, a variety of different substances are added to foods for a variety of different reasons. Some act as preservatives to prevent microbial contamination (typically, bacteria and fungi are of concern). Others enhance the flavor, texture, or appearance of the food. There are literally thousands of these additives in regular use. We will examine some of the common categories of additives, along with examples of some substances of possible toxicological interest.

PRESERVATIVES AND ANTIOXIDANTS

Preservation of foods has long been an issue for humans, as the supply of many foodstuffs is not always perfectly synchronized with demand. It is also an issue when food must be transported some distance between the point of production and the point of use. Thus, the importance of developing additives which can increase the length of time that food can be safely stored. The two major problems that must be combatted are the growth of microbes and the tendency of food constituents (such as fats) to break down and become unpalatable.

Historical measures for food preservation have included physical methods which slow or stop growth of microorganisms and/or reduce the enzymatic activities which lead to food breakdown (*spoilage*). These physical methods include cooling, freezing, and drying (water, of course, being necessary for microbial growth). *Salt curing* and *sugar curing* were two of the earliest chemical preservation techniques to have been developed. In these processes, salt or sugar is added to foods, pulling water out of tissues and creating an osmotically unfavorable environment for microbial growth. *Pickling* combines salt curing with the addition of acids which also inhibit microbial growth and activity. These acids may either be added directly, or produced by non-pathogenic bacteria (often grouped together under the term *lactic acid bacteria or LAB*) through *fermentation*. Many of these bacteria also manufacture their own antimicrobial compounds, known as *bacteriocins*. Finally, the process of canning blocks microbial contamination by vacuum sealing microbe-free foods into sterilized containers.

Many of these processes, of course, continue to be employed; however, in addition, several chemical additives which inhibit microbial growth and food spoilage have also been developed. *Propionates, benzoates,* and *sorbates* are weak organic acids with antimicrobial properties. *BHA* and *BHT* are antioxidants which protect against the tendency for unsaturated fatty acids in foods to oxidize and produce the unpleasant products generally referred to as *rancid* fats and oils. Ironically, the trend toward using more of the healthier, polyunsaturated fats in foods has also resulted in foods with a higher susceptibility to becoming rancid. Other preservatives with antioxidant properties include ascorbic and citric acids.

Antioxidants and some other preservatives can also prevent further ripening or other chemical reactions in foods. *Sulfites*, for example, can prevent browning in fruits, and before a 1986 ban were commonly used on fresh fruits and vegetables (at salad bars, for example). Their use is, however, still permitted in processed foods. The problematic issue with sulfites is not so much direct toxicity as it is that some individuals can suffer allergic reactions to them. Individuals with pre-existing asthma seem to be more prone to this; however, as is typical of hypersensitivity reactions, the problem can develop in anyone at any time.

One final group of preservative compounds are the *nitrites* and *nitrates*. These have long been used in the preservation of meats, partially because of their effectiveness in protection against *Clostridium botulinum* (the microorganism responsible for botulism) and partly because they contribute to the flavor and appearance of the cured meat. However, during the 1970s, concerns were raised about the potential for nitrites and nitrates to interact with secondary amines also found in food (and in the stomach) to produce carcinogenic nitrosamines. The issue is complicated by the fact, though, that nitrites and nitrates are common constituents of many foods, and, in fact, nitrates from vegetables account for the vast majority (85%–90%) of nitrates in the average diet. Most ingested nitrite actually comes from saliva, where it is produced through reduction of nitrate by bacteria in the oral cavity. In the time period since the concerns were first raised, however, regulations have, in fact, been tightened and amounts of nitrites and nitrates in cured meats have been reduced.

SWEETENERS AND FLAVOR ENHANCERS

It is clear that many, if not most, humans have an affinity for sweet foods. Unfortunately, there is also strong evidence that consumption of high levels of dietary sugars can lead to adverse health consequences such as obesity, diabetes, cardiovascular, and kidney disease. Thus, there has been significant interest in the development of *non-nutritive sweeteners* (NNSs), compounds which can supply a sensation of sweetness but without the high caloric content. This is possible, due to the fact that most of these substances are tens to thousands of times sweeter than sucrose, and thus much smaller amounts are necessary in order to provide the desired sweetening effects.

The earliest of the NNSs to be discovered and developed was *saccharin*. Introduced in the early 1900s, it was controversial from the start, with arguments raging over whether or not it was an allowable sugar substitute. Sugar shortages during World War I and World War II accelerated its acceptance, but in the 1970s controversy ignited again when a few studies claimed an association between saccharin exposure and bladder cancer in rats. It was later determined that significant species differences between rats and humans rendered those studies inapplicable to humans, and, in fact, neither studies in other primates nor epidemiological studies have seen any evidence of a saccharin-related cancer risk. Thus, saccharin remains on the market (as Sweet 'N Low©, to give one brand example).

Other NNSs currently on the market include *aspartame*, an amino acid derivative marketed as Nutra Sweet©, and Equal©, among other brands. There are many anecdotal claims for adverse effects produced by aspartame, but the only well-documented risk involves individuals with the disease phenylketonuria. These individuals have defects in the metabolic pathway by which phenylalanine, a breakdown product of aspartame, is itself broken down and thus must limit consumption of phenylalanine. Acesulfame-K, sucralose (Splenda©), and the sugar alcohols (sorbitol, mannitol, and xylitol) are also in common use as sweeteners.

Besides increasing sweetness, additives can also be used to enhance other aspects of flavor. There are literally thousands of flavorings in use in the food industry, so to assess their safety poses a significant challenge. The approach chosen by the Joint FAO/WHO Expert Committee on Food Additives (JECFA), an international committee of experts who conducts risk assessment for food additives and contaminants, was to first sort flavorings out into groups on the basis of structural similarities (for example, phenol and phenol derivatives, sulfur-containing heterocyclic compounds, aliphatic and aromatic ethers, and so forth). Then, *no observed effect levels (NOELs)* for the compounds were derived using toxicological data, or where no data was available, were extrapolated from data for structurally related compounds. *Maximized survey-derived daily intakes (MSDIs)* and a measure called the *possible average daily intake (PADI)* (which is deliberately weighted toward higher estimates) were then used to estimate exposures. Then, a *margin of safety*, defined as the NOEL dose divided by the estimated exposure, was calculated. In their most recent testing, the Committee determined that over 99% of compounds they tested had a margin of safety of over 1000; only one compound had a margin of safety below 100.

One flavor enhancer, *monosodium glutamate* or *MSG*, has, over the years, been anecdotally identified as a cause of a hypersensitivity reaction in some individuals. However, this effect has been difficult to replicate in controlled studies and the additive remains approved for use.

FOOD COLORS

One of the most controversial categories of food additives is the category which includes the substances which modify the color of foods. *Color additives* used in the United States can be grouped into two categories: additives which require FDA certification (including most synthetic organic colors) and those that do not (including colors derived from natural sources). Certified colors are typically named with the prefix "FD&C" followed by the color (e.g., "blue") and a numerical designation. In the European Union, color additives are designated with an *E number*.

The certified color additives currently in use include:

- FD&C Blue No. 1 (Brilliant Blue) and No. 2 (Indigotine)
- FD&C Green No. 3 (Fast Green)
- FD&C Red No. 3 (Erythrosine) and No. 40 (Allura Red)
- FD&C Yellow No. 5 (Tartrazine) No. 6 (Sunset Yellow)
- Orange B
- Citrus Red No. 2

Structures for some of these are shown in Figure 19.1. Most of these are aromatic amines or azo compounds, which are normally a very reactive type of compound. However, chemical substitutions (including sulfonation), along with the large molecular size of many color additives tend to reduce gastrointestinal absorption and thus reduce toxicity. Also, because color additives are generally present in foods in small amounts, *estimated daily intakes* (in units of mg/person/day) are typically low (20 mg/person/day or lower). For most color additives, this is an order of magnitude or more below the *acceptable daily intake*, either for adults or children.

There are a number of color additives that have been approved or provisionally approved and have later been removed from the approved list. Perhaps the most notorious of these is FD&C Red No. 2 which had its provisional approval revoked in the United States in response to studies that indicated a possible weak link between exposure and development of cancer. Public concern was high, causing some companies to withdraw red foods from the market even if they did not contain the additive (most notably red M&Ms©, which were withdrawn by the Mars Corporation and not re-introduced until 1986).

Current concern over color additives has tended to focus on the possible connection between the additives and development of attention-deficit hyperactivity disorder (ADHD) in children. There are several studies that have offered evidence of a potential connection; however, the complex nature of diets and the abundance of confounding factors for behavior have made it difficult to sort out the specific dietary

FD & C Blue No. 1   FD & C Green No. 3

FD & C Red No. 40

FIGURE 19.1   The chemical structures of some of the certified color additives.

components that may be factors. More recent studies have taken a more holistic look at diet, including possible individual food sensitivities, as a possible contributing factor to a number of behavioral issues in children.

OTHER ADDITIVES, INCLUDING INDIRECT ADDITIVES

There are a variety of other types of food additives including additives that improve texture, retain moisture, control pH, act as emulsifiers, or even add nutrients to foods. *Indirect additives* are those which enter food from packaging or other materials that come into contact with foods at some point prior to reaching the consumer. The FDA uses a complex method of estimating exposure to these substances based on the amount of the substance in use and the propensity for the substance to migrate into foods (a measure which is based on extraction studies performed with a variety of solvents).

Plastics

See also:
   Case Study: Plastic Debris in
      the Marine Environment
                     Chapter 14

   Other Organic Compounds
                     Chapter 17

Endocrine Disruption

See also:
  *Endocrine Disrupters
  and Interference with
  Hormonal Controls*
                    Chapter 7

One group of indirect additives which have been of concern is *plasticizers*, chemicals used to enhance durability and flexibility of plastic packaging materials. Because they are typically of low molecular weight, are present in relatively high concentrations, and are not an integral part of the polymer structure of the plastics itself, these molecules are potentially free to migrate out of the plastic and into the foods which they contain. This is particularly an issue with foods with high fat content, as many plasticizers are quite lipophilic. Examples of plasticizers include *phthalate esters*, used in polyvinyl chloride plastics, *alkylphenols*, and *BPA*, used in polycarbonate plastics. Several of these compounds are endocrine disrupters and thus have the potential to impact not only human health, but ecosystem health, as well. The FDA has determined that levels measured in foods are well below those necessary to produce health effects; however, the potential for ecosystem level effects due to leaching from discarded packaging is a concern, with negative effects on reproduction in fish, amphibians, and insects noted in a number of studies.

Nanoparticles

See also:
  *Deposition of Particulates
  Case Study: Nanoparticles*
                    Chapter 8

Another relatively new issue is the use of nanoparticles (particles with diameters in the 1–100 nm range) in foods. Many new food-related uses are proposed for nanoparticles, including enhancing the stability of emulsions (requiring less fat content in foods), improving nutrient delivery and absorption, and using nanoencapsulation to mask components with unpalatable tastes. Silver and gold nanoparticles have also been shown to decrease microbial growth in foods when incorporated into packaging, although it should be mentioned that the mechanism for this antimicrobial action is not entirely clear. To go one step further, the incorporation of nanoparticles into *polymer nanocomposites* could even lead the way to packaging which could act as a monitor for freshness or an indicator of contamination of the food it contains.

However, some caution is in order. There are many unanswered questions about the behavior of nanoparticles in the gastrointestinal tract, and the potential ease of absorption of nanoparticles through that route is clearly of concern. Also, because many materials behave much differently at the nanoparticle scale than at normal scale, even previously approved additives would need to be re-tested and re-approved if scaled down to nanoparticle size.

CHEMICAL CONTAMINANTS IN FOODS

Pesticides

See also:
  Pesticides
                    Chapter 17

One type of chemical contaminant which is of concern in foods is, of course, pesticides. Pesticide use is regulated by the EPA, and in order to be used on food crops, a pesticide must be specifically registered for use on that crop. Regulations are designed to minimize residues at time of harvest; nonetheless, one must expect that at least some dietary exposure is likely to occur. What is at question is the significance of those exposures with respect to effects on human health. This is not trivial to determine, in part because multiple avenues for pesticide exposure are present for most individuals, and, of course, dietary exposures themselves are difficult to estimate. Also, processing (frying, blanching, washing, etc.) can alter pesticide concentrations in foods, in most cases leading to significant reductions.

However, in general, most studies have concluded that overall health risks of pesticide residues on food are generally low. In one study involving life-cycle impact assessment of pesticides in the diet (using residue data from the USDA Pesticide Data Program), an average of 3.2 min of life per person in the United States was estimated as lost due to negative health effects from pesticide residues. Another study (also using USDA data) examined 12 of the fruits and vegetables identified as most likely to harbor residues, as well as the levels of the 10 most frequently detected pesticides on each. These data were then combined with estimates of mean exposure with the result that in every case, estimated exposures were much lower than the *chronic reference doses* (*RfD*), the maximum allowable dose according to the EPA. In fact, the highest estimate (for methamidophos on bell peppers) was 2% of the RfD, with 75% of the estimates falling below 0.01% of the RfD.

Another category of potential food contaminants is the heavy metals. They may enter the food chain through bioaccumulation (several plants are prone to accumulate metals from soils and water), from packaging (lead solder on cans, for example), or even from dinnerware (some pottery glazes are known to contain lead). One of the most notorious examples of contamination of foods by metals was the incident in Minamata, Japan, where the accumulation of methylmercury in an aquatic food chain led to significant effects both on wildlife and on human life. In another incident in Japan, cadmium accumulation in rice grown downstream from a mining operation led to the development of itai-itai disease, a condition characterized by kidney and bone damage.

Synthetic chemicals such as PCBs and PBBs have also entered the food supply on occasion. In an incident which began in 1973 in Michigan, a farmer began to note unusual health problems in his livestock. Suspecting a problem with the feed, he submitted it for a chemical analysis, which ultimately revealed the presence of PBBs. Upon investigation,

Mercury

See also:
   Vascular Spasms and Blood Pressure
                                    Chapter 9

   Other Neurotoxicants
                                    Chapter 10

   Damage to the Proximal Tubule
                                    Chapter 12

   Metals
                                    Chapter 17

   Mercury
                                    Appendix

Lead

See also:
   Interference with Cell Division
                                    Chapter 7

   Anemias, Hemolysis, and Related Disorders
                                    Chapter 9

   Myelinopathies
   Other Neurotoxicants
                                    Chapter 10

   Toxicant-Induced Immunosuppression
                                    Chapter 13

   Metals
                                    Chapter 17

   Lead
                                    Appendix

Cadmium

See also:
Interference with Cell Division
Chapter 7

Vascular Spasms and Blood Pressure
Chapter 9

Damage to the Proximal Tubule
Chapter 12

Metals
Chapter 17

Cadmium
Appendix

PCBs and PBBs

See also:
Endocrine Disrupters and Interference
with Hormonal Controls
Chapter 7

Toxicant-Induced Immunosuppression
Chapter 13

Other Organic Compounds
Chapter 17

it was discovered that Michigan Chemical had mixed up the packaging of two of their products, FireMaster (PBBs) and NutriMaster (MgO), substituting somewhere in the order of 500–1000 pounds of the PBB product for the MgO feed supplement. And to complicate matters, not only had equipment used to mix and process feed passed the contamination on to other feeds, but tissues from animals now known to have been exposed to PBBs had also been added to other feeds. Thousands of residents were exposed through dairy, beef, poultry, and egg products; epidemiological studies have not yet reported any significant associations between exposure and disease, although monitoring of exposed individuals continues. There have also been a number of other incidents in both the United States and Europe in which similar compounds have been found to contaminate the food supply. These would include a 1997 incident in which milk in Germany was found to contain dioxins originating in contaminated citrus pulp from Brazil, as well as a 1999 episode in Belgium in which PCB-contaminated recycled fat also found its way into animal feed.

Finally, there have also been some concerns over the presence of antibiotics and other drugs in animal feeds. Antibiotics have been used not only to prevent disease, but also to produce enhanced weight gains in cows, pigs, chickens, and other livestock, and have even found uses in aquaculture (particularly in less-developed countries). One issue with this practice has been that the use of antibiotics in this fashion (low doses over long periods of time) encourages the development of resistant strains of microbes much more effectively than short-term targeted therapy for a particular medical condition does. Resistant bacteria can then be passed on to humans through the food chain, an event that has, in fact, already been documented in multiple cases. Resistance can also pass between microbial species, compounding the problem. There has been a ban in Europe using antibiotics for growth promotion since 1999, and, encouragingly, there is now some evidence for declines in resistance to tetracycline, vancomycin (a structural relative of the growth promoter avoparcin), and other drugs. In the United States, the FDA has requested that manufacturers of animal pharmaceuticals voluntarily alter labels on antibiotics to eliminate growth production uses and has also moved to require veterinary approval for purchase of the drugs.

CONTAMINANTS OF BIOLOGICAL ORIGIN

Examples of food contaminants of biological origin include mycotoxins (such as aflatoxin or ergot alkaloids), dinoflagellate toxins (red tides), and bacterial toxins (such as botulinum toxin). These are covered elsewhere in the textbook and so will not be addressed here.

Mycotoxins
See also:
Fungal Toxins
Chapter 20

Prions are another example of contaminants of biological origin. These misfolded variations of normal cell glycoproteins (the normal PrP protein takes the form of an alpha helix, the disease-causing form is folded into a beta pleated sheet) seem to be able to convert the normal proteins into the pathogenic form. Unfortunately, the pathogenic form is also much more resistant to both enzymatic and nonenzymatic degradation, facilitating its accumulation in tissues, where it causes diseases such as *scrapie* in sheep, *bovine spongiform encephalopathy (BSE)* in cows, and *Creutzfeldt–Jakob Disease (CJD)* in humans. These diseases are characterized by the development of vacuoles, loss of neurons, and often development of PrP plaques. Symptoms include demetia, ataxia, and other neuromotor symptoms.

Red Tide
See also:
Dinoflagellates and Paralytic Shellfish Toxins
Chapter 20

Botulinum Toxin
See also:
Pharmacokinetics and Toxicokinetics
Chapter 2
Acetylcholine Case Study: Botulinum Toxin
Chapter 10
Bacterial Toxins
Chapter 20

Most cases of CJD are sporadic, however some cases (variant CJD or vCJD) can be acquired through the diet. One notable example of this is *kuru*, a human prion disease which was found to be endemic among some populations of New Guinea in the 1950s, and which was subsequently determined to be transmitted through ritual cannibalism. More recently, approximately 229 cases of vCJD have been identified primarily in the United Kingdom and France which were most likely caused by consumption of meat products from cows suffering from BSE.

TRANSGENIC FOODS

Ever since the dawn of agriculture, humans have been attempting the genetic modification of crops and livestock raised for food purposes. Until recently, however, this has taken the form of selective breeding programs designed to select for desired attributes, or even deliberate production of mutations combined with assessment and selection of the resulting offspring for desirable traits. With the tools available now, however, direct manipulation of the genome has become not only possible, but routine. This typically takes the form of the insertion of genes from one species directly into the genome of another. The resulting organisms are commonly referred to as *genetically modified organisms (GMOs)*.

There are many common GMO food crops currently being grown. Many of these have been altered to contain genes that confer pest resistance. The first GMOs to be developed, for example, were crops containing *cry* genes from the organism *Bacillus thuringiensis* (*Bt*) which code for insecticidal endotoxins. Corn, cotton, and canola varieties containing these genes are available. Other genes confer herbicide tolerance, such as the resistance to glyphosate which is found in Monsanto's Roundup Ready soybeans. *Stacked* crops contain multiple modifications.

There are several concerns which have been raised about GMO crops, including questions over the likelihood of the interbreeding of these crops with unaltered crops, allowing the "escape" of the genetic modification into the environment. Of course, the issue of whether any changes in the biochemistry of a crop plant might affect the food products made from that plant is also a concern. Typically, analytical testing is done on GMO crops in order to compare composition with non-GMO varieties; safety studies are also done to ensure that no unexpected adverse health effects are associated with the new variety.

REGULATIONS AND REGULATORY AGENCIES

Regulation of food began with the Pure Food and Drug Act of 1906, which was one of a number of health and safety-related acts passed during the early years of the century. These laws were motivated, in part, by public reaction to the sensational contents of Upton Sinclair's novel, *The Jungle*, describing conditions in the meat-packing industry. The *Food and Drug Administration*, created out of an administrative reorganization in 1930, recommended revisions in the 1906 act, which led to the passage of the *Federal Food, Drugs, and Cosmetics Act of 1938*. This act contained the first provision for requiring approval of products and ingredients prior to marketing them, but laid the burden of proof on the FDA to demonstrate that a product was unsafe.

By the 1950s, there was a veritable explosion of ingredients (literally thousands) in common use in foods, most with little or no formal safety evaluation to back up their use. At this point, it was recognized that to try and require retroactive comprehensive testing of each one would be prohibitively difficult, and so the idea of the Generally Recognized As Safe (GRAS) list was born. Ingredients can be added to the GRAS list based on scientific testing or history of usage (if the ingredient was introduced prior to 1958). Inclusion on the GRAS list means that, technically, the ingredient is no longer defined as being a food additive. By amendment, however, some substances (food colors, for example) have been declared ineligible for inclusion on the GRAS list.

Food contaminants are dealt with separately, with the FDA also possessing the authority to set allowable limits for what are considered unavoidable contaminants. Foods containing avoidable contaminants which may pose a health risk are considered to be adulterated and are subject to recall, with the magnitude of the recall effort (there are Class I, II, and III recalls) determined by the potential severity of the risk.

TESTING

Determination of the safety of food additives and contaminants is a complicated, multistep process. One key factor to be evaluated has already been mentioned: the estimated

TABLE 19.1

Types of Testing Required for the FDA's Concern Levels I, II, and III

Concern Level	Type of Testing
I	Genetic toxicity testing
	Short-term rodent toxicity testing
II	Genetic toxicity testing
	Short-term rodent toxicity testing
	Subchronic toxicity testing with rodents and non-rodent species
	Reproductive toxicity testing
	Developmental toxicity testing
	Metabolism and pharmacokinetic studies
III	Genetic toxicity testing
	Short-term rodent toxicity testing
	Subchronic toxicity testing with rodents and non-rodent species
	Reproductive toxicity testing
	Developmental toxicity testing
	Metabolism and pharmacokinetic studies
	Chronic toxicity/carcinogenicity studies in rodents
	One-year toxicity studies with non-rodent species
	Human studies

daily intake. To calculate this, one must not only know the concentration of the additive in any foods in which it appears, but also the average daily intake of each of those foods. There are a number of databases available in which consumption of various categories of food has been recorded, often separated out by demographic groups. Exposure to the additive from nonfood sources, if applicable, must also be considered.

In terms of toxicity testing, additives are assigned to categories based on structure. The FDA has identified three categories: A for structures with low potential for toxicity, B for structures with unknown or intermediate potential, and C for structures with higher potential. These categories, along with estimates of level of exposure, are used to assign additives to *concern levels*, which then dictate appropriate testing. For example, a compound in category A with low estimated exposure would be assigned to concern level I, with the least stringent testing requirements. A compound in category B with the identical estimated exposure, on the other hand, might be assigned to the intermediate concern level II. Concern level III corresponds to the most rigorous testing requirements. Some of the FDA's recommended tests for each of the levels are shown in Table 19.1. Of course, positive results at any level might indicate the need for further testing beyond the initial requirements of that level.

CARCINOGENICITY

The issue of potential carcinogenicity of food additives has always been controversial. Back in 1958, a series of hearings on food and drug regulation was held. Out of those hearings, a piece of legislation was introduced by Congressman James Delaney

(D-NY) which stated, in part, that "... no additive shall be deemed to be safe if it is found to induce cancer when ingested by man or animal, or if it is found, after tests which are appropriate for the evaluation of the safety of food additives, to induce cancer in man or animal ..." (FD&C Act, section 409(c)(3)(a)). Two years later, the clause was extended to specifically include color additives.

The Delaney Clause, as it has come to be known, was, of course based on the science of the time and does not take into account the concepts of dose–response or risk–benefit analysis. It applies neither to GRAS substances (which, as you may recall, are defined not to be additives) nor to unavoidable contaminants. Loopholes in the strict interpretation have been created, however. In 1962, residues from drugs used in livestock were allowed if no residues could be "found" in edible tissue. The question for the FDA, of course, is how hard to require producers to look. The solution was that producers were required to use a method showing that residues were not found above a concentration corresponding to an increased lifetime risk for cancer of 1 in a million. The clause was also interpreted by the FDA as not applying to constituents of additives, but only to the additive itself. Thus, if an additive contains low levels of a carcinogenic constituent, but does not test positively for carcinogenicity itself, it is allowable. Also, in 1996, the status of pesticide residues was clarified to allow levels not exceeding those allowable in the raw foods, regardless of carcinogenicity. The FDA does retain the authority to interpret exactly what it means to "induce cancer," a process that is clearly much more complex than was understood in the 1950s.

THE IMPORTANCE OF LABELING

Wherever there are potential individual sensitivities and idiosyncratic reactions to products, accurate labeling is critical. This is certainly the case with food products. The Food Allergen Labeling and Consumer Protection Act was passed in 2004 and requires that products containing any of eight *major food allergens* be labeled as such. These foods (milk, eggs, fish, crustaceans, walnuts and pecans, and other nuts from trees, peanuts, wheat, and soybeans) were chosen because they are jointly responsible for an estimated 90% of all adverse food-related allergic reactions. The FDA also regulates and monitors "gluten-free" labeling.

Another area in which labeling has become an issue is GMO foods. Although a number of countries mandate identification of foods from GMOs, the FDA considers GMO and non-GMO foods to be equivalent and does not require labeling. However, many consumers beg to differ, leading to a fair amount of controversy, with some states stepping in to mandate labeling of GMO-derived foods. Manufacturers, on the other hand, argue that labeling will imply that there are somehow significant safety differences between GMO and non-GMO versions of the same foods, something that studies to this point have not indicated to be the case.

BIBLIOGRAPHY

Bellisle, F., Glutamate and the UMAMI taste: Sensory, metabolic, nutritional and behavioural considerations. A review of the literature published in the last 10 years, *Neurosci. Biobehav. Rev.* 23, 423, 1999.

Brul, S. and Coote, P., Preservative agents in foods. Mode of action and microbial resistance mechanisms, *Int. J. Food Microbiol.*, 50, 1, 1999.

Burdock, G.A. and Carabin, I.G., Generally recognized as safe (GRAS): History and description, *Toxicol. Lett.*, 150, 3, 2004.

Cabello, F.C., Heavy use of prophylactic antibiotics in aquaculture: A growing problem for human and animal health and for the environment, *Environ. Microbiol.*, 8, 1137, 2006.

Cushen, M., Kerry, J., Morris, M., Cruz-Romero, M., and Cummins, E., Nanotechnologies in the food industry—Recent developments, risks and regulation, *Trends Food Sci. Technol.*, 24, 30, 2012.

Dalton, L., Food preservatives, *Chem. Eng. News Arch.*, 80, 40, 2002.

Duncan, T.V., Applications of nanotechnology in food packaging and food safety: Barrier materials, antimicrobials and sensors, *J. Colloid Interf. Sci.*, 363, 1, 2011.

Fasano, E., Bono-Blay, F., Cirillo, T., Montuori, P., and Lacorte, S., Migration of phthalates, alkylphenols, bisphenol A and di(2-ethylhexyl)adipate from food packaging, *Food Control*, 27, 132, 2012.

Fries, G.F., The PBB episode in Michigan: An overall appraisal, *Crit. Rev. Toxicol.*, 16, 105, 1985.

Haik, S. and Brandel, J.-P., Infectious prion diseases in humans: Cannibalism, iatrogenicity and zoonoses, *Infect. Genet. Evol.*, 26, 303, 2014.

Hattersley, S., Ward, R., Baka, A., and Crevel, R.W.R., Advances in the risk management of unintended presence of allergenic foods in manufactured food products—An overview, *Food Chem. Toxicol.*, 67, 255, 2014.

Inaba, T., Kobayashi, E., Suwazono, Y., Uetani, M., Oishi, M., Nakagawa, H., and Nogawa, K., Estimation of cumulative cadmium intake causing Itai-itai disease, *Toxicol. Lett.*, 159, 192, 2005.

Juraske, R., Mytel, C.L., Stoessel, F., and Hellweg, S., Life cycle human toxicity assessment of pesticides: Comparing fruit and vegetable diets in Switzerland and the United States, *Chemosphere*, 77, 939, 2009.

Kamle, S. and Ali, S., Genetically modified crops: Detection strategies and biosafety issues, *Gene*, 522, 123, 2013.

Keikotlhaile, B.M., Spanoghe, P., and Steurbaut, W., Effects of food processing on pesticide residues in fruits and vegetables: A meta-analysis approach, *Food Chem. Toxicol.*, 48, 1, 2010.

Kotsonis, F.N. and Burdock, G.A., Food toxicology, in *Casarett and Doull's Toxicology*, Klaassen, C.D., Ed., McGraw-Hill, New York, 2013, chap. 31.

Malisch, R. and Kotz, A., Dioxins and PCBs in feed and food—Review from European perspective, *Sci. Total Environ.*, 491, 2, 2014.

Marinovich, M., Galli, C.L., Bosetti, C., Gallus, S., and LaVecchia, C., Aspartame, low-calorie sweeteners and disease: Regulatory safety and epidemiological issues, *Food Chem. Toxicol.*, 60, 109, 2013.

Marshall, B.M. and Levy, S.B., Food animals and antimicrobials: Impacts on human health, *Clin. Microbiol. Rev.*, 24, 718, 2011.

McCann, D., Barrett, A., Cooper, A., Crumpler, D., Dalen, L., Grimshaw, K., Kitchin, E. et al., Food additives and hyperactive behavior in 3-year-old and 8/9-year-old children in the community: A randomized, double-blinded, placebo-controlled trial, *Lancet*, 370, 1560, 2007.

Merrill, R.A., Food safety regulation: Reforming the Delaney Clause, *Annu. Rev. Public Health*, 18, 313, 1997.

Munro, I.C. and Danielewska-Nikiel, B., Comparison of estimated daily intakes of flavouring substances with no-observed-effect levels, *Food Chem. Toxicol.*, 44, 758, 2006.

Rangan, C. and Barceloux, D.G., Food additives and sensitivities, in *Medical Toxicology of Natural Substances: Foods, Fungi, Medicinal Herbs, Toxic Plants, and Venomous Animals*, Barceloux, D.G., Ed., John Wiley and Sons, Hoboken, NJ, 2008.

Reich, M.R., Environmental politics and science: The case of PBB contamination in Michigan, *AJPH*, 73, 302, 1983.

Ross, R.P., Morgan, S., and Hill, C., Preservation and fermentation: Past, present and future, *Int. J. Food Microbiol.*, 79, 3, 2002.

Rulis, A.M. and Levitt, J.A., FDA's food ingredient approval process. Safety assurance based on scientific assessment, *Regul. Toxicol. Pharmacol.*, 53, 20, 2009.

Shankar, P., Ahuja, S., and Sriram, K., Non-nutritive sweeteners: Review and update, *Nutrition*, 29, 1293, 2013.

Sindelar, J.J. and Milkowski, A.L., Human safety controversies surrounding nitrate and nitrite in the diet, *Nitric Oxide*, 26, 259, 2012.

Sinn, N., Nutritional and dietary influences on attention deficit hyperactivity disorder, *Nutr. Rev.*, 66, 558, 2008.

Snell, C.L., Bernheim, A., Berge, J.-B., Kuntz, M., Pascal, G., Paris, A., and Ricroch, A.E., Assessment of the health impact of GM plant diets in long-term and multigenerational animal feeding trials: A literature review, *Food Chem. Toxicol.*, 50, 1134, 2012.

Takayama, S., Sieber, S.M., Adamson, R.H., Thorgeirsson, U.P., Dalgard, D.W., Arnold, L.L., Cano, M., Eklund, S., and Cohen, S.M., Long-term feeding of sodium saccharin to non-human primates: Implications for urinary tract cancer, *J. Natl. Cancer Inst.* 90, 10, 1998.

Winter, C.K. and Katz, J.M., Dietary exposure to pesticide residues from commodities alleged to contain the highest contamination levels, *J. Toxicol.*, 2011, Article ID 589674, doi:10.1155/2011/589674.

20 Applications
Toxins

TOXINS

It should certainly be no surprise to anyone that living organisms contain biologically active molecules which can affect other living organisms. Nor should it be surprising that some of those biologically active molecules can produce adverse (toxic) effects. The term for these toxicants of biological origin is *toxins*, and the world is full of them. Toxins can be found in a wide variety of organisms from prokaryotes to eukaryotes, a fact which, in and of itself, is a strong indicator that there must be a significant survival value to producing such a substance. In this chapter, we will examine the wide range of toxins found in living organisms as well as the evolutionary pressures which contributed to their development.

The study of the way organisms interact through exchange of chemical products is called *chemical ecology*. Briefly, some organisms appear to use toxins in their own defense, either preventing attacks by predators or discouraging competition. Many plants, for example, manufacture bitter compounds which seem to play a role in discouraging insects from consuming their leaves and other tissues. Other plants manufacture chemicals which discourage competition through inhibiting germination of potential competing plants in the nearby soil. Some animals also rely on toxins to discourage predators, whereas some, of course, rely on toxins to help capture prey. Either way, it is clear that there can be strong evolutionary selection pressures promoting development of toxins as tools for survival.

BACTERIAL TOXINS

There are a wide variety of bacterial toxins, many of which are likely to have evolved to promote successful survival and reproduction of the bacterium. *Exotoxins* are toxins which are secreted by the bacterium which manufactures them, while *endotoxins* are part of the structural component of the bacterial cell itself.

BOTULINUM AND TETANUS TOXINS

One of the best known bacterial exotoxins is botulinum toxin. This complex molecule is a two-chain polypeptide linked by disulfide bonds with three functional domains and has the distinction of being perhaps the most toxic substance known based on

Botulinum Toxin

See also:
  Acetylcholine
  Case Study: Botulinum
    Toxin
                Chapter 10

an estimated LD_{50} in humans of 1–3 ng/kg. Produced by the bacterium *Clostridium botulinum*, a gram-positive, anaerobic bacterium, seven different variations (serotypes) of the toxin have been identified, with types A, B, and E responsible for most cases of human poisoning (serotype A is shown in Figure 20.1). The most common route of exposure for humans is absorption through the gastrointestinal tract, where the toxin is protected from gastric proteases by auxiliary proteins which are released by the bacterium along with it.

The mechanism of action of botulinum toxin involves block of acetylcholine release, thus, symptoms of botulism include muscle paralysis, as well as interference with parasympathetic function. Typical patients might present with nausea and vomiting, visual disturbances, dryness of the mouth and throat, weakness, and ataxia. The most serious complication is the risk of paralysis of the diaphragm. Generally, confirmation of diagnosis is made through identification of the toxin itself in serum or feces. Treatment involves support of respiration and, if deemed necessary, administration of *anti-toxin* (an antibody against the toxin produced from serum of immunized horses). Botulinum toxin is discussed in more detail in the case study in Chapter 10.

A toxin which is closely related to botulinum toxin structurally is *tetanus toxin*. This toxin is produced by the bacterium *Clostridium tetani*, which lives in soil and is found worldwide. Typically, human exposure occurs when wounds are contaminated with *C. tetani* spores. These then germinate and produce the toxin, which is taken up by peripheral nerves in the vicinity. Retrograde axonal transport then moves the toxin to dendrites, where it is released and is taken up by presynaptic inhibitory

FIGURE 20.1 Botulinum toxin molecule, showing three functional domains. (Reprinted from *Trends in Biochemical Sciences*, 27, 552, Turton, K., Chaddock, J.A., and Acharya, K.R., Botulinum and tetanus neurotoxins: structure, function and therapeutic utility, Copyright 2002, with permission from Elsevier.)

neurons. There it blocks vesicle formation and release of inhibitory neurotransmitters such as glycine and GABA. The result is overstimulation of muscle contraction leading to tetanic paralysis. Unlike the case with botulinum toxin, however, an effective vaccine against tetanus toxin exists.

CYANOBACTERIAL TOXINS

Cyanobacteria, also known as "blue-green algae," are not actually algae, but are a ubiquitous group of photosynthetic prokaryotes named for a pigment (phycocyanin) found in many species. The fossil record indicates that they were one of the earliest organisms on earth, and it is likely that their photosynthetic activities contributed significantly to the formation of the early atmosphere. Today they are found in freshwater, marine, and terrestrial ecosystems and are particularly abundant under eutrophic (nutrient-rich) conditions. In fact, under the proper conditions (typically warm, nutrient-rich, still, surface waters), they are able to undergo rapid proliferation, dramatically increasing their biomass in a short time in a phenomenon known as a *bloom.*

Not all cyanobacteria produce toxins, and even in those that do the amount of toxins produced may vary greatly with environmental conditions. Cyanotoxins are generally grouped according to their site of action into hepatotoxins, neurotoxins, and dermatotoxins. The most widespread of the hepatotoxins are the *microcystins*, which are produced by a number of cyanobacterial species including species in the *Microcystis, Nostoc, Oscillatoria,* and *Anabaena* genera. Microcystins produce liver damage through binding to and inhibiting serine/threonine phosphatases, leading to hyperphosphorylation of proteins and disruption of hepatocyte structure and function. Death is caused by hemorrhagic shock which follows the breakdown of liver circulation and accumulation of blood in the liver. At lower, chronic exposures, microcystins can act as promoters and can also lead to chronic inflammation.

Eutrophication

See also:
   Freshwater Ecosystems
      Chapter 14

   Phosphorus and Nitrogen
      Chapter 17

Neurotoxins include *anatoxin-a*, which is an agonist at nicotinic acetylcholine receptors, and the structurally unrelated *anatoxin-s* which is an acetylcholinesterase inhibitor. *BMAA*, another neurotoxic cyanotoxin, stimulates glutamate receptors and has been hypothesized to be a causative agent in lytico-bodig, the neurodegenerative disease endemic to Guam. *Saxitoxin,* a sodium channel blocker, is also produced by some cyanobacteria (as well as by some dinoflagellates).

Cyanotoxins have been determined to be responsible for toxicity to both wildlife and to domestic livestock. Growth of cyanobacteria has been observed to reduce growth of green algae as

Acetylcholinesterase

See also:
   Effects of Toxicants on
     Enzymes
      Chapter 4

   Acetylcholine
      Chapter 10

   Organophosphates
      Appendix

well as aquatic plants such as duckweed and elodea. Interestingly, some bivalves appear to be able to tolerate the presence of cyanotoxins, possibly due to increased activity of the p-glycoprotein pump, which may act to excrete the toxins. Fish, however, appear susceptible, and numerous reports of fish kills associated with cyanobacterial blooms have been reported. Of course, there are also concerns over human exposure to these toxins, either through drinking water or through the food chain. And because of the possible association of BMAA exposure with the development of lytico-bodig, hypotheses have been advanced that link exposure to this and other cyanobacterial toxins to development of other neurodegenerative diseases including Parkinson's disease, Alzheimer's disease, and amyotrophic lateral sclerosis. Additional research will be necessary in order to assess the validity of this potential risk.

Neurotransmitter Receptors

See also:
  Effects of Toxicants on
  Receptors and Ion
  Channels
                Chapter 4

  Effects of Toxicants on
  Synaptic Function
                Chapter 10

BMAA

See also:
  Excitotoxicity
                Chapter 10

OTHER BACTERIAL TOXINS

There are a host of other examples of toxins produced by bacteria. Perhaps one of the best studied is *diphtheria toxin*, produced by *Corynebacterium diptheriae*, which exerts its effects through inhibition of protein synthesis. *Bacillus anthracis*, the bacterium which causes anthrax, produces three components (*protective antigen*, *lethal factor*, and *edema factor*) which combine to form the two toxins *lethal toxin* (PA + LF) and *edema toxin* (PA + EF). The PA component mediates binding and entry into cells, while the LF and EF components have enzymatic activity (metalloproteinase and adenylyl cyclase activity, respectively.) Dermal anthrax infection produces localized sores and edema; gastrointestinal or respiratory exposure can lead to a lethal systemic infection in which bacteria initially infiltrate and produce massive damage to immune cells, and later spread to most major organ systems leading to cardiovascular collapse. Toxins are also produced by *Vibrio cholera* (the agent which causes cholera), *Streptococcus pyogenes* (responsible for strep throat, scarlet fever, and toxic shock), and many other bacterial species.

PROTIST TOXINS

The term "protist" is often used to refer to organisms which are eukaryotic and generally unicellular or colonial in structure. It is sometimes said that these are the organisms that, for one reason or another, do not seem to fit in with any of the other eukaryotes (fungi, plants, or animals). Certainly, it is a diverse group, including as it does protozoans (amoebas, paramecia), algae (green, brown, and red), and slime molds and water molds.

Dinoflagellates and Paralytic Shellfish Toxins

Probably the most notorious protists in terms of toxin production are the *dinoflagellates* (members of the algae). There are probably around 2000 species of dinoflagellates, some of which are autotrophic, some of which are heterotrophic, and some of which can utilize both mechanisms for nutrient incorporation. Around 70 dinoflagellate species have been found to produce toxins. Just as cyanobacteria can proliferate rapidly under the right conditions, so can dinoflagellates. The resulting phenomenon is often referred to as a red tide, as the coloration of the organisms involved can dramatically discolor the water. Red tides and related blooms can occur in many coastal areas, as can be seen in the WHOI/NOAA map of harmful algal bloom-related events which have occurred in the United States (Figure 20.2).

Several of the toxins produced by dinoflagellates can be grouped together under the term *paralytic shellfish toxins (PSTs)*, a grouping which includes saxitoxin and related compounds. Saxitoxin is a sodium channel blocker and thus blocks neurotransmission. Another group of toxins, the *brevetoxins*, are produced by the organism *Karenia brevis*; a few dinoflagellates also produce the neurotoxin *domoic acid*. Brevetoxins also act on sodium channels, locking them in an open position, leading to rapid depolarization.

Tetrodotoxin, Saxitoxin

See also:
Effects of Toxicants on Voltage-Activated Ion Channels
Chapter 4

Effects of Toxicants on Electrical Conduction
Chapter 10

Mollusks
Fish, Birds, and Mammals
Chapter 20

TTX, STX
Appendix

P-Glycoprotein

See also:
The Blood–Brain Barrier
Chapter 10

Bioaccumulation

See also:
Material Cycling in Ecosystems
Chapter 14

Dinoflagellates from the genus *Pfiesteria* have been reported to be responsible for fish kills in estuarine areas and have been reported to produce a number of hydrophilic, metal-containing toxins, although both the toxins and the mechanisms through which they may work have yet to be completely characterized.

The high lipid solubility of these toxins can lead to accumulation in filter-feeding shellfish and scavenging gastropods, and ultimately to consumption by humans. Symptoms include paralysis (thus leading to the descriptor "paralytic" shellfish toxin) as well as numbness and paresthesias. Death would follow paralysis of the diaphragm. Unfortunately, the toxin is heat-stable and thus is not inactivated by cooking (indeed, upon boiling some of the toxin may, in fact, be transferred to the cooking water). Red tides and PSP are not new phenomena; reports of paralytic shellfish poisoning (PSP) can be found in historical records as far back as the late 1700s. Deaths from PSP are now fairly rare, however, as most developed countries have monitoring programs which prevent harvesting of shellfish from waters during dinoflagellate blooms. Fish can also be affected by dinoflagellate blooms, as can turtles, dolphins, manatees, and even seabirds.

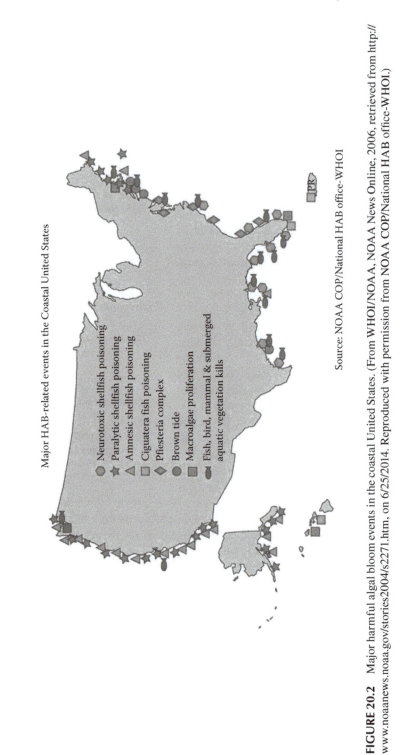

FIGURE 20.2 Major harmful algal bloom events in the coastal United States. (From WHOI/NOAA. NOAA News Online, 2006, retrieved from http://www.noaanews.noaa.gov/stories2004/s2271.htm, on 6/25/2014. Reproduced with permission from NOAA COP/National HAB office-WHOI.)

FUNGAL TOXINS

There are a number of fungi which produce toxins (generally known as *mycotoxins*), many of which have a long history of impact on human health. For example, the first records of the fungus *Claviceps purpurea* contaminating grain date back to around 600 BC. The common name for this fungus, *ergot*, comes from the French word "argot" which means "cock's spur" and refers to the purple or black "spurs" of fungus that can develop on harvested grain, particularly under moist conditions. Consumption of bread made from contaminated grain (usually rye) could lead to *ergotism*, a condition characterized by either swollen extremities, jaundice, and gangrene (gangrenous ergotism) or redness of the skin, diarrhea, vomiting, peripheral pain, mania, and convulsions (convulsive ergotism).

In the middle ages, epidemics of ergotism were reportedly responsible for tens of thousands of deaths, especially amongst the poor. The burning sensations in the limbs experienced by some sufferers caused the disease to be referred to as "St. Anthony's Fire," a name which also refers to the religious order who ran hospitals where afflicted sufferers came for treatment. By the 1600s, the cause of the malady was known; however, it took much longer for effective preventative measures to become established. The toxins responsible for ergotism are the *ergot alkaloids*, all of which are derivatives of lysergic acid.

Aflatoxins, to cite another example of mycotoxins, are produced by fungi from the genus *Aspergillus* and are found in foods which have been contaminated by those fungi. Originally discovered when moldy peanuts killed hundreds of thousands of turkeys in the 1960s, aflatoxins turn out to have a ubiquitous presence in a wide variety of foods including corn, peanuts, and other nut and cereal products. The most toxic aflatoxin is aflatoxin B_1, which is metabolized to an epoxide by cytochrome P450. It is both carcinogenic and hepatotoxic, with effects on the reproductive and immune systems, as well. Interestingly, aflatoxins have been detected in the liver of children with Reye's syndrome, also. Other mycotoxins which are potential contaminants of food include *ochratoxin A* (nephrotoxic; a probable human carcinogen), and a variety of toxins produced by *Fusarium* species.

Other concerns about exposures to fungal toxins have centered on respiratory exposures. Pulmonary exposures to molds (either outdoors or indoors) may produce allergic reactions, and exposure to mold has been implicated in the development of asthma in children. Mycotoxins, however, are not volatile molecules, and thus toxin-containing mold spores themselves would need to be inhaled in order to produce effects. Whether or not this scenario is actually capable of delivering mycotoxin levels sufficient to produce toxicity has been debated, particularly with respect to some *Stachybotrys* species (commonly referred to as "black mold"). Molds can, however, produce both volatile organic compounds and particulates which may under some circumstances serve as respiratory irritants.

Reye's Syndrome

See also:
Case Study: Reye's
Syndrome
Chapter 11

Finally, of course, mushrooms belong to the fungal kingdom. There are a number of mushrooms which contain toxins, but perhaps the best known are the *Amanita* species

FIGURE 20.3 Death cap mushrooms, *A. phalloides*. (Reprinted from *Medicine*, 40, 135, Persson, H., Copyright 2012, with permission from Elsevier.)

(including *Amanita phalloides*, the "death cap" or "destroying angel"; Figure 20.3). Found throughout the world (it is native to Europe, but has been introduced into the United States), this mushroom is responsible for more lethal poisonings worldwide than any other. This is likely to be at least in part due to its similarity to other, nontoxic, mushrooms, including some edible *Amanitas*. The death cap and its relatives contain *amatoxins*, heat-stable peptides which remain intact throughout the cooking process. The mechanism of action of amatoxins is the inhibition of RNA polymerase, leading to a blockade of protein synthesis. The gastrointestinal system is the primary target, with initial symptoms (severe gastrointestinal distress), typically beginning a few hours after exposure, and signs of liver damage following a few days later. The only treatment is supportive care; mortality is high and generally results from hepatic damage.

Other mushrooms that contain toxins include *Gyromitra* species (false morels) which contain *gyromitrin*, an inhibitor of GABA synthesis. The cholinergic agonist *muscarine* can be found in several mushrooms, as can the hallucinogenic compound *psilocybin*. Many other mushrooms contain toxins that are gastric irritants.

PLANT TOXINS

Scientists have been aware of the existence of *secondary plant products* for many years. These are chemicals which are manufactured by the plant but do not seem to play a role in basic metabolism, growth, reproduction, or other fundamental life processes. One of the first secondary plant products to be isolated and studied was morphine, in the early 1800s. Since then, literally hundreds of thousands of secondary plant products have been identified. Many of them have proved to have toxic effects on humans, an observation which should not be too surprising given that many secondary plant products are likely to have evolved as defense mechanisms against herbivorous grazers.

Initial characterization of secondary plant products focused primarily on categorization by structure, although there are a number of different ways to sort them out.

Some of the major groups of products of toxicological interest include *alkaloids* (a group which includes not only morphine, but also cocaine, caffeine, and atropine), *terpenes* and *terpenoids* (including steroids and saponins), and *glycosides*. While a full review of toxin-containing plants is obviously far beyond the scope of this chapter, we can, however, look at a few representative examples of these toxins, the plants which produce them, and their effects on human health.

PLANT ALKALOIDS

The term "alkaloid" is generally used to refer to a diverse group of chemicals of botanical origin which contain heterocyclic nitrogen and are generally, but not always, derived from amino acids. Examples of some alkaloids and their structures are found in Table 20.1. The table also contains information on the general mechanism of action of each, along with the chapter where additional information on its effects can be found.

One alkaloid whose actions are not described in detail elsewhere in this textbook is caffeine. *Caffeine*, along with its relatives *theophylline* and *theobromine*, is a xanthine alkaloid. The physiological effects of these compounds have been known literally for centuries, and probably were first deduced from observations of the effects of coffee berries on livestock. Xanthines are found in multiple plants, including coffee plants (*Coffea* sp.), where they probably act to discourage pests from feeding on the plant and to inhibit germination of other plants in the surrounding soil. All three xanthines are also found in the tea plant (*Camellia sinensis*), and in *Theobroma cacao*, the plant whose seeds are used to produce chocolate.

The pharmacological actions of caffeine and the other two xanthines appear to relate primarily to the blockade of adenosine receptors, which play a role in modulating release of several neurotransmitters (including acetylcholine, dopamine, serotonin, glutamate, and GABA). Adenosine receptors are G-protein-coupled receptors and come in four subtypes. Caffeine has the highest affinity for A_1 and A_2 receptors, activation of which lead to suppression of neuronal activity, among other actions. The ubiquity of adenosine receptors in the brain and elsewhere leads to the wide diversity of effects of caffeine, which has been reported to increase alertness, decrease reaction times, and improve mood. On the other side of the coin, caffeine can also increase anxiety. Chronic consumption of caffeine has been linked, in some studies, to reduction in risk for development of dementia, and adenosine receptor blockade has been shown in some animal models to be protective against ischemic damage in adults. However, other studies have indicated exactly the opposite in developing animals.

Discontinuation of chronic caffeine intake can cause withdrawal symptoms, as many individuals who have cut back on their coffee habit have discovered. This is due to the upregulation of adenosine receptor numbers that occurs in response to chronic caffeine-induced adenosine receptor blockade. Once exposure ceases, there is a period of time before compensatory downregulation restores receptor balance. In that interim, the overabundance of adenosine receptors can lead to, among other effects, the increase in cerebral vasodilation which produces the classic caffeine withdrawal headache.

TABLE 20.1

Structures and Mechanisms of Action of Some Common Plant Alkaloids

Alkaloid	Structure	Mechanism of Action	Additional Information
Cocaine		Blocks reuptake of catecholamines	Ch. 16, Forensic Toxicology
Caffeine		Adrenergic receptor antagonist	This chapter
Nicotine		Nicotinic cholinergic receptor agonist	Ch. 10, Neurotoxicology
Morphine		Opioid receptor agonist	Ch. 10, Neurotoxicology Ch. 16, Forensic Toxicology
Atropine		Muscarinic cholinergic receptor antagonist	Ch. 10, Neurotoxicology
Strychnine		Glycine receptor antagonist	Ch. 10, Neurotoxicology
Capsaicin		Transient receptor potential vanilloid 1 (TRPV1) receptor agonist	This chapter

Another interesting alkaloid is *capsaicin*. Found in plants of the genus *Capsicum*, it is believed to act as a defense against herbivores. Research has discovered that capsaicin is an agonist at a receptor called the *transient receptor potential vanilloid 1 receptor (TRPV1)*. This receptor is found in a variety of areas of the mammalian brain and plays a role in heat and pain pathways. Thus, binding of capsaicin to the receptor produces a sensation which simulates the sensation of heat. But because binding of capsaicin to the receptor also induces a refractory state, it is used in a variety of topical analgesic preparations (although it is generally used in combination with another topical anesthetic to block any unpleasant irritant sensations). Interestingly, avian TRPV1 is sufficiently different in structure from mammalian TRPV1 that capsaicin does not bind effectively. Thus, birds can consume capsaicin-containing fruits and seeds without adverse effect.

PLANT TERPENES AND TERPENOIDS

One important defense mechanism used by many plants is the production of *latex*, a complex fluid released upon injury and containing a variety of compounds including terpenes and terpenoids. One notorious example of a latex-producing plant is the poinsettia (*Euphorbia pulcherrima*). This common holiday plant rocketed into notoriety on the basis of a 1944 book reporting an unconfirmed death of a child following ingestion of a leaf. In truth, a study of 22,793 cases of poinsettia ingestion and/or dermal exposure between 1985 and 1992 found no fatalities, and generally no effects, or minor effects following exposure. The terpenoids in poinsettias, however, do have the potential to irritate human skin, producing a condition termed *contact dermatitis*.

PLANT GLYCOSIDES

Chemically, a *glycoside* consists of a sugar bonded to a non-sugar (the *aglycone*) through a glycosidic bond. Types of glycosides are typically sorted out by the chemical nature of the aglycone. There are a number of toxicologically interesting plant glycosides. *Amygdalin* (found in almonds) as well as *linamarin* (found in the cassava plant) are classified as cyanogenic glycosides, with cyanide as the aglycone. These compounds probably serve a defensive function, in that if the plant is attacked, the cyanogenic glycoside is broken down to release hydrogen cyanide. In spite of this, cassava root is a major source of food in areas of Africa, but it must be processed properly in order to remove all traces of the toxic glycoside.

Digoxin and digitoxin are steroidal glycosides found in the foxglove, *Digitalis purpurea*. Often referred to as the cardiac glycosides, these compounds are Na^+/K^+ ATPase blockers and have been used to treat a variety of cardiovascular conditions due to their effects on myocardial contraction and conductivity.

OTHER IRRITANTS

As mentioned earlier, numerous plants produce chemicals which can lead to either dermal or gastrointestinal irritation. One example of plants capable of causing irritation is the group consisting of *Dieffenbachia* species. Colloquially known as "dumb cane," these plants, if ingested, can produce pain and swelling of the mouth, tongue, and oral cavity. This effect has long been attributed to the crystals of calcium oxalate (*raphides*) which are found in the plant; however, other chemical constituents may be involved, as well. Raphides are also found in daffodils (*Ranunculus* spp.), and, along with a variety of alkaloids produce the "daffodil itch" which is a common complaint of florists. The sap of stinging nettles (*Urtica spp.*) contains a complex mix of chemicals including formic acid, acetylcholine, and histamine, which its stinging hairs (*trichomes*) can inject into organisms unfortunate enough to brush up against it.

POISON IVY AND ALLERGIC CONTACT DERMATITIS

Poison Ivy

See also:
    Toxicant-Induced Allergies
        Chapter 13

Finally, *allergic contact dermatitis* is a different type of dermatitis which involves interaction of plant constituents with the immune system to produce an allergic reaction. The plant most widely known for this is poison ivy (Figure 20.4), which belongs to the genus *Toxicodendron* along with other "poison" plants such as poison oak and poison sumac. Common in North America, these plants are responsible for millions of allergic contact dermatitis cases a year, causing a significant number of lost-time injuries in many outdoor occupations. Sensitivity to the chemical responsible for the dermatitis, *urushiol*, seems to be limited to a few species of primates and clearly has some genetic basis. Sensitization to the chemical can occur at any time, and it is estimated that one-half to three-quarters of the adults in the United States are sensitive.

Urushiol is found in leaves, roots, and stems of *Toxicodendron* species and is released following damage (such as by insects, for example). Slight differences

FIGURE 20.4   Poison ivy. (Modified with permission from http://www.shutterstock.com/pic-195121376/stock-photo-poison-ivy-toxicodendron-radicans-vine-growing-on-a-tree.html?src=&ws=1. Accessed on August 9, 2014.)

exist between forms of urushiol found in different plant species, but generally a person who is sensitive to one form will also be sensitive to the others. Upon contact with the chemical, a rash typically develops within a day or two of exposure. Although normally nonvolatile, urushiol can be volatilized by fire and can pose a significant respiratory hazard to forest firefighters. The rash will generally resolve within 2–3 weeks. Persistent folk information holds that the spread of the fluid contents of rash vesicles can spread the irritation; this is not true. It is true, however, that due to the stability of urushiol, re-exposure can occur following contact with objects which have been contaminated with the chemical. Washing the chemical off following contact is only effective if done in the first few minutes of exposure.

ANIMAL TOXINS

Toxins are also employed by a wide number of animal species, either in protection or in predation. In general, animals that can deliver a poison through a bite or sting are described as *venomous*. Venoms are typically complex mixtures containing proteins and polypeptides as well as multiple other organic and inorganic components. Venom components vary greatly, even, in some cases, between populations of the same species and are likely to have evolved via gene duplication and modification. Absorption of venom components, as well as delivery to target tissues, would follow the same principles as with any other toxicant and would depend on chemical structure and properties. Smaller, lipid-soluble components can cross cell membranes; larger molecules may enter the circulation through more permeable lymphatic vessels.

CNIDARIANS

One phylum which is well represented in terms of toxin production is the Cnidaria, a group which contains sea anemones, corals, and jellyfish. The common structure possessed by cnidarians is the *nematocyst*, a complex organelle typically found in epithelial cells known as *nematocytes*. When triggered, these organelles (which have a very high internal osmotic pressure) discharge rapidly, shooting out a spine-tipped, coiled tubule and thus delivering toxins directly into target tissues.

While both sea anemones and corals contain relatively potent toxins, it is jellyfish toxins that most people are both most familiar with and most likely to encounter in person. Perhaps the most notorious jellyfish is the box jellyfish, *Chironex fleckeri*. Sometimes described as the most dangerous animal in the world, this Australian jellyfish has been responsible for around 70 deaths (although most envenomed victims do survive). The sting is extremely painful and leads rapidly to cardiovascular complications, with initial hypertension followed by hypotension and arrhythmias. Respiratory arrest completes the picture. The mechanisms behind these physiological effects are not completely clear, but may involve effects on ion transport, as *Chironex* venom contains proteins CfTX-1 and CfTX-2 which share structural features with pore-forming proteins. The severe pain associated with the sting may well involve stimulation of TRPV1 receptors (the same receptors that capsaicin interacts with). There is an antivenom available; however, it must be given early to be effective.

Other jellyfish which are of clinical significance in terms of human exposures include *Carukia barnesi*, the cause of *irukandji syndrome*. This syndrome is characterized by headache, backache, abdominal pain, nausea and vomiting, hypertension, and severe anxiety. It is extremely painful, but only rarely lethal. Stings delivered by the well-known man o'war jellyfish (*Physalia spp.*) also tend to be painful rather than lethal. Finally, the larvae of the thimble jellyfish, *Linuche unguiculata*, are the cause of "seabather's eruption," an allergic dermatitis caused when these tiny organisms become trapped in clothing and fire their nematocysts, particularly after the swimmer exits the water. Although there is little initial effect, the ensuing dermatitis can last for days, if not weeks.

ARTHROPODS: ARACHNIDS, INSECTS, AND MORE

There are a variety of venomous scorpions in the world, and death and serious injury from encounters with scorpions is higher than those from spiders or snakes (well over a million incidents involving scorpions per year, according to estimates). Many known scorpion toxins interact with Na^+ or K^+ ion channels; some are specific for mammalian channels, some for insect channels, and some bind to both. Some peptides found in scorpion venom also bind to receptors involved in intracellular calcium storage and release. In North, Central, and South America, scorpions from the genus *Centruroides* are the most likely to be involved in human envenomations. Symptoms of *Centruroides* stings include local pain, tachycardia and hypertension, and ataxia. In most cases, symptoms resolve within 12–24 h, with fatalities most common in untreated children.

Spiders, also, have evolved complex venoms as well as venom glands. The clinical significance of spider bites in humans, however (with a few clear exceptions) is debatable, as a relatively small percentage of spider species have been reported to cause serious complications. Among the species known to be dangerous to humans are *Atrax spp.* (Australian funnel web spiders), *Phoneutria spp.* (Brazilian wandering spiders), *Latrodectus spp.* (including the black widow spider), and *Loxoscales spp.* (including the brown recluse spider).

Spider venoms serve the joint purposes of paralyzing (and killing) prey, initiating digestion, and also providing a defensive weapon. They have been found to contain glutamate receptor antagonists, calcium, sodium, and potassium channel blockers, and other neurotoxins. *Latrodectus spp.* venoms contain α-*latrotoxin*, which causes massive neurotransmitter release, probably through facilitating calcium influx (through insertion into the membrane and subsequent pore formation) and triggering neurotransmitter vesicle exocytosis. α-Latrotoxin is specific for vertebrates; related compounds show specificity for insects and crustaceans. Signs and symptoms of black widow envenomation may include muscle cramping, vomiting, abdominal pain, hypertension, and tachycardia. Fatalities are rare.

Another spider whose bite can be clinically significant is the brown recluse, *Loxosceles reclusa* (Figure 20.5). *Loxosceles*, of which there are over a hundred species worldwide, are small, nocturnal spiders that live under bark and beneath rocks

FIGURE 20.5   Brown recluse spider. (Reprinted from *Clin. Dermatol.*, 24, 213, Swanson, D.L. and Vetter R.S., Loxoscelism, Copyright 2006, with permission from Elsevier.)

in the natural environment, as well as in the hidden nooks and crannies of human habitations. The venom of this spider is rich in enzymes, including the phospholipase sphingomyelinase D, proteases, and hyaluronidases. The sphingomyelinase D appears to be primarily responsible for the local reaction to *Loxosceles* envenomations, which includes pain, itching, swelling, and edema generally occurring a few hours after the bite. However, several days later a necrotic area may form around the bite, which can then take several weeks to heal. Systemic reactions to the bite are relatively rare and can involve hemolysis which may then lead to renal failure.

The clinical significance of brown recluse bites in the United States has been debated. The issue is problematic partly because there is generally little if any pain associated with a brown recluse bite, so patients typically present several hours later with a lesion but with no evidence of the causative agent. There is some evidence, however, that physicians may be diagnosing a variety of non-arachnid-related dermatological lesions (bacterial and fungal infections, contact dermatitis, other insect bites) as brown recluse bites. Diagnoses, for example, are not infrequently made in regions in which populations of the spider do not occur.

One example of a problematic venomous insect is the fire ant (*Solenopsis invicta*). Fire ant venom contains a piperidine derivative that stimulates release of histamine from mast cells, causing the local burning sensation, followed by development of a raised lesion. Bees and yellow jackets (family Apidae), and wasps and hornets (family Vespidae) are also, of course, venomous. With all of these insects, the risk exists for development of allergic responses to venom proteins such as phospholipases, hyaluronidase, and melittin (a peptide found in bee venom). Because several of these stinging species have venom components in common, cross-reactivity to different venoms is not uncommon. Incidentally, the major difference in terms of threat between common honeybees and the Africanized "killer bees" is not in venom composition or volume of delivery, but in aggressiveness of attack. (The Africanized bees respond in greater numbers and at a greater distance from the hive than their counterparts.)

MOLLUSKS

This chapter would not be complete without at least a mention of venomous mollusks. Two examples come immediately to mind: cone snails (*Conus spp.*) and the blue ringed octopus (*Hapalochlaena spp.*). Cone snails are predatory marine snails that use their venom to prey on marine worms, fish, and other mollusks. The venoms are complex, containing a group of peptides collec-

tively known as *conotoxins*, as well as other components. Conotoxins, which are rich in disulfide bonds, can be separated into gene families and superfamilies, and tend to target ion channels and receptors important in the neurotransmission process, including sodium, potassium, and calcium channels, the nicotinic acetylcholine receptor, serotonin receptors, and glutamate receptors. It has also been found that many conotoxins tend to be highly specific for binding to targets only in the type of prey (fish, mollusk, etc.) that the snail which produces the toxin preys upon. It is no wonder that there is great interest in investigating the potential therapeutic applications of conotoxins. One conotoxin, Ziconotide, is currently in clinical trials for use in pain management.

The blue-ringed octopus is a relatively small marine mollusk (measuring roughly 6 in across) whose beautiful coloration occasionally leads to it being picked up out of tide pools by hapless tourists (the typical way in which envenomation occurs). The main component of the venom is *tetrodotoxin*, a relatively ubiquitous toxin which is also found in a variety of other organisms (including puffer fish and newts). The toxin is produced in salivary glands and injected through a bite from the sharp beak of the octopus. The bite is often not noticed, and realization that a bite has occurred comes only with the advent of symptoms such as weakness, numbness, ataxia, and ultimately respiratory arrest. The only treatment is supportive care. Along with its presence in the venom, TTX has also been found in the skin of the octopus, indicating that it may serve a defensive function as well as functioning in prey capture.

The occurrence of tetrodotoxin in such a variety of species has always been somewhat of a mystery. It is possible, of course, that the biosynthetic pathways that produce tetrodotoxin in all these species were basically derived from a pathway inherited from a common ancestor long in the evolutionary past. It is also possible, however, that tetrodotoxin is not being manufactured in these species at all, but is, in fact, being absorbed from some other (possibly bacterial) environmental source. This question is still being debated, with some evidence existing for both points of view. The existence of symbiotic bacteria which produce compounds of ecological value to the host is well documented, and, in fact TTX-producing bacteria (including some *Vibrio*, *Pseudomonas*, and *Bacillus* species) have been isolated from puffer fish and blue-ringed octopi. However, these bacteria have not been able to be isolated from all known TTX-producing animals, such as newts, for example. The inability of many captive organisms to produce TTX also argues for an environmental source, although it could also be simply that a necessary dietary item is lacking in order for the organisms to make TTX. At any rate, this is a question that continues to be studied and will hopefully soon be definitively answered.

AMPHIBIANS

Toxins in amphibians almost certainly have evolved as a defensive mechanism against predation. Perhaps the best known examples of toxic amphibians are the neotropical poison dart frogs (*Dendrobates spp.* and *Phyllobates spp.*). These small, brightly colored frogs secrete alkaloids including *batrachotoxin*, which locks sodium channels in an open position. Toad secretions can contain epinephrine, norepinephrine, GABA, and serotonin, as well as *bufotenin*, a hallucinogenic compound. Some newt tissues, as already discussed, contain tetrodotoxin.

REPTILES

There are, of course, a large number of venomous and/or toxic reptiles, including snakes. The four families to which most venomous snakes belong are listed in Table 20.2, along with examples from each family.

Snakes from these families are found worldwide, but envenomation is a significant health problem primarily in tropical areas where rural workers are the primary victims. In Europe, most venomous snakes are of the genus *Vipera*, and include the asp viper, common viper, and blunt-nosed viper. In Africa, venomous natives include cobras (*Naja spp.*), mambas (*Dendroaspis spp.*), and saw-scaled vipers (*Echis spp.*). In Asia, cobras and kraits (*Bungarus spp.*) are found, but the most dangerous Asian snake is the Russell's viper (*Daboia spp.*). Australia is home to a strikingly high number of venomous snakes (some 70% of native snakes fall into that category) including brown snakes (*Pseudonaja spp.*) and the taipan (*Oxyuranus scutellatus*). In Central and South America, lance-headed vipers (of a variety of different species) predominate and include the fer de lance or terciepelo (*Bothrops atrox*). The largest venomous snake in the region is the Bushmaster (*Lachesis muta*). In North America, venomous snakes include a variety of rattlesnakes (*Crotalus spp.*), water moccasins and copperheads (*Agkistrodon spp.*), and coral snakes (varieties of all these species are also found in Central and South America).

Snake venom is produced in venom glands in the upper jaws and is delivered through ducts to the base of enlarged teeth (fangs). These teeth often contain grooves,

TABLE 20.2
Families which Contain Venomous Snakes

Family	Characteristics/Habitat	Examples
Viperidae	Slender bodies, triangular heads. Primarily terrestrial	Puff adders, rattlesnakes
Atractaspididae	"Stiletto snakes." Burrow in the ground	
Elapidae	Smooth scales. Terrestrial, arboreal, marine	Cobras, mambas, coral snakes, sea snakes
Colubridae	Many harmless; a few venomous	African boomslang

which may be open or closed, and through which the venom flows and is delivered upon biting. In some snakes (Viperidae), the fangs are basically hollow, allowing particularly efficient venom delivery. In addition, the Viperidae fangs can be retracted and are regularly replaced (every 6–10 weeks in rattlesnakes, for example). Actual injection of venom is produced by action of a muscle on the venom gland and is under the control of the snake. Because there is a metabolic cost to venom manufacture, the ability to meter (measure out) venom may provide survival advantages. Nonetheless, snakes typically deliver doses of venom that may be in the range of 100 times the lethal dose for their prey.

Snakes swallow their prey whole, and in order to facilitate this, it is preferable for the prey to be dead, or at least immobilized prior to consumption. To achieve this, snake venom is a complex mixture which includes proteins, peptides, lipids, amines, and inorganic components. Although many different venoms share components in common, there is also wide variation in composition between species and even populations of the same species. Still, components seem to be targeted towards disrupting two major systems: the nervous system and the cardiovascular system.

Some components of snake venoms have enzymatic activity. Table 20.3 lists some examples of types of enzymes found in snake venoms.

Another group of compounds found in snake venoms is the non-enzymatic polypeptides, of which hundreds have been identified. Examples of these include *dendrotoxins* (from mamba venoms) which are potassium channel blockers, *cardiotoxins* (from cobra venoms) which produce red cell hemolysis, *erabutoxins* (from sea snake venom) and *cobratoxins* which block the nicotinic acetylcholine receptor, and *crotamine* (from rattlesnake venom) which acts on sodium channels. Several venom polypeptides have found usefulness as tools in physiology research, and others are being investigated for potential therapeutic value.

Treatment for snakebite depends, of course, on the species involved and the specifics of its venom. Generally, supportive care is in order and may or may not be supplemented by administration of the appropriate antivenom, if one is available and

---

TABLE 20.3
Some Enzymatic Components of Snake Venoms

Venom Component	Mechanism of Action	Examples of Physiological Targets and Effects
Phospholipase A_2 enzymes	Hydrolysis of phospholipids; may act alone or form complexes with other venom components	Block presynaptic neurotransmitter release Myotoxic (effects on muscle) Hemorrhagic (liver, kidney, lungs)
Hyaluronidases	Breakdown of hyaluronic acid (interstitial component)	"Spreading factor," breaks down tissues allows venom to penetrate
Acetylcholinesterases	Hydrolysis of acetylcholine	Disruption of neurotransmission
Phosphatases	Hydrolysis of DNA and RNA	Unclear
Peptidases and proteases	Hydrolysis of peptides and proteins	Hemorrhagic (interfere with clotting process) Digestion of prey

can be administered within a few hours of envenomation. Antivenoms may be monovalent (specific for one species) or polyvalent (active against several species, typically from the same region). There is some risk to the administration of antivenom, as the patient may suffer a hypersensitivity reaction.

Other venomous reptiles include the Gila monster and the Mexican beaded lizard.

FISH, BIRDS, AND MAMMALS

The puffer fish and its toxin, tetrodotoxin, have already been discussed earlier in this book, but there are other toxic fish that are worth a mention. Several species of stingrays possess barbed and serrated spines which also feature toxin-producing cells. These cells sustain damage when the spine is thrust into target tissue, and some of the toxin enters the wound. The function of the spine and toxin system is believed to be defensive. There are also several toxic species of scorpionfish and lionfish. Typically, these fish have venom glands located in grooves on the dorsal, anal, or ventral fins. However, they are rarely a threat to humans, and most incidents involve removal of scorpionfish which have been caught in nets or attempts by divers to grab lionfish on a reef. Similarly, stonefish typically cause injury to humans only when they are stepped on as they are lying camouflaged with dorsal spines erect.

Tetrodotoxin

See also:
Effects of Toxicants on
Voltage-Activated Ion
Channels
Chapter 4

Effects of Toxicants on
Electrical Conduction
Chapter 10

Cyanobacterial Toxins
Dinoflagellates and
Paralytic Shellfish Toxins
Mollusks
Chapter 20

TTX, STX
Appendix

The list of known toxic birds is very short and their existence was only discovered in 1990. The hooded pitohui (*Pitohui dichrous*) from Papua New Guinea was discovered to be toxic when a graduate student was pecked by a mist netted bird, reflexively brought the wound to his mouth, and was surprised to experience a burning sensation, as well as pain and numbness in his mouth and tongue. Upon analysis, the feathers, skin, and muscles of the pitohui were found to contain *homobatrachotoxin*, a toxin closely related to batrachotoxin (as also found, of course, in poison dart frogs). There is evidence that the pitohui uses the toxin for defense, as it advertises its unpalatability with bright colors and a strong odor. Natives only rarely eat the bird and call it the "rubbish bird" because of its bitter taste. Another New Guinea bird, the blue-capped ifrita, has also been found to contain batrachotoxin. There is some evidence that the pitohui may accumulate the toxin through the food chain, as a beetle eaten by the bird has been found to contain the toxin. Of course, if this is the case, then the question merely shifts from how the bird acquires the toxin to how the beetle acquires the toxin. It is entirely possible that other bird species may also either manufacture or accumulate toxins, however relatively little research has been done in this area.

Finally, there are also relatively few venomous or toxic mammals. Most of these are *shrews* which produce venom from enlarged submaxillary salivary glands. The major toxin that has been identified from shrew venom is BLTX, a protease. The

FIGURE 20.6   Platypus (*Ornithorhynchus anatinus*). (Reprinted from *Toxicon*, 59, 680, Ligabue-Braun, R., Verli, H., and Carlini, C.R., Venomous mammals: A review, Copyright 2012, with permission from Elsevier.)

function of the venom is unclear, although it may aid the insectivorous shrews in collecting live, yet comatose, prey. The common vampire bat also produces toxic saliva.

Perhaps the best known venomous mammal, however, is the platypus (Figure 20.6), which is a native of Australia. The adult males of this species possess spurs on their hind legs with venom-producing glands at the base. Because the glands increase in size and output during the mating season, it is hypothesized that the spurs and venom play a role in competition for mating. At one time, the platypus was hunted, and as a result human envenomation was much more common than it is now. Platypus venom is a complex mixture of peptides, proteases, and other compounds, and envenomation in humans is accompanied by excruciating pain and swelling in the affected area, and can last for weeks.

BIBLIOGRAPHY

Araoz, R., Molgo, J., and de Marsac, N.T., Neurotoxi cyanobacterial toxins, *Toxicon*, 56, 813, 2010.

Batoreu, M.C.C., Dias, E., Pereira, P., and Franca, S., Risk of human exposure to paralytic toxins of algal origin, *Environ. Toxicol. Pharmacol.*, 19, 401, 2005.

Beckmann, A. and Ozbek, S., The nematocyst: A molecular map of the Cnidarian stinging organelle, *Int. J. Dev. Biol.* 56, 577, 2012.

Brinkman, D. and Burnell, J., Biochemical and molecular characterization of cubozoan protein toxins, *Toxicon,* 54, 1162, 2009.

Burkholder, J.M. and Marshall, H.G., Toxigenic *Pfiesteria* species—Updates on biology, ecology, toxins, and impacts, *Harmful Algae*, 14, 196, 2012.

Bush, R.K., Portnoy, J.M., Saxon, A., Terr, A.I., and Wood, R.A., The medical effects of mold exposure, *J. Allergy Clin. Immunol.*, 117, 326, 2006.

Casewell, N.R., Wuster, W., Vonk, F.J., Harrison, R.A., and Fry, B.G., Complex cocktails: The evolutionary novelty of venoms, *Trends Ecol. Evol.*, 28, 219, 2013.

Caya, J.G., Agni, R., and Miller, J.E. *Clostridium botulinum* and the clinical laboratorian, *Arch. Pathol. Lab. Med.*, 128, 653, 2004.

Chau, R., Kalaitzis, J.A., and Neilan, B.A., On the origins and biosynthesis of tetrodotoxin, *Aquat. Toxicol.*, 104, 61, 2011.

Comella, C.L. and Pullman, S.L., Botulinum toxins in neurological disease, *Muscle Nerve*, 29, 628, 2004.

Contardo-Jara, V., Pfulgmacher, S., and Wiegand, C., Multi-xenobiotic-resistance a possible explanation for the insensitivity of bivalves towards cyanobacterial toxins, *Toxicon*, 53, 936, 2008.

Cox, P.A., Banack, S.A., Murch, S.J., Rasmussen, U., Tien, G., Bidigare, R.R., Metcalf, J.S., Morrison, L.F., Codd, G.A., and Bergman, B., Diverse taxa of cyanobacteria produce beta-*N*-methylamino-L-alanine, a neurotoxic amino acid, *PNAS*, 102, 5074, 2005.

Dumbacher, J.P., Wako, A., Derrickson, S.R., Samuelson, A., Spande, T.F., and Daly, J.W., Melyrid beetles (*Choresine*): A putative source for the batrachotoxin alkaloids found in poison-dart frogs and toxic passerine birds, *PNAS*, 101, 15857, 2004.

Etheridge, S.M., Paralytic shellfish poisoning: Seafood safety and human health perspectives, *Toxicon*, 56, 108, 2010.

Fautin, D.G., Structural diversity, systematics, and evolution of cnidae, *Toxicon*, 54, 1054, 2009.

Gladman, A.C., Toxicodendron dermatitis: Poison ivy, oak, and sumac, *Wilderness Environ. Med.*, 17, 120, 2006.

Gremski, L.H., Trevisan-Silva, D., Ferrer, V.P., Matsubara, F.H., Meissner, G.O., Wille, A.C.M., Vuitika, L. et al., Recent advances in the understanding of brown spider venoms: From the biology of spiders to the molecular mechanisms of toxins, *Toxicon*, 83, 91, 2014.

Guichard, A., Nizet, V., and Bier, E., New insights into the biological effects of anthrax toxins: Linking cellular to organismal responses, *Microbes Infect.*, 14, 97, 2012.

Hartmann, T., From waste products to ecochemicals: Fifty years research of plant secondary metabolism, *Phytochemistry*, 68, 2831, 2007.

Hoagland, P., Jin, D., Beet, A., Kirkpatrick, B., Reich, A., Ullmann, S., Fleming, L.E., and Kirkpatrick, G., The human health effects of Florida Red Tide (FRT) blooms: An expanded analysis, *Environ. Int.*, 68, 144, 2014.

Holmes, R.K., Biology and molecular epidemiology of diphtheria toxin and the *tox* gene, *J. Infect. Dis.*, 181, S156, 2000.

Hossain, M.A., Ahmed, M.S., and Ghannoum, M.A., Attributes of *Stachybotrys chartarum* and its association with human disease, J. Allergy Clin. Immunol., 113, 200, 2004.

Karlson-Stiber, C. and Persson, H., Cytotoxic fungi—An overview, *Toxicon*, 42, 339, 2003.

Kini, R.M., Excitement ahead: Structure, function and mechanism of snake venom phospholipase A_2 enzymes, *Toxicon*, 42, 827, 2003.

Konno, K., Plant latex and other exudates as plant defense systems: Roles of various defense chemicals and proteins contained therein, *Phytochemistry*, 72, 1510, 2011.

Krenzelok, E.P., Jacobsen, T.D., and Aronis, J.M., Poinsettia exposures have good outcomes … just as we thought, *Am. J. Emerg. Med.*, 14, 671, 1996.

Landsberg, J.H., Flewelling, L.J., and Naar, J., *Karenia brevis* red tides, brevetoxins in the food web, and impacts on natural resources: Decadal advancements, *Harmful Algae*, 8, 598, 2009.

Ligabue-Braun, R., Verli, H., and Carlini, C.R., Venomous mammals: A review, *Toxicon*, 59, 680, 2012.

Liu, S., Moayeri, M., and Leppla, S.H., Anthrax lethal and edema toxins in anthrax pathogenesis, *Trends Microbiol.*, 22, 317, 2014.

Marin, S., Ramos, A.J., Cano-Sancho, G., and Sanchis, V., Mycotoxins: Occurrence, toxicology, and exposure assessment, *Food Chem. Toxicol.*, 60, 218, 2013.

Marroquin-Cardona, A.G., Johnson, N.M., Phillips, T.D., and Hayes, A.W., Mycotoxins in a changing global environment—A review, *Food Chem. Toxicol.*, 69, 220, 2014.

McCleary, R.J.R. and Kini, R.M., Non-enzymatic proteins from snake venoms: A gold mine of pharmacological tools and drug leads, *Toxicon*, 62, 56, 2013.

Mebs, D., *Venomous and Poisonous Animals*, CRC Press, Boca Raton, FL, 2002.

Merel, S., Walker, D., Chicana, R., Snyder, S., Baures, E., and Thomas, O., State of knowledge and concerns on cyanobacterial blooms and cyanotoxins, *Environ. Int.*, 59, 303, 2013.

Persson, H., Mushrooms, *Medicine*, 40, 135, 2012.

Porciuncula, L.O., Sallaberry, C., Mioranzza, S., Botton, P.H.S., and Rosemberg, D.B., The Janus face of caffeine, *Neurochem. Int.*, 63, 594, 2013.

Pratheepa, V. and Vasconcelos, V., Microbial diversity associated with tetrodotoxin production in marine organisms, *Environ. Toxicol. Pharmacol.*, 36, 1046, 2013.

Quintero-Hernandez, V., Jimenez-Vargas, J.M., Gurrola, G.B., Valdivia, H.H., and Possani, L.D., Scorpion venom components that affect ion-channels function, *Toxicon*, 76, 328, 2013.

Rash, L.D. and Hodgson, W.C., Pharmacology and biochemistry of spider venoms, *Toxicon*, 40, 225, 2002.

Rossetto, O., Megighian, A., Scorzeto, M., and Montecucco, C., Botulinum neurotoxins, *Toxicon*, 67, 31, 2013a.

Rossetto, O., Megighian, A., Scorzeto, M., and Montecucco, C., Tetanus neurotoxin, *Toxicon*, 67, 59, 2013b.

Sharma, S.K., Vij, A.S., and Sharma, M., Mechanisms and clinical uses of capsaicin, *Eur. J. Pharmacol.*, 720, 55, 2013.

Simpson, L., The life history of a botulinum toxin molecule, *Toxicon*, 68, 40, 2013.

Suput, D., In vivo effects of cnidarian toxins and venoms, *Toxicon*, 54, 1190, 2009.

Swanson, D.L. and Vetter R. S., Loxoscelism, *Clin. Dermatol.*, 24, 213, 2006.

Thapa, P., Espiritu, M.J., Cabalteja, C.C., and Bingham, J.-P., Conotoxins and their regulatory considerations, *Regul. Toxicol. Pharmacol.*, 70, 197, 2014.

Tibbals, J., Australian venomous jellyfish, envenomation syndromes, toxins and therapy, *Toxicon*, 48, 830, 2006.

Tibbals, J., Yanagihara, A.A., Turner, H.C., and Winkel, K., Immunological and toxinological responses to jellyfish stings, *Inflamm. Allergy Drug Targets*, 10, 438, 2011.

Tighe, A.P. and Schiavo, G., Botulinum neurotoxins: Mechanism of action, *Toxicon*, 67, 87, 2013.

Turton, K., Chaddock, J.A., and Acharya, K.R., Botulinum and tetanus neurotoxins: Structure, function and therapeutic utility, *Trends Biochem. Sci.*, 27, 552, 2002.

Van Dongen, P.W.J. and de Groot, A.N.J.A., History of ergot alkaloids from ergotism to ergometrine, *Eur. J. Obstet. Gynecol. Reprod. Biol.*, 60, 109, 1995.

Vetter, R.S., Cushing, P.E., Crawford, R.L., and Royce, L.A., Diagnoses of brown recluse spider bites (loxoscelism) greatly outnumber actual verifications of the spider in four western American states, *Toxicon*, 42, 413, 2003.

Vila-Luna, S., Cabrera-Isidoro, S., Vila-Luna, L., Juarez-Diaz, I., Bata-Garcia, J.L., Alvarez-Cervera, F.J. et al., Chronic caffeine consumption prevents cognitive decline from young to middle age in rats, and is associated with increased length, branching, and spine density of basal dendrites in CA1 hippocampal neurons, *Neuroscience*, 202, 384, 2012.

Walker, J.P., Morrison, R., Stewart, R., and Gore, D., Venomous bites and stings, *Curr. Prob. Surg.*, 50, 9, 2013.

Watkins, J.B., III, Toxic effects of plants and animals, in *Casarett and Doull's Toxicology*, Klaassen, C.D., Ed., McGraw-Hill, New York, 2013, chap. 26.

Wei, C.J., Li, W., and Chen, J.-F., Normal and abnormal functions of adenosine receptors in the central nervous system revealed by genetic knockout studies, *Biochim. Biophys. Acta*, 1808, 1358, 2011.

Wheeler, A. and Smith, H.S., Botulinum toxins: Mechanisms of action, antinociception, and clinical applications, *Toxicology*, 306, 124, 2013.

Wiegand, C. and Pflugmacher, S., Ecotoxicological effects of selected cyanobacterial secondary metabolites. A short review, *Toxicol. Appl. Pharmacol.*, 203, 201, 2005.

Williams, B.L., Stark, M.R., and Caldwell, R.L., Microdistribution of tetrodotoxin in two species of blue-ringed octopuses (*Hapalochlaena lunulata* and *Hapalochlaena fasciata*) detected by fluoresecent immunolabeling, *Toxicon*, 60, 1307, 2012.

Appendix: List of Selected Toxicants

Name: Acetaminophen

Chemical formula: *p*-Hydroxyacetanilide, $C_8H_9NO_2$

Physical properties: Crystals

Sources and uses: Synthetic compound used to treat headache, inflammation, fever

Toxicity: LD_{50} (mice, oral), 338 mg/kg

See also:
Glutathione Conjugation
Chapter 3

Liver Cell Death: Necrosis and Apoptosis
Chapter 11

Acetaminophen

Name: Aflatoxins

Chemical formula: Aflatoxin B1 (2,3,6a,9a-tetrahydro-4-methoxycyclopenta[c]furo[3′,2′:4,5]furo[2,3-h][1]benzopyran-1,11-dione), $C_{17}H_{12}O_6$

Physical properties: Crystals

Sources and uses: Produced by *Aspergillus* sp. (fungi); can contaminate food products under proper conditions

Toxicity: Hepatotoxicants, carcinogens; LD_{50} (mouse, i.p.), 9.5 mg/kg

See also:
Fungal Toxins
Chapter 20

Aflatoxin B₁

Name: Arsenic

Chemical formula: As, Atomic wt. 74.92159

Physical properties: Gray-black, metallic crystal; many different species, most toxic

Sources and uses: Naturally occurring element; used in manufacture of glass, metal working; contaminant of precious metal ore

Toxicity: Gastrointestinal irritant, nephrotoxicant, hepatotoxicant; arsenic trioxide LD_{50} (mouse, oral), 39.4 mg/kg

See also:
Criminal Poisonings
Chapter 16

Airborne Toxicants
Categories of Waste
Chapter 17

As_2O_3
Arsenic trioxide (arsenous acid)

Name: Asbestos

Chemical formula: Chrysotile (the most common form of asbestos), $Mg_6(Si_4O_{10})(OH)_8$

Physical properties: Fibrous silicate

Source and uses: Naturally occurring; used as a heat- and fire-resistant material

Toxicity: Respiratory toxicant, carcinogen

See also:
Fibrosis and
Pneumoconioses
Chapter 8

$[Mg_6(Si_4O_{10})(OH)_8]$
Asbestos (chrysotile)

Name: Benzene

Chemical formula: C_6H_6

Physical properties: Clear, colorless, flammable liquid

Source and uses: Can be purified from coal or synthesized; used in manufacture of many compounds

Toxicity: Acute exposures lead to nervous system depression with an LD_{50} (rat, oral) of 3.8 mL/kg; chronic exposure has been associated with bone marrow depression and leukemia

See also:
Anemias, Hemolysis, and Related Disorders
Chapter 9

Toxicant-Induced Immunosuppression
Chapter 13

Benzene

Name: Cadmium
  Chemical formula: Cd, Atomic wt. 112.411
  Physical properties: Heavy metal
  Source and uses: Naturally occurring; used in
    making of metal alloys, batteries, and other
    products
  Toxicity: Reproductive toxicant, cardiovascular
    toxicant, renal toxicant

See also:
  Damage to the Proximal
    Tubule
                     Chapter 12

Metals
                     Chapter 17

Name: Carbamate pesticides
  Chemical formula: Carbaryl(1-naphthalenol
    methylcarbamate), $C_{12}H_{11}NO_2$; Aldicarb
    (2-methyl-2(methylthio)propanal
    O-[(methylamino)carbonyl]oxime),
    $C_7H_{14}N_2O_2S$
  Physical properties: Crystals
  Sources and uses: Synthetic pesticides, insecticides
  Toxicity: Neurotoxicity; carbaryl LD_{50} (rat, oral), 250 mg/kg; aldicarb (rat, oral),
    1 mg/kg

See also:
  Pesticides
                     Chapter 17

Carbaryl
(Sevin®)

Aldicarb

Name: Carbon disulfide
  Chemical formula: CS_2
  Physical properties: Flammable liquid
  Source and uses: Synthesized; used in
    manufacture of rayon and as cleaning
    solvent
  Toxicity: Cardiovascular toxicant,
    neurotoxicant

See also:
  Atherosclerosis
                     Chapter 9

Distal Axonopathies
                     Chapter 10

Name: Carbon monoxide
  Chemical formula: CO
  Physical properties: Colorless, odorless gas
  Source and uses: Produced during combus-
    tion of organic materials; toxic by-prod-
    uct of combustion
  Toxicity: Cardiovascular toxicant (com-
    bines with hemoglobin)

See also:
  Effects of Toxicants on Transport
    Proteins
                                    Chapter 4

  Atherosclerosis
  Effects of Toxicants on
    Hemoglobin
                                    Chapter 9

  Carbon Oxides
                                    Chapter 17

Name: Chlorofluorocarbons
  Chemical formula: CFC-11 = CCl_3F,
    CFC-12 = CCl_2F_2,
    CFC-113 = CCl_2FCClF_2
  Physical properties: Stable, nontoxic
    compounds
  Sources and uses: Synthetic chemicals used as refrigerants, propellants, among
    other uses
  Toxicity: Nontoxic; influences health indirectly through damage to the ozone
    layer

See also:
  Chlorofluorocarbons
                                    Chapter 17

Name: Hydrogen cyanide
  Chemical formula: HCN
  Physical properties: Liquid or gas; odor of
    bitter almonds
  Source and uses: Industrial processes,
    chemical weapon
  Toxicity: LD_{50} (mouse, oral), 3.7 mg/kg

See also:
  Effects of Toxicants on Enzymes
                                    Chapter 4

  Criminal Poisonings
                                    Chapter 16

Name: 2,4-D; 2,4,5-T

Chemical formula: 2,4-D (2,4-dichlorophenoxy) acetic acid, $C_8H_6Cl_2O_3$; 2,4,5-T (2,4,5-trichlorophenoxy) acetic acid, $C_8H_5Cl_3O_3$

See also:
Pesticides
Chapter 17

Physical properties: Crystalline powder
Source and uses: Synthetic herbicides
Toxicity: Irritants; 2,4-D LD_{50} (rat, oral), 375 mg/kg; 2,4,5-T LD_{50} (rat, oral), 500 mg/kg

2,4-D
((2,4-dichlorophenoxy)acetic acid)

2,4,5-T
((2,4,5-trichlorophenoxy)acetic acid)

Name: Diethylstilbestrol (DES)

Chemical formula: 4,4′-(1,2-diethyl-1,2ethenediyl)bisphenol, $C_{18}H_{20}O_2$
Physical properties: Crystalline powder
Source and uses: Synthetic compound with estrogenic activity; used at one time in prevention of miscarriage; also used in livestock feed to promote growth
Toxicity: Carcinogen

See also:
Examples of Teratogens
Chapter 7

Diethylstilbestrol

Name: Ethanol
Chemical formula: C_2H_6O
Physical properties: Colorless liquid
Source and uses: Produced through the
process of fermentation; used as recreational drug, as solvent in foods and
medicines, and as industrial solvent
Toxicity: Acute exposures lead to nervous
system depression; chronic exposures
lead to increased risks of neurologic and
hepatic disease; carcinogen, teratogen

See also:
   Examples of Teratogens
               Chapter 7

   Other Neurotoxicants
               Chapter 10

   Fatty Liver
   Liver Cell Death: Necrosis and
    Apoptosis
   Fibrosis and Cirrhosis
               Chapter 11

   Forensic Toxicology and Alcohol
   Use
               Chapter 16

Name: Ethylene dibromide (EDB)
Chemical formula: 1,2-dibromoethane,
   $C_2H_4Br_2$
Physical properties: Heavy liquid
Sources and uses: Synthetic; used as fumigant
Toxicity: Carcinogen, hepatotoxicant, nephrotoxicant

EDB
1,2-dibromoethane

Name: Formaldehyde
Chemical formula: CH_2O
Physical properties: Colorless irritant gas
Source and uses: By-product of combustion;
also manufactured synthetically; used in
production of resins, textiles, particleboard,
and other products
Toxicity: Respiratory irritant, possible
carcinogen

See also:
   Immediate Responses:
    Upper Airway Effects
               Chapter 8

   Toxicant-Induced Allergies
               Chapter 13

Formaldehyde

Name: Halogenated hydrocarbon solvents

Chemical formulas: Carbon tetrachloride, CCl_4

Chloroform, $CHCl_3$

Dichloromethane, CH_2Cl_2

Trichloroethylene, C_2HCl_3

Physical properties: Colorless, heavy liquids

Source and uses: Synthesized; used as industrial solvents, cleaners, in synthesis of many organic compounds

Toxicity: Neurotoxic, hepatotoxic, toxic to cardiovascular and renal systems, carcinogenic

Carbon tetrachloride

See also:
Primary Biotransformation
(Phase I Reactions): Reduction
Chapter 3

Arrhythmias
Chapter 9

Other Neurotoxicants
Chapter 10

Liver Cell Death: Necrosis and
Apoptosis
Fibrosis and Cirrhosis
Chapter 11

Damage to the Proximal Tubule
Chapter 12

Name: Lead

Chemical formula: Pb, Atomic wt. 207.2

Physical properties: Heavy metal; also may form organolead compounds

Source and uses: Naturally occurring; used in manufacture of alloys, batteries, pipes, pigments, and many other products

Toxicity: Neurotoxic, reproductive toxin

See also:
Anemias, Hemolysis, and Related
Disorders
Chapter 9

Myelinopathies
Other Neurotoxicants
Chapter 10

Toxicant-Induced
Immunosuppression
Chapter 13

Metals
Chapter 17

Name: Mercury

Chemical formula: Hg, Atomic wt. 200.59
Physical properties: Heavy metal; also may
   form organomercury compounds
Source and uses: Naturally occurring
Toxicity: Teratogenic, neurotoxicant, renal
   toxicant, irritant

$CH_3\text{-}Hg^+$
Methylmercury

See also:
   Other Neurotoxicants
                             Chapter 10

   Damage to the Proximal Tubule
                             Chapter 12

   Metals
                             Chapter 17

Name: Nitrates, nitrites

Chemical formula: xNO_3, xNO_2
Physical properties: Usually white or yellow powder or granules
Source and uses: Occur naturally; used in
   manufacturing, preservation of meats,
   and fertilizers
Toxicity: Cardiovascular toxicant; possible
   role in carcinogenesis; sodium nitrate
   LD_{50} (rabbits, oral), 2 g/kg; sodium
   nitrite LD_{50} (rat, oral), 180 mg/kg

$NaNO_2$
Sodium nitrite
$NaNO_3$
Sodium nitrate

See also:
   Effects of Toxicants on Transport
    Proteins
                             Chapter 4

   Vascular Spasms and Blood
    Pressure
   Effects of Toxicants on
    Hemoglobin
                             Chapter 9

   Phosphorus and Nitrogen
                             Chapter 17

Name: Nitrogen dioxide

Chemical formula: NO_2
Physical properties: Irritant gas
Source and uses: By-product of combustion
Toxicity: Respiratory toxicant

See also:
   Immediate Responses:
    Lower Airway Effects
                             Chapter 8

   Sulfur Oxides and Nitrogen
    Oxides
                             Chapter 17

Name: Organochlorine pesticides

Chemical formula: DDT (dichlorodiphenyltri-chloroethane), $C_{14}H_9Cl_5$

Chlordane (1,2,4,5,6,7,8,8-octochloro-2,3,3a,4,7,7a-hexahydro-4,7-methano-1H-indene), $C_{10}H_6Cl_8$

Aldrin (1,2,3,4,10,10-hexachloro-1,4,4a,5,8,8a-hexahydro-1,4:5,8-dimethanonaphthalene), $C_{12}H_8Cl_6$

Dieldrin (3,4,5,6,9,9-hexachloro-1a,2,2a,3,6,6a,7,7a-octahydro-2,7:3,6-dimethanonapth[2,3-b]oxirene), $C_{12}H_8Cl_6O$

Mirex (1,1a,2,2,3,3a,4,5,5,5a,5b,6-dodeca-chloro-octahydro-1,3,4-metheno-1H-cyclobuta[cd]pentalene), $C_{10}Cl_{12}$

Chlordecone (kepone) (decachlorooctahydro-1,3,4-metheno-2H-cyclobuta[cd] pentalen-2-one), $C_{10}Cl_{10}O$

Physical properties: Crystals; insoluble in water

Source and uses: Synthetic insecticides; persistent in the environment

Toxicity: Neurotoxic; chronic exposure may lead to hepatotoxicity; DDT LD_{50} (human), 500 mg/kg; chlordane LD_{50} (rat, i.p.), 343 mg/kg; aldrin LD_{50} (rats, oral), 30–60 mg/kg; dieldrin LD_{50} (rat, oral), 46 mg/kg; mirex LD_{50} (rat, oral), 600 mg/kg; chlordecone LD_{50} (rat, oral), 125 mg/kg

See also:

Effects of Toxicants on Electrical Conduction

Chapter 10

Material Cycling in Ecosystems

Chapter 14

Pesticides

Chapter 17

DDT
(Dichlorodiphenyltrichloroethane)

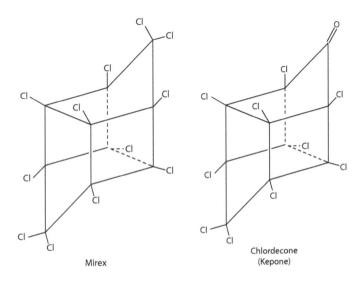

Mirex

Chlordecone
(Kepone)

Name: Organophosphates

Chemical formula: Parathion (phos-
phorothioic acid *O,O*-diethyl *O*-(4-
nitrophenyl)ester), $C_{10}H_{14}NO_5PS$
Malathion ([(dimethoxypnosphino-
thioyl)thio]buta-nedioic acid diethyl
ester), $C_{10}H_{19}O_6PS_2$
Physical properties: Pale liquid
Source and uses: Synthetic insecticide
Toxicity: Neurotoxicants; parathion LD_{50}
(rat, oral), 3–10 mg/kg; malathion LD_{50}
(rat, oral), 1000–1400 mg/kg

See also:
  Serine Hydrolases
                                          Chapter 3

  Effects of Toxicants on Enzymes
                                          Chapter 4

  Acetylcholine
  Distal Axonopathies
                                          Chapter 10

  Pesticides
                                          Chapter 17

Methyl parathion

Malathion

Name: Ozone

Chemical formula: O_3
Physical properties: Irritant gas
Source and uses: Produced as a by-product
of combustion; air pollutant
Toxicity: Respiratory irritant

See also:
  Immediate Responses: Lower
  Airway Effects
                                          Chapter 8

  Hydrocarbons and the Formation
  of Secondary Pollutants
  (Including Ozone)
                                          Chapter 17

Ozone

Name: Paraquat

Chemical formula: 1,1′-Dimethyl-4,4′-
bipyridinium, $[C_{12}H_{14}N_2]^{2+}$
Physical properties: Crystals
Source and uses: Synthetic herbicide
Toxicity: Irritant, respiratory toxicant

See also:
  Immediate Responses: Lower
  Airway Effects
                                          Chapter 8

  Pesticides
                                          Chapter 17

Paraquat

Name: Petroleum products

Chemical formula: Mixture of aliphatic and cyclic hydrocarbons, some aromatic hydrocarbons, and other compounds

Physical properties: Oily liquid

Source and uses: Naturally occurring; petroleum or crude oil is distilled and separated into components that are typically used for fuels or lubricants

Toxicity: Some components are neurotoxic; others are potentially carcinogenic

See also:

Marine Ecosystems

Chapter 14

Petroleum Products as Water Pollutants

Chapter 17

Isooctane
(2,2,4-trimethylpentane)

H₃CCH₃
Ethane

$CH_3CH_2CH_2CH_2CH_2CH_2CH_3$
n-Heptane

Cyclopentane

Name: Polychlorinated biphenyls (PCBs) and polybrominated biphenyls (PBBs)

Chemical formula: C_{12} with varying amounts of H and Cl

Physical properties: Usually liquids

Source and uses: Synthetic compounds used in manufacture of electrical equipment

Toxicity: Hepatotoxic, immunotoxic, potential carcinogens; PCBs LD_{50} (rat, oral), 1500 mg/kg

See also:

Endocrine Disrupters and Interference with Hormonal Controls

Chapter 7

Toxicant-Induced Immunosuppression

Chapter 13

Other Organic Compounds

Chapter 17

PCB or PBB
(Polychlorinatedbiphenyl; x=Cl)
(Polybrominatedbiphenyl; x=Br)

Name: Polycyclic aromatic hydrocarbons (PAHs)

Chemical formula: Anthracene $C_{14}H_{10}$

Benzo[a]pyrene, $C_{20}H_{12}$

3-Methylcholanthrene (3-MC), $C_{21}H_{16}$

Physical properties: Clear or yellowish powder

Source and uses: Produced during incomplete combustion of organic materials

Toxicity: Carcinogenic

See also:
The Role of Cytochrome P450
Chapter 3
Genetic Carcinogens
Chapter 6

Polycyclic aromatic hydrocarbon

3-methylcholanthrene

Name: Pyrethroids

Chemical formula: Allethrin (2,2-dimethyl-3-(2-methyl-1-propenyl) cyclopropanecarboxylic acid 2-methyl-4-oxo-3-(2-propenyl)-2-cyclopenten-1-yl ester), $C_{19}H_{26}O_3$

Permethrin (3-(2,2-dichloroethenyl)-2,2-dimethylcyclopropanecarboxylic acid (3-phenoxyphenyl) methyl ester), $C_{21}H_{20}Cl_2O_3$

Cypermethrin is the -cyano derivative of permethrin

Physical properties: Liquid

Source and uses: Synthetic insecticide

Toxicity: Neurotoxic

See also:
Pesticides
Chapter 17

Cypermethrin

Name: Strychnine

Chemical formula: $C_{21}H_{22}N_2O_2$
Physical properties: Crystalline powder
Source and uses: From plant *Strychnos nux-vomica*; used as pesticide
Toxicity: Neurotoxicant

See also:
Amino Acid Neurotransmitters
Chapter 10

Criminal Poisonings
Chapter 16

Strychnine

Name: Sulfur dioxide

Chemical formula: SO_2
Physical properties: Irritant gas
Source and uses: By-product of combustion
Toxicity: Respiratory irritant

See also:
Immediate Responses: Upper Airway Effects
Chapter 8

Sulfur Oxides and Nitrogen Oxides
Chapter 17

Name: Tetrachlorodibenzodioxin (TCDD)

Chemical formula:
2,3,7,8-Tetrachlorodibenzo[b,e][1,4] dioxin, $C_{12}H_4Cl_4O_2$
Physical properties: Crystalline solid
Source: Contaminant formed during manufacture of certain herbicides
Toxicity: Hepatotoxic, immunotoxic, teratogenic, carcinogenic; LD_{50} (rat, oral), 0.02–0.04 mg/kg

See also:
The Role of Cytochrome P450
Chapter 3

Chemical Carcinogenesis: Promotion
Chapter 6

Endocrine Disrupters and Interference with Hormonal Controls
Chapter 7

Toxicant-Induced Immunosuppression
Chapter 13

Pesticides
Chapter 17

TCDD
(2,4,7,8-tetrachlorodibenzodioxin)

Name: Tetrodotoxin (TTX), saxitoxin (STX)

Chemical formula: TTX, $C_{11}H_{17}N_3O_8$

STX, $[C_{10}H_{17}N_7O_4]^{2+}$

Physical properties: Crystalline solid

Sources and uses: Naturally occurring
toxins produced by fish (TTX) and
dinoflagellates (STX)

Toxicity: Neurotoxic; TTX LD_{50} (mice,
i.p.), 10 mg/kg; STX LD_{50} (mice, oral),
263 mg/kg

See also:
Effects of Toxicants on Voltage-
    Activated Ion Channels
                                        Chapter 4

Effects of Toxicants on Electrical
    Conduction
                                        Chapter 10

Mollusks
Cyanobacterial Toxins
Protist Toxins
Fish, Birds, and Mammals
                                        Chapter 20

Tetrodotoxin

Saxitoxin

Name: Thalidomide

Chemical formula: 2-(2,6-Dioxo-3-
piperidinyl)-1*H*-isoindole-1,3(2*H*)-
dione, $C_{13}H_{10}N_2O_4$

Physical properties: Powder

Source and uses: Synthetic drug used to
treat morning sickness

Toxicity: Teratogen

See also:
Examples of Teratogens
                                        Chapter 7

Thalidomide

Name: Tobacco

Chemical formula: Complex mixture of compounds, including polycyclic aromatic hydrocarbons, phenols, nicotine, nitrosamines

Physical properties: Leaf of tobacco plant

Source and uses: Tobacco plant, recreational use (smoking)

Toxicity: Respiratory toxicant, carcinogenic

See also:
Genetic Carcinogens
Chapter 6

Lung Cancer
Chronic Obstructive Pulmonary
  Disease: Bronchitis and
  Emphysema
Chapter 8

Atherosclerosis
Effects of Toxicants on
  Hemoglobin
Chapter 9

Nicotine

Name: Toluene diisocyanate

Chemical formula: 2,4-Diisocyanatotoluene, $C_9H_6N_2O_2$

Physical properties: Liquid

Source and uses: Synthetic compound used in manufacturing

Toxicity: Respiratory toxicant, immunotoxic

See also:
Asthma and Immune-
  Related Chronic
  Conditions
Chapter 8

Toxicant-Induced Allergies
Chapter 13

Toluene 2,4-diisocyanate

BIBLIOGRAPHY

Most toxicity data in this appendix were derived from the same sources as cited in the chapters on that subject. Physical data, chemical data, and some LD$_{50}$ figures were obtained from O'Neil, M.J., Smith, A., Heckelman, P.E., Obenchain, J.R., Gallipeau, J.R., D'Arecca, M.A., and Budavari, S., Eds., *Merck Index*, 13th ed., Merck and Co., Whitehouse Station, NJ, 2001.

The following books are textbooks that would serve as good general references for background information on basic biology, biochemistry, cell biology, anatomy and physiology, ecology, and environmental science:

Alberts, B., Johnson, A., Lewis, J., Raff, M., Roberts, K., and Walter, P., *Molecular Biology of the Cell*, 5th ed., Garland Science, New York, 2007.

Begon, M., *Ecology: From Individuals to Ecosystems*, 4th ed., Blackwell Publishing, Malden, MA, 2006.

Berg, J.M., Tymoczko, J.L., and Stryer, L., Biochemistry, 7th ed., W.H. Freeman and Company, San Francisco, 2010.

Berne, R.M. and Levy, M.N., *Physiology*, 6th ed., C.V. Mosby, St. Louis, 2009.

Freeman, S., Quillin, K., and Allison, L., *Biological Science*, 5th ed., Benjamin Cummings, San Francisco, 2013.

Hall, J.E., *Guyton and Hall Textbook of Medical Physiology*, 12th ed., W.B. Saunders, Philadelphia, 2010.

Martini, F.H., Nath, J.L., and Bartholomew, E.F, *Fundamentals of Anatomy and Physiology*, 9th ed., Benjamin Cummings, San Francisco, 2011.

Miller, G.T., Jr. and Spoolman, S., *Living in the Environment*, 18th ed., Cengage Learning, Stamford, CT, 2014.

National Library of Medicine, PubChem, 2009. Available online at https://pubchem.ncbi.nlm.nih.gov/.

Reece, J.B., Urry, L.A., Cain, M.L., Wasserman, S.A., Minorsky, P.V., and Jackson, R.B., *Campbell Biology*, 10th ed., Benjamin Cummings, San Francisco, 2013.

Saladin, K., *Anatomy and Physiology: The Unity of Form and Function*, 7th ed., McGraw-Hill, New York, 2014.

Index

A

Ab, *see* Antibodies (Ab)
ABC transporters, *see* ATP binding cassette
transporters (ABC transporters)
Accutane, 146
Acetaminophen, 46, 232–233, 397
Acetylation, 48, 55; *see also* Biotransformation
Acetylcholine (ACh), 199; *see also* Neurotoxicants;
Neurotransmitter; Receptors
atropine, 202–203
autonomic nervous system, 199–200
cholinergic agonists, 202
curare, 202
fight-or-flight response, 200
hydrolase, *see* Acetylcholinesterase
hydrolysis, 55, 203
locations in nervous system, 200
receptors, 61, 201
synthesis, 201
toxicant site of action, 201, 202
Acetylcholinesterase (AChase), 33, 54, 203; *see*
also Acetylcholine (ACh); Enzymes;
Hydrolysis; Organophosphates
acetylcholine hydrolysis, 55
aging, 56
irreversible inhibition, 56
phosphorylation of, 56
reaction catalyzed by, 55
ACGIH, *see* American Conference of
Governmental and Industrial
Hygienists (ACGIH)
ACh, *see* Acetylcholine (ACh)
AChase, *see* Acetylcholinesterase (AChase)
Acid rain, 319–320
Acinar zones, 225; *see also* Liver
Acinus, 225; *see also* Liver
Aconitine, 173
Acoustic startle chamber, 220
Acquired immunodeficiency syndrome (AIDS),
266; *see also* Immunotoxicology
Acrolein, 160
Acrylamide, 212; *see also* Axonopathies
Action potential, 63, 195, 196
Active transport, 16, 17; *see also* Toxicant
absorption
Acutely transforming retroviruses, 104
Acute radiation syndrome, 352
Acute toxicity, 5; *see also* Toxicity
Adduct, 109

Adenomatous polyposis coli (APC), 108
Adenosine tri phosphate (ATP), 149
Adenylate cyclase, 63
ADH, *see* Alcohol dehydrogenase (ADH);
Antidiuretic hormone (ADH)
ADHD, *see* Attention-deficit hyperactivity
disorder (ADHD)
Adult stem cells, 138; *see also* Reproductive system
Aerosol, 158
Aflatoxins, 235, 381, 397; *see also* Toxicant-
induced liver injury
Aging, 56
Agonists, 61; *see also* Toxicants
AH, *see* Aryl hydrocarbon hydroxylase (AH)
AHR, *see* Aryl hydrocarbon receptor (AHR)
Ah receptor, 36
AIDS, *see* Acquired immunodeficiency syndrome
(AIDS)
AIF, 71
Airborne toxicants, 321
Air pollution, 315
acid rain, 319–320
airborne toxicants, 321
carbon oxides, 317–318
chlorofluorocarbons, 321
control of, 323, 324
effects of air pollutants, 316–317
gaseous air pollutants, 315
global warming, 317
greenhouse effect, 317, 318
indoor, 322–323
Kyoto protocol, 318
methyl isocyanate, 322
Montreal protocol, 321
nitrogen oxides, 319
ozone, 320, 321, 322
particulate pollutants, 315
particulates, 321
primary pollutants, 315
secondary pollutants, 315
secondary pollutants, 320
sulfur oxides, 319
TCDD, 322
thermal inversion, 316
Air toxics, *see* Airborne toxicants
Alcohol dehydrogenase (ADH), 42; *see also*
Oxidation
Alcoholic cardiomyopathy, 175; *see also* Heart
Aldehyde dehydrogenase (ALDH), 42, 175;
see also Oxidation

413